GLOBAL
PHYSICAL
CLIMATOLOGY

SECOND EDITION

GLOBAL PHYSICAL CLIMATOLOGY

SECOND EDITION

DENNIS L. HARTMANN
Department of Atmospheric Sciences,
University of Washington, Seattle, WA, USA

AMSTERDAM • BOSTON • HEIDELBERG • LONDON
NEW YORK • OXFORD • PARIS • SAN DIEGO
SAN FRANCISCO • SINGAPORE • SYDNEY • TOKYO

Elsevier
Radarweg 29, PO Box 211, 1000 AE Amsterdam, Netherlands
The Boulevard, Langford Lane, Kidlington, Oxford OX5 1GB, UK
225 Wyman Street, Waltham, MA 02451, USA

Notices
Knowledge and best practice in this field are constantly changing. As new research and experience broaden our understanding, changes in research methods, professional practices, or medical treatment may become necessary.

Practitioners and researchers must always rely on their own experience and knowledge in evaluating and using any information, methods, compounds, or experiments described herein. In using such information or methods they should be mindful of their own safety and the safety of others, including parties for whom they have a professional responsibility.

To the fullest extent of the law, neither the Publisher nor the authors, contributors, or editors, assume any liability for any injury and/or damage to persons or property as a matter of products liability, negligence or otherwise, or from any use or operation of any methods, products, instructions, or ideas contained in the material herein.

British Library Cataloguing-in-Publication Data
A catalogue record for this book is available from the British Library

Library of Congress Cataloging-in-Publication Data
A catalog record for this book is available from the Library of Congress

ISBN: 978-0-12-328531-7

For information on all Elsevier publications
visit our website at http://store.elsevier.com/

Working together
to grow libraries in
developing countries

www.elsevier.com • www.bookaid.org

Contents

Preface to the Second Edition

I had not intended that 21 years should pass between the first edition of this book and the second, but much else has intervened to command my attention and time. The timing of this second edition is nonetheless propitious, as I started in earnest just after finishing participation in the Intergovernmental Panel on Climate Change (IPCC) Working Group I Fifth Assessment in late 2013. So I had good familiarity with a recent summary of the state of knowledge as a starting point. Another advantage of waiting so long is that the easy availability of data sets and modern tools to manipulate them made it possible for me to make most of the new figures myself. The second edition is thus heavy with color figures of data from observations and models.

The basic outline of the book is the same as that of the first edition, except that I have added a new Chapter 8 on natural internal variability that is captured by the instrumental record. This chapter makes use of some statistical and mathematical techniques that I do not explain in detail, but I feel that by looking at the pictures and trusting me, the student will gain an intuitive sense of the structures of atmospheric and oceanic variability. The first seven chapters are suitable for a course for undergraduate science majors, and the final six chapters are more appropriate for an introductory overview course for graduate students. When I use the book to teach third-year atmospheric sciences majors, I go through the first seven chapters and mix in material from Chapters 9, 12, and 13 to add spice and relevance.

Many people contributed to the second edition. Marc L. Michelsen made many of the figures in Chapter 6 and on many occasions rescued me when my own computer skills led me into blind alleys. Bryce E. Harrop did the radiative–convective equilibrium calculations in Chapters 3 and 13. I wanted to update Manabe and Wetherald (1967) with more physical clouds, but I have to say that modern calculations are not very different from theirs. Paulo Ceppi provided me with multimodel mean data from CMIP5. Mark D. Zelinka generated some key figures in Chapter 11. I have made liberal use of figures from the IPCC Fifth Assessment, and I am grateful to the IPCC authors for producing them and the IPCC for letting me use them here. I thank all the people who created data sets and put them on the web in handy format. Some of these are acknowledged in figure captions.

A number of people have used the first edition over the years and provided feedback, and some have used the draft of second edition and provided editorial and content suggestions. I will forget to mention some of them, and for that I apologize. David W. J. Thompson used the draft of second edition in a class and provided a particularly insightful critique of the new Chapter 8. Donald J. Wuebbles also used the draft and provided suggestions and corrections. Tsubasa Kohyama did a very careful edit of most chapters of the second edition. Hansi K. A. Singh read Chapter 7 and provided comments that improved it. Casey Wall provided useful suggestions on the later chapters. Sara Berry, Paulo Ceppi, Bryce Harrop, Tsubasa Kohyama, Daniel McCoy, and Casey Wall helped with proof correction. Many students helped me involuntarily by using the book in class, and I thank them also for the insights I gained from their work. My present and former graduate students have taught me a lot. It is also a great boon for me to work in the Department of Atmospheric Sciences at the University of Washington, where excellence is combined with collegiality to good effect.

I am grateful for continued climate research support from the federal government through the National Science Foundation, National Aeronautics and Space Administration, National Oceanographic and Atmospheric Administration, and the Department of Energy. Climate research is a public good activity and without the support of the US taxpayer we would know much less about climate and the particular challenges that climate change presents to the world. Of course, other nations have also contributed greatly to what we have learned, and US citizens have benefited from their efforts too.

The dedication of the second edition remains the same as the first – to my family – who share their love and strengthen my spirit. It is also good to have great friends and colleagues. Thanks.

Dennis L. Hartmann
Seattle, October 2015

Preface to the First Edition

The science of climatology began to evolve rapidly in the last third of the twentieth century. This rapid development arose from several causes. During this period, the view of Earth from its moon made people more aware of the exceptional nature of their planetary home at about the same time that it became widely understood that humans could alter our global environment. Scientific and technological developments gave us new and quantitative information on past climate variations, global observations of climate parameters from space, and computer models with which we could simulate the global climate system. These new tools together with concern about global environmental change and its consequences for humanity caused an increase in the intensity of scientific research about climate.

Modern study of the Earth's climate system has become an interdisciplinary science incorporating the atmosphere, the ocean, and the land surface, which interact through physical, chemical, and biological processes. A fully general treatment of this system is as yet impossible, because the understanding of it is just beginning to develop. This textbook provides an introduction to the physical interactions in the climate system, viewed from a global perspective. Even this endeavor is a difficult one, since many earth science subdisciplines must be incorporated, such as dynamic meteorology, physical oceanography, radiative transfer, glaciology, hydrology, boundary-layer meteorology, and paleoclimatology. To make a book of manageable size about such a complex topic requires many difficult choices. I have endeavored to provide a sense of the complexity and interconnectedness of the climate problem without going into excessive detail in any one area. Although the modern approach to climatology has arisen out of diverse disciplines, a coherent collection of concepts is emerging that defines a starting point for a distinct science. This textbook is my attempt to present the physical elements of that beginning with occasional references to where the chemical and biological elements are connected.

This book is intended as a text for upper-division undergraduate physical science majors and, especially in the later chapters, graduate students. I have used the first seven chapters as the basis for a 10-week undergraduate course for atmospheric sciences majors. A graduate course can be fashioned by supplementing the text with readings from the current literature. Most climatology textbooks are descriptive and written from the perspective of geographers, but this one is written from the perspective of a physicist. I have attempted to convey an intuition for the workings of the climate system that is based on physical principles. When faced with a choice between providing easy access to an important concept and providing a rigorous and comprehensive treatment, I have chosen easy access. This approach should allow students to acquire the main ideas without great pain. Instructors may choose to elaborate on the presentation where their personal interests and experience make it desirable to do so.

This book could not have been produced without the assistance of many people. It evolved from 15 years of teaching undergraduate and graduate students, and I thank the ATMS 321 and ATMS 571 students at the University of Washington who have endured my experimentation and provided comments on early drafts of this book. Professor Steve Esbensen and his AtS 630 class at Oregon State University provided commentary on a near final draft of Chapters 1–7 in the spring of 1993. Valuable comments and suggestions on specific chapters were also provided by David S. Battisti, Robert J. Charlson, James R. Holton, Conway B. Leovy, Gary A. Maykut, Stephen G. Porter, Edward S. Sarachik, J. Michael Wallace, and Stephen G. Warren. The encouragement and advice given by James R. Holton were critical for the completion of this book. Many people contributed graphics, and I am particularly grateful for the special efforts given by Otis Brown, Frank Carsey, Jim Coakley, Joey Comiso, Scott Katz, Gary Maykut, Pat McCormick, Robert Pincus, Norbert Untersteiner, and Stephen Warren.

Grace C. Gudmundson applied her professional editorial skills to this project with patience, dedication, and good humor. Her efforts greatly improved the quality of the end product. Similarly, Kay M. Dewar's artistic and computer skills produced some of the more appealing figures. Marc L. Michelsen's genius with the computer extracted data from many digital archives and converted them into attractive and informative computer graphics. Luanna Huynh and Christine Rice were especially helpful with the appendices and tables.

My efforts to understand the climate system have been generously supported over the years by research grants and contracts from the US government. I am particularly happy to acknowledge support from the Climate Dynamics Program in the Atmospheric Sciences Division of the National Science Foundation, and the Earth Radiation Budget Experiment and Earth Observing System programs of the National Aeronautics and Space Administration. I also thank all of my colleagues from whom I have learned, who have shared their ideas with me, and who have given me the respect of serious argument.

This book is dedicated to my family, especially my wife, Lorraine, and my children, Alan and Jennifer, whose love and sacrifice were essential to its completion. I hope this book will help to explain why I spend so many evenings and weekends in my study. I thank my parents, Alfred and Angeline, for a good start in life and support along the way toward happy employment.

Dennis L. Hartmann

1

Introduction to
the Climate System

1.1 ATMOSPHERE, OCEAN, AND LAND SURFACE

Climate is the synthesis of the weather in a particular region. It can be defined quantitatively by using the expected values of the meteorological elements at a location during a certain month or season. The expected values of the meteorological elements can be called the climatic elements and include variables such as the average temperature, precipitation, wind, pressure, cloudiness, and humidity. In defining the climate, we usually employ the values of these elements at the surface of Earth. Thus, one can characterize the climate of Seattle by stating that the average annual mean precipitation is 38 in. and the annual mean temperature is 52°F. However, one might need a great deal more information than the annual means. For example, a farmer would also like to know how the precipitation is distributed through the year and how much rain would fall during the critical summer months. A hydroelectric plant engineer needs to know how much interannual variability in rainfall and snow accumulation to expect. A homebuilder should know how much insulation to be installed and the size of the heating or cooling unit needed to provide for the weather in the region. Sailors might like to know that the wind blows in the winter, but not so much in the summer.

The importance of climate is so basic that we sometimes overlook it. If the climate were not more or less as it is, life and civilization on this planet would not have developed as they have. The distribution of vegetation and soil type over the land areas is determined primarily by the local climate. Climate affects human lives in many ways; for example, climate influences the type of clothing and housing that people have developed. In the modern world, with the great technological advances of the past century, one might think that climate no longer constitutes a force capable of changing the course of human history. It is apparent, on the contrary,

Global Physical Climatology. http://dx.doi.org/10.1016/B978-0-12-328531-7.00001-3

that we are as sensitive now as we have ever been to climate fluctuations and climate change.

Because food, water, and energy supply systems are strained to meet demand and are optimized to the current average climatic conditions, fluctuations or trends in climate can cause serious difficulties for humanity. Moreover, since the population has grown to absorb the maximum agricultural productivity in much of the world, the absolute number of human lives at risk of starvation during climatic anomalies has never been greater. In addition to natural year-to-year fluctuations in the weather, which are an important aspect of climate, we must be concerned with the effects of human activities in producing long-term trends in the climate. It is now clear that humans are affecting the global climate, and this influence is growing. The actions of humankind that can change the global climate include altering the composition of the atmosphere and the nature of Earth's surface.

The surface climate of Earth varies greatly with location, ranging from the heat of the tropics to the cold of the polar regions, and from the drought of a desert to the moisture of a rain forest. Nonetheless, the climate of Earth is favorable for life, and living creatures exist in every climatic extreme. The climate of a region depends on latitude, altitude, and orientation in relation to water bodies, mountains, and the prevailing wind direction. In this book, we are concerned primarily with the global climate and its geographic variation on scales of hundreds to thousands of kilometers. In order to focus on these global issues, climate variations on horizontal spatial scales smaller than several tens of kilometers are given only minimal discussion.

The climate of Earth is defined in terms of measurable weather elements. The weather elements of most interest are temperature and precipitation. These two factors together largely determine the species of plants and animals that survive and prosper in a particular location. Other variables are also important, of course. The *humidity*, the amount of water vapor in the air, is a critical climate factor that is related closely to the temperature and precipitation. Condensation of water in the atmosphere produces clouds of water droplets or ice particles that greatly change the radiative properties of the atmosphere. The occurrence of clouds is important in itself for aviation and other activities, but clouds also play a role in determining both precipitation and surface temperature. Cloudiness influences the transmission of terrestrial radiation through the atmosphere and the amount of solar radiation that reaches the surface. The mean wind speed and direction are important considerations for local climate, air-pollution dispersion, aviation, navigation, wind energy, and many other purposes. The climate system of Earth determines the distribution of energy and water near the surface and consists primarily of the atmosphere, the oceans, and the land surface. The workings of this global system are the topic of this book (Fig. 1.1).

FIGURE 1.1 **Earth as seen on July 6, 2015 by the NASA Earth Polychromatic Imaging Camera aboard the NOAA Deep Space Climate Observatory spacecraft one million miles from Earth.**

1.2 ATMOSPHERIC TEMPERATURE

Temperature is the most widely recognized climatic variable. The global average temperature at the surface of Earth is about 288 K, 15°C, or 59°F. The range of temperatures encountered at the surface is favorable for the life forms that have developed on Earth. The extremes of recorded surface temperature range from the coldest temperature of –89.2°C (–128.6°F) at Vostok, Antarctica to the warmest temperature of 56.7°C (134°F) at Furnace Creek Ranch in Death Valley, California. These temperature extremes reflect the well-known decrease of temperature from the tropics, where the warmest temperatures occur, to the polar regions that are much colder. Both the warm temperature in Death Valley and the cold temperature at Vostok also

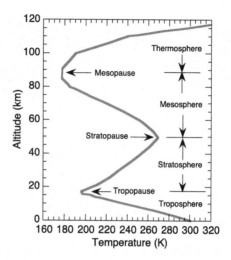

FIGURE 1.2 **The main zones of the atmosphere defined according to the temperature profile of the standard atmosphere profile at 15°N for annual-mean conditions.** *Data from U.S. Standard Atmosphere Supplements (1966).*

result partly from the decrease of temperature with altitude because Death Valley is below sea level and Vostok is 3450 m above sea level.

An important feature of the temperature distribution is the decline of temperature with height above the surface in the lowest 10–15 km of the atmosphere (Fig. 1.2). This rate of decline, called the *lapse rate*, is defined by

$$\Gamma \equiv -\frac{\partial T}{\partial z} \tag{1.1}$$

where T is the temperature and z is altitude and the deltas indicate a partial derivative. The global mean tropospheric lapse rate is about 6.5 K km^{-1}, but the lapse rate varies with altitude, season, and latitude. In the upper *stratosphere*, the temperature increases with height up to about 50 km. The increase of temperature with height that characterizes the stratosphere is caused by the absorption of solar radiation by ozone. Above the stratopause at about 50 km the temperature begins to decrease with height in the *mesosphere*. The temperature of the atmosphere increases rapidly above about 100 km because of heating produced by absorption of ultraviolet radiation from the sun, which dissociates oxygen and nitrogen molecules and ionizes atmospheric gases in the *thermosphere*.

The decrease of temperature with altitude in the *troposphere* is crucial to many of the mechanisms whereby the warmth of the surface temperature of Earth is maintained. The lapse rate in the troposphere and the mechanisms that maintain it are also central to the determination of

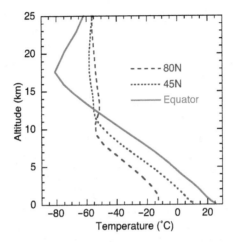

FIGURE 1.3 Annual mean temperature profiles for the lowest 25 km of the atmosphere in three latitude bands. *Data from ERA-Interim.*

climate sensitivity, as discussed in Chapter 10. The lapse rate and temperature in the troposphere are determined primarily by a balance between radiative cooling and convection of heat from the surface. The vertical distribution of temperature varies with latitude and season. At the equator, the temperature decreases with altitude up to about 17 km (Fig. 1.3). The tropical *tropopause* is the coldest part of the lowest 20 km of the atmosphere in the annual mean. In middle and high latitudes, the temperature of the lower stratosphere is almost independent of height. The tropospheric lapse rate in polar latitudes is less than it is nearer the equator. At high latitudes, the temperature actually increases with altitude in the lower troposphere in the winter and spring (Fig. 1.4). A region of negative lapse rate is called a *temperature inversion*. The polar temperature inversion has important implications for the climate of the polar regions. It arises because the surface cools very efficiently through emission of infrared radiation in the absence of insolation during the winter darkness. The air does not emit radiation as efficiently as the surface, and heat transported poleward in the atmosphere keeps the air in the lower troposphere warmer than the surface.

The variation of the zonal mean temperature with latitude and altitude is shown in Fig. 1.5. In Southern Hemisphere winter (June, July, and August, JJA), the polar stratosphere is colder than 180 K, and is the coldest place in the atmosphere, even colder than the tropical tropopause. The Northern Hemisphere stratosphere does not get as cold, on average, because planetary Rossby waves generated by surface topography and east–west surface temperature variations transport heat to the pole during *sudden stratospheric warming* events.

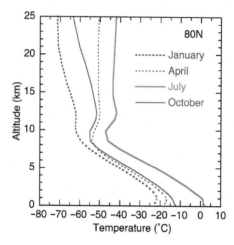

FIGURE 1.4   **Season variation profiles at 80°N.** *Data from ERA-Interim.*

The geographic and seasonal variation of surface temperature is shown in Fig. 1.6. The surface temperature is greatest near the equator, where it exceeds 296 K (23°C) across a broad band of latitudes in all seasons. Outside this belt, surface temperature decreases steadily toward both poles. The interiors of the northern continents become very cold during winter, but they are warmer than ocean areas at the same latitude during summer. Seasonal variations of surface temperature in the interiors of North America and Asia are very large (Fig. 1.6). Seasonal variation in the Southern Hemisphere is much smaller because of the greater fraction of the surface covered by ocean. The smaller seasonal variation of air temperature in mid-latitudes of the Southern Hemisphere is associated with the larger fraction of ocean-covered surface there. The ocean stores heat very effectively. During the summer season it stores the heat provided by the sun. Because a large amount of heat is required to raise the surface temperature of the oceans, the summer insolation raises the surface temperature by only a small amount. During winter, a large amount of heat is released to the atmosphere with a relatively small change in sea surface temperature. Land areas heat up and cool down much more quickly than oceans (see Chapter 4).

1.3 ATMOSPHERIC COMPOSITION

The composition of the atmosphere is a key determinant of Earth's climate. The interaction of atmospheric gases with radiant energy modulates the flow of energy through the climate system. The atmosphere has a mass of about 5.14×10^{18} kg, which is small compared to the mass of the ocean, 1.39×10^{21} kg, and the solid Earth, 5.98×10^{24} kg. Dry atmospheric air

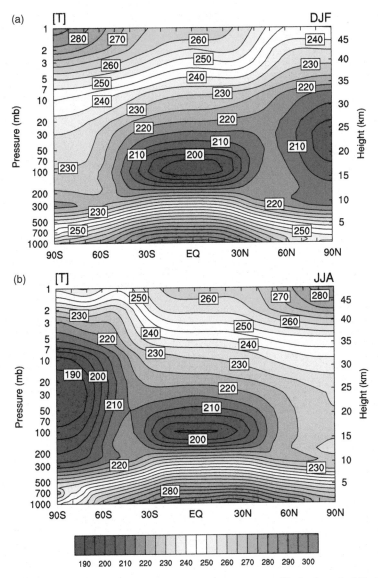

FIGURE 1.5 **Zonal average temperature (K) as a function of latitude and altitude for the (a) December, January, February (DJF) and (b) June, July, August (JJA) Seasons.** *Data from ERA 40 reanalysis.*

is composed mostly of molecular nitrogen (78%) and molecular oxygen (21%). The next most abundant gas in the atmosphere is argon (1%), an inert noble gas. The atmospheric gases that are important for the absorption and emission of radiant energy comprise less than 1% of the atmosphere's mass. These include water vapor (3.3×10^{-3} of the atmosphere's total mass), carbon dioxide (5.3×10^{-7}), and ozone (6.42×10^{-7}), in order of

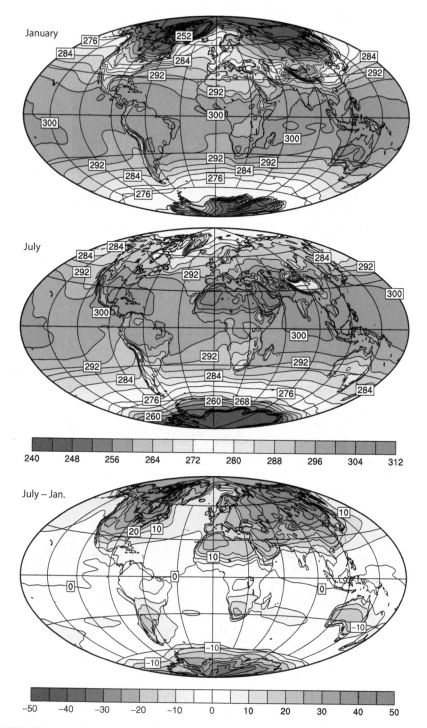

FIGURE 1.6　**Global map of the January and July surface temperature and July minus January.** *Data from ERA-Interim reanalysis.*

TABLE 1.1   Composition of the Atmosphere

Constituent	Chemical formula	Molecular weight ($^{12}C = 12$)	Fraction by volume in dry air	Total mass (g)
Total atmosphere		28.97		5.136×10^{21}
Dry air		28.96	100.0 %	5.119×10^{21}
Nitrogen	N_2	28.01	78.08 %	3.87×10^{21}
Oxygen	O_2	31.99	20.95 %	1.185×10^{21}
Argon	Ar	39.95	0.934 %	6.59×10^{19}
Water vapor	H_2O	18.02	Variable	1.7×10^{19}
Carbon dioxide	CO_2	44.0	391 ppmv*	$\sim 2.76 \times 10^{18}$
Neon	Ne	20.18	18.18 ppmv	6.48×10^{16}
Krypton	Kr	83.8	1.14 ppmv	1.69×10^{16}
Helium	He	4.00	5.24 ppmv	3.71×10^{15}
Methane	CH_4	16.04	1.8 ppmv*	$\sim 4.9 \times 10^{15}$
Xenon	Xe	131.3	87 ppbv	2.02×10^{15}
Ozone	O_3	47.99	Variable	$\sim 3.3 \times 10^{15}$
Nitrous oxide	N_2O	44.01	324 ppbv*	$\sim 2.3 \times 10^{15}$
Carbon monoxide	CO	28.0	120 ppbv	$\sim 5.9 \times 10^{15}$
Hydrogen	H_2	2.02	500 ppbv	$\sim 1.8 \times 10^{14}$
Ammonia	NH_3	17.0	100 ppbv	$\sim 3.0 \times 10^{13}$
Nitrogen dioxide	NO_2	46.0	1 ppbv	$\sim 8.1 \times 10^{12}$
Sulfur dioxide	SO_2	64.1	200 pptv	$\sim 2.3 \times 10^{12}$
Hydrogen sulfide	H_2S	34.1	200 pptv	$\sim 1.2 \times 10^{12}$
CFC-12	CCl_2F_2	120.9	528 pptv*	$\sim 1.0 \times 10^{13}$
CFC-11	CCl_3F	137.4	238 pptv*	$\sim 6.8 \times 10^{12}$

* Values of trace constituents valid in 2011 (ppmv = 10^{-6}, ppbv= 10^{-9}, pptv = 10^{-12}).

importance for surface temperature, followed by methane, nitrous oxide, and a host of other minor species (Table 1.1).

1.4 HYDROSTATIC BALANCE

The atmosphere is composed of gases held close to the surface of the planet by gravity. The vertical forces acting on the atmosphere at rest are gravity, which pulls the air molecules toward the center of the planet, and the pressure gradient force, which tries to push the atmosphere out into space. These forces are in balance to a very good approximation, and by equating the pressure gradient force and the gravity force one obtains the *hydrostatic balance*. Since force is mass times acceleration, we may express the vertical force balance per unit mass as an equation between the downward acceleration of gravity, g, and the upward acceleration that would be caused by the increase of pressure toward the ground, if gravity were not present to oppose it.

$$g = -\frac{1}{\rho}\frac{dp}{dz} \tag{1.2}$$

For an ideal gas, pressure (p), density (ρ), and temperature (T) are related by the formula

$$p = \rho R T \tag{1.3}$$

where R is the gas constant. After some rearrangement, (1.2) and (1.3) yield

$$\frac{dp}{p} = -\frac{dz}{H} \tag{1.4}$$

where

$$H = \frac{RT}{g} = \text{scale height.} \tag{1.5}$$

If the atmosphere is *isothermal*, with temperature ~260 K, then the temperature and scale height are constant and the hydrostatic equation may be integrated from the surface, where $p = p_s = 1.01325 \times 10^5$ Pa, to an arbitrary height, z, yielding an expression for the distribution of pressure with height.

$$p = p_s e^{-z/H} \tag{1.6}$$

The pressure thus decreases exponentially away from the surface, declining by a factor of $e^{-1} = (2.71828)^{-1} = 0.368$ every scale height. The scale height for the mean temperature of Earth's atmosphere is about 7.6 km. Figure 1.7 shows the distribution of atmospheric pressure with altitude. The pressure is largest at the surface and decreases rapidly with altitude in accord with the exponential decline given by (1.6). We can rearrange (1.2) to read

$$dm \equiv \rho dz = \frac{-dp}{g} \tag{1.7}$$

The mass between two altitudes, dm, is related to the pressure change between those two levels. Because of hydrostatic balance, the total mass of the atmosphere may be related to the global mean surface pressure.

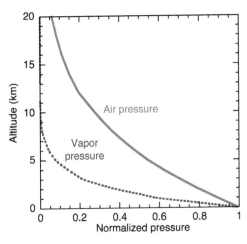

FIGURE 1.7 **Vertical distributions of air pressure and partial pressure of water vapor as functions of altitude for globally and annually averaged conditions.** Values have been normalized by dividing by the surface values of 1013.25 and 17.5 hPa, respectively.

$$\text{Atmospheric mass} = \frac{p_s}{g} = 1.03 \times 10^4 \text{ kg m}^{-2} \tag{1.8}$$

The vertical column above every square meter of Earth's surface contains about 10,000 kg of air.

Because the surface climate is of primary interest, and because the mass of the atmosphere is confined to within a few scale heights of the surface, or several tens of kilometers, it is the lower atmosphere that is of most importance for climate. For this reason, most of this book will be devoted to processes taking place in the troposphere, at the surface, or in the ocean. The stratosphere has some important effects on climate, however, and these will be described where appropriate.

1.5 ATMOSPHERIC HUMIDITY

Atmospheric humidity is the amount of water vapor carried in the air. It can be measured as vapor pressure, mixing ratio or specific humidity. Specific humidity is the ratio of vapor mass to total air mass, whereas mixing ratio is the ratio of the mass of vapor to the mass of dry air. The atmosphere must carry away the water evaporated from the surface and supply water to areas of rainfall. Water that flows from the land to the oceans in rivers was brought to the land areas by transport in the atmosphere as vapor. Atmospheric water vapor is also the most important greenhouse gas in the atmosphere. Water vapor condenses to form clouds,

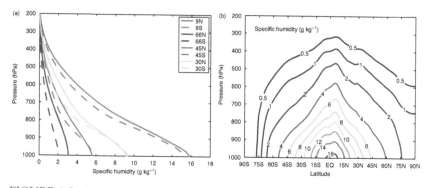

FIGURE 1.8   **Profiles of specific humidity as a function of pressure, for annual-mean conditions as (a) Line plots and (b) Contour plot.** *Data from ERA-Interim.*

which can release rainfall and are also extremely important in both reflecting solar radiation and reducing the infrared radiation emitted by Earth.

The partial pressure of water vapor in the atmosphere decreases very rapidly with altitude (Fig. 1.7). The partial pressure of water vapor decreases to half of its surface value by 2 km above the surface and to less than 10% of its surface value at 5 km. Atmospheric water vapor also decreases rapidly with latitude (Fig. 1.8). The amount of water vapor in the atmosphere at the equator is nearly 10 times that at the poles.

The rapid upward and poleward decline in water vapor abundance in the atmosphere is associated with the strong temperature dependence of the saturation vapor pressure. The vapor pressure in equilibrium with a wet surface increases very rapidly with temperature. The temperature dependence of saturation pressure of water vapor over a water surface is governed by the *Clausius–Clapeyron relationship.*

$$\frac{de_s}{dT} = \frac{L}{T(\alpha_v - \alpha_1)} \tag{1.9}$$

In (1.9), e_s is the saturation vapor pressure above a flat liquid surface, L is the latent heat of vaporization, T is the temperature in K, and α represents the specific volume of the vapor α_v and liquid α_1 forms of water. The Clausius–Clapeyron relation can be manipulated to express the fractional change of saturation vapor pressure $\Delta e_s / e_s$, and thereby the specific humidity at saturation, q^*, to the fractional change of temperature. The specific humidity is related to the water vapor pressure approximately as $q \approx 0.622 \dfrac{e}{p}$.

$$\frac{\Delta q^*}{q^*} = \frac{\Delta e_s}{e_s} \approx \left(\frac{L}{R_v T}\right)\frac{\Delta T}{T} = r\frac{\Delta T}{T} \tag{1.10}$$

where R_v is the gas constant for water vapor.

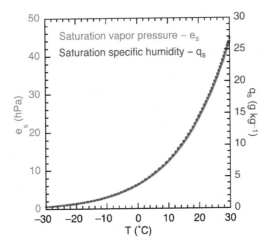

FIGURE 1.9 Saturation vapor pressure and specific humidity as functions of temperature at standard pressure.

For terrestrial conditions, $T \sim 260$ K, and the factor r is approximately 20. This means that a 1% change in temperature of about 3 K will result in a 20% change in the saturation vapor pressure, or about 7% for 1 K. If the relative humidity (the ratio of the actual specific humidity to the saturation specific humidity) remains fixed, then the actual water vapor in the atmosphere will increase by 7% for every 1 K temperature increase. This rapid exponential increase of saturation pressure with temperature can be seen more explicitly if we consider the approximate solution to (1.9) valid near standard pressure and temperature of 1013.25 hectoPascals (hPa) and 273 K.

$$e_s \cong 6.11 \cdot \exp\left\{\frac{L}{R_v}\left(\frac{1}{273} - \frac{1}{T}\right)\right\} \qquad (1.11)$$

The exponential dependence of the saturation vapor pressure on temperature expressed by (1.11) is shown in Fig. 1.9.

1.6 ATMOSPHERIC THERMODYNAMICS, VERTICAL STABILITY AND LAPSE RATE

Conservation of energy is a central constraint on climate, and when combined with the hydrostatic relationship, it determines much about the vertical structure of the atmosphere, including the lapse rate.

1.6.1 First Law of Thermodynamics

The first law of thermodynamics states that energy is conserved, so that for a unit mass of gas, the applied heat δQ is equal to the sum of the change in internal energy dU and the work done dW. If we assume that the external work done by air is only that associated with volume changes, $dW = p\,d\alpha$, and we use the definition of the specific heat at constant volume $c_v = (dU / dT)_v$, we obtain a useful form of the first law of thermodynamics.

$$\delta Q = c_v\,dT + p\,d\alpha \tag{1.12}$$

Here p is pressure, dT is the change in temperature, and $d\alpha$ is the change in specific volume, α. Using the ideal gas law,

$$p\alpha = RT \tag{1.13}$$

we may write that,

$$p\,d\alpha = R\,dT - \alpha\,dp \tag{1.14}$$

Using (1.14), and

$$R = c_p - c_v \tag{1.15}$$

(1.12) becomes,

$$\delta Q = c_p\,dT - \alpha\,dp \tag{1.16}$$

or using (1.13) again,

$$\delta Q = c_p\,dT - \frac{RT}{p}\,dp \tag{1.17}$$

1.6.2 Potential Temperature

An adiabatic process is one for which no heat is added or taken away so, $\delta Q = 0$, and (1.17) can be rearranged to read,

$$\frac{dT}{T} - \frac{R}{c_p}\frac{dp}{p} = d\ln T - \frac{R}{c_p}d\ln p = d\ln\left(T p^{-R/c_p}\right) = 0 \tag{1.18}$$

so that for a parcel of gas that undergoes an adiabatic process,

$$T p^{-R/c_p} = \text{constant} \tag{1.19}$$

If we define Θ to be the temperature at some reference pressure p_o, which is usually taken to be 1000 hPa, then the potential temperature Θ is the temperature a parcel of air would have if it were brought adiabatically to the reference pressure.

$$\Theta = T \left(\frac{p_o}{p} \right)^{R/c_p} \tag{1.20}$$

1.6.3 Static Stability and the Adiabatic Lapse Rate

The potential temperature is useful because it remains constant as a parcel undergoes an adiabatic change of pressure. The vertical gradient of potential temperature determines the dry static stability of the atmosphere. If the potential temperature increases with height, then parcels raised adiabatically from their initial height will always be colder and thus denser than their environment and will sink back to their original pressure. If the potential temperature decreases with height, then parcels raised up will be warmer than their environment and will be accelerated upward by buoyancy; therefore, if the potential temperature decreases with height, the temperature profile is unstable.

$$\frac{d\Theta}{dz} > 0 \quad \text{Stable}$$
$$\frac{d\Theta}{dz} < 0 \quad \text{Unstable} \tag{1.21}$$

The rate at which temperature changes as a parcel moves up or down in the atmosphere without heating can be derived by using the hydrostatic equation,

$$\alpha \, dp = -g \, dz \tag{1.22}$$

in (1.16), and setting $\delta Q = 0$, to give

$$c_p dT + g dz = 0 \tag{1.23}$$

or,

$$-\left(\frac{\partial T}{\partial z} \right)_{\text{adiabatic}} = \frac{g}{c_p} = \Gamma_d = 9.8 \, \text{K km}^{-1} \tag{1.24}$$

1.6.4 Moist Processes and Equivalent Potential Temperature

When moisture is present in air and an air parcel is raised adiabatically, the parcel can become supersaturated such that the water vapor condenses and latent heat is released. One can incorporate the latent heat release as heating in the first law of thermodynamics by writing the heat release in terms of the change in saturation water vapor mixing ratio, dq^*. The saturated adiabatic lapse rate can then be derived (Wallace and Hobbs; 2006).

$$\Gamma_s = \frac{\Gamma_d}{1 + \dfrac{L}{c_p}\dfrac{dq^*}{dT}} \tag{1.25}$$

The saturated adiabatic lapse rate is generally less than the dry adiabatic lapse rate, and becomes smaller as the temperature rises. As a saturated parcel rises, water condenses, latent heat is released and the parcel cools more slowly with increasing altitude than an unsaturated or dry parcel.

Another useful quantity is the equivalent potential temperature, which is the temperature that would be obtained by a moist air parcel if it were first raised moist-adiabatically until all of its water condensed out, and then brought adiabatically back to a reference surface pressure.

$$\Theta_e = \Theta \exp\left(\frac{Lq^*}{c_p T}\right) \tag{1.26}$$

Equivalent potential temperature incorporates the moist static energy of air parcels in a hydrostatic atmosphere. The moist static energy includes the sensible, potential, and latent energy per unit mass (1.27).

$$\text{Moist static energy} = c_p T + gz + Lq \tag{1.27}$$

Potential temperature includes the dry static energy, from which the latent energy in (1.27) is excluded. If the equivalent potential temperature decreases with height, then the air parcel is only conditionally unstable. It is unstable only if it becomes saturated. If the equivalent potential temperature increases with height, then the parcel is absolutely stable.

The important difference between dry adiabatic ascent and moist ascent can be illustrated by plotting the dry and moist adiabats for a few representative cases. The adiabats are the temperature profiles that would be experienced by parcels as they are raised upward adiabatically from the surface. Examples of some dry and saturated adiabats are plotted in Fig. 1.10. Because of the release of latent heat, the temperatures of moist adiabats decrease less rapidly with height in the lower troposphere, but become parallel to the dry adiabats at low temperatures where latent heating is nearly zero, because the saturation vapor pressure is very small. This difference is particularly evident at high temperatures, where the saturated parcel starts out at the surface with much more latent energy and therefore its temperature drops less rapidly with altitude. Because the curvature of the saturated adiabats increases with temperature, when the temperature of the saturated parcel at the surface is increased, its temperature when it arrives at any layer higher in the troposphere is increased by a larger amount than the surface temperature increase. For example, parcels started from the surface at 20 and 30°C have temperatures of −45.6 and −15.5°C when they reach 10 km altitude. The difference of 30°C at 10 km

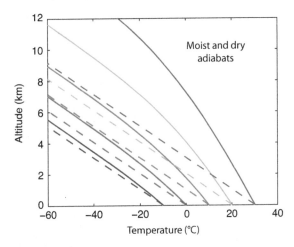

FIGURE 1.10 **Moist and dry adiabatic temperature profiles.** Parcels of air start at 1000 hPa either saturated with water vapor (solid) or completely dry (dashed), and are raised adiabatically while conserving moist static energy. Starting temperatures of −10, 0, 10, 20 and 30°C are shown.

is three times their difference at the surface. This basic mechanism would indicate that significant changes in lapse rate and dry static stability should be expected when the climate changes, especially in the tropics where moist convection strongly controls the temperature profile.

1.7 THE WORLD OCEAN

The atmosphere contains a tiny fraction of the total water in the climate system, about one part in 10^5. Most of the surface water of Earth is contained in the oceans and in ice sheets (Table 1.2). Earth contains

TABLE 1.2 Water on Earth

Water reservoir	Depth if spread over the entire surface of Earth (m)	Total (%)
Oceans	2650	97
Icecaps and glaciers	60	2.2
Groundwater	20	0.7
Lakes and streams	0.35	0.013
Soil moisture	0.12	0.013
Atmosphere	0.025	0.0009
Total	2730	100

about 1.35×10^9 km³ of water, of which about 97% is seawater. Since all the oceans are connected to some degree, we can think of them collectively as the world ocean. The world ocean is a key element of the physical climate system. Ocean covers about 71% of Earth's surface to an average depth of 3730 m. The ocean has tremendous capability to store and release heat and chemicals on time scales of seasons to centuries. Ocean currents move heat poleward to cool the tropics and warm the extratropics. The world ocean is the reservoir of water that supplies atmospheric water vapor for rain and snowfall over land. The ocean plays a key role in determining the composition of the atmosphere through the exchange of gases and particles across the air–sea interface. The ocean removes carbon dioxide from the atmosphere and produces molecular oxygen, and participates in other key geochemical cycles that regulate the surface environment of Earth.

Temperature in the ocean generally decreases with depth from a temperature very near that of the surface air temperature to a value near the freezing point of water in the deep ocean (Fig. 1.11). A thin, well-mixed surface layer is stirred by winds and waves so efficiently that its temperature and salinity are almost independent of depth. Most of the temperature change occurs in the *thermocline*, a region of rapid temperature change with depth in the first kilometer or so of the ocean. Below the thermocline is a deep layer of almost uniform temperature. In middle and high latitudes, the mixed layer is thin in summer and deep in winter (e.g., 45°N in Fig. 1.11).

Salinity of seawater is defined as the number of grams of dissolved salts in a kilogram of seawater. Salinity in the open ocean ranges from about $33 \, g \, kg^{-1}$ to $38 \, g \, kg^{-1}$. In seawater with a salinity of 35 g kg⁻¹, about 30 g kg⁻¹ are composed of sodium and chloride (Table 1.3). Salinity is an important

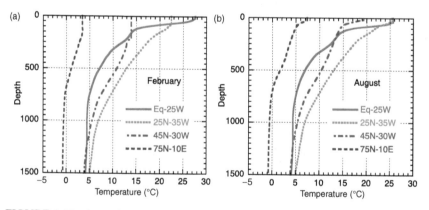

FIGURE 1.11   **Annual-mean ocean potential temperature profiles for various latitudes and as a function of depth in meters for (a) February and (b) August.** *MIMOC data.*

TABLE 1.3   Concentrations of the Major Components of Sea Water with a Salinity of 35‰

Component	Grams per kilogram
Chloride	19.4
Sodium	10.8
Sulfur	0.91
Magnesium	1.29
Calcium	0.41
Potassium	0.39
Bicarbonate	0.14
Bromide	0.067
Strontium	0.008
Boron	0.004
Fluoride	0.001

contributor to variations in the density of seawater at all latitudes and is the most important factor in high latitudes and in the deep ocean, where the temperature is close to the freezing point of water. Variations in the density of seawater drive the deep-ocean circulation, which is critical for heat storage and transport and for the recirculation of nutrients necessary for life. Salinity of the global ocean varies systematically with latitude in the upper layers of the ocean (Fig. 1.12). In subtropical latitudes (10–30°), the surface salinity is large because evaporation exceeds precipitation and leaves the seawater enriched in salt. In middle and high

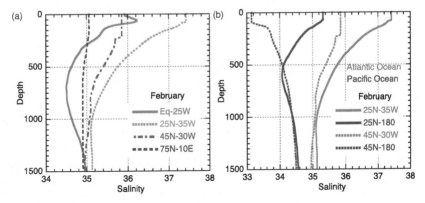

FIGURE 1.12   **Profiles of salinity during February for various locations.** *MIMOC data.*

latitudes, precipitation of freshwater exceeds evaporation, and so surface salinities are quite low. Near the equator, a thin layer of fresh water from precipitation sits atop more saline water below. In the deep ocean, salinity variations are much smaller than near the surface, because the sources and sinks of freshwater are at the surface and the deep water comes from a few areas in high latitudes. The Atlantic is much saltier than the Pacific at nearly all latitudes and for this reason the formation of cold, salty water that can sink to the bottom of the ocean is much more prevalent in the Atlantic than the Pacific (see Chapter 7).

1.8 THE CRYOSPHERE

All of the ice near the surface of Earth is called the *cryosphere*. About 2% of the water on Earth is frozen, and this frozen water constitutes about 80% of the freshwater. Most of the mass of ice is contained in the great ice sheets of Antarctica (89%) and Greenland (8.6%) (Table 1.4). For climate, it is often not the mass of ice that is of primary importance, but rather the surface area that is covered by ice of any depth. This is because surface ice of any depth generally is a much more effective reflector of solar radiation than the underlying surface. Also, sea ice is a good insulator and allows air temperature to be very different from that of the seawater under just a few meters of sea ice. Currently, year-round (perennial) ice covers about 11% of the land area and 7% of the world ocean. During some seasons, the amount of land covered by seasonal snow cover exceeds the surface area covered by perennial ice cover. The surface areas covered by ice sheets, seasonal snow, and sea ice are comparable. Ice sheets cover about 16×10^6 km^2, seasonal snow about 50×10^6 km^2, and sea ice up to 23×10^6 km^2.

1.9 THE LAND SURFACE

Although the land surface covers only 29% of Earth, the climate over the land surface is extremely important to us because humans are land-dwelling creatures. Cereal grains are the world's most important food source, and supply about half of the world's calories and much of the protein. About 80% of the animal protein consumed by humans comes from meat, eggs, and dairy products, and only 20% from seafood.

Over the land surface, temperature and soil moisture are key determinants of natural vegetation and the agricultural potential of a given area. Vegetation, snow cover, and soil conditions also affect the local and

TABLE 1.4 Estimated Global Inventory of Land and Sea Ice

			Area (km^2)	Volume (km^3)	Total ice mass (%)
Land ice	Antarctic ice sheet		13.9×10^6	30.1×10^6	89.3
	Greenland ice sheet		1.7×10^6	2.6×10^6	8.6
	Mountain glaciers		0.5×10^6	0.3×10^6	0.76
	Permafrost	Continuous	8×10^6	(Ice content) 0.2–0.5×10^6	0.95
		Discontinuous	17×10^6		
	Seasonal snow (avg. max)	Eurasia	30×10^6	2–3×10^3	
		America	17×10^6		
Sea ice	Southern Ocean	Max	18×10^6	2×10^4	
		Min	3×10^6	6×10^3	
	Arctic Ocean	Max	15×10^6	4×10^4	
		Min	8×10^6	2×10^4	

The volume of water in the ground that annually freezes and thaws at the surface of permafrost (active layer), and in regions without permafrost but with subfreezing winter temperatures is not included in this table.

After Untersteiner (1984); printed with permission from Cambridge University Press.

global climate, so that local climate and land surface conditions participate in a two-way relationship. Land topography plays an important role
in modifying regional climates, and weathering of rocks on land is a key
component of the carbon cycle that controls the carbon dioxide content of
the atmosphere on millennial time scales.

The arrangement of land and ocean areas on Earth plays a role in determining global climate. The arrangement of land and ocean varies on time
scales of millions of years as the continents drift about. At the present time,
about 68% of Earth's land area is in the Northern Hemisphere (Fig. 1.13).
The fact that the Northern Hemisphere has most of the land area causes
significant differences in the climates of the Northern and Southern Hemispheres, and plays an important role in climate change. The Northern
Hemisphere has much more dramatic east–west variations in continental
elevation, especially in middle latitudes where the Himalaya and Rocky
Mountains are prominent features (Fig. 1.14). The topography of the land
surface and the arrangement and orientation of mountain ranges are key
determinants of climate.

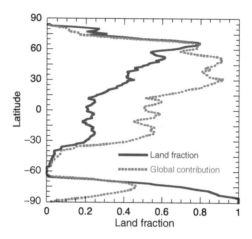

FIGURE 1.13   Fraction of surface area covered by land as a function of latitude (solid line) and contribution of each latitude belt to the global land surface area (dashed line).

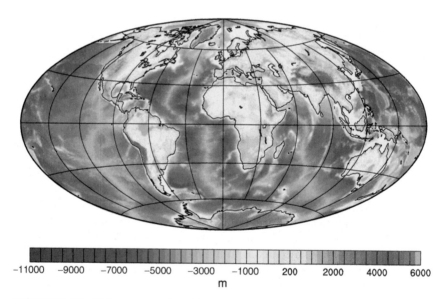

FIGURE 1.14   Color contour plot of the topography of Earth relative to sea level. Scale is in meters.

EXERCISES

1. Give several reasons why the amplitude of the annual variation of surface temperature is greatest in Siberia (Fig. 1.6).
2. If you are standing atop Mount Everest at 8848 m, about what fraction of mass of the atmosphere is below you? (Use eq. (1.6).)
3. An airplane is flying at 10,000 m above the surface. What is the pressure outside the airplane in hectoPascals? What is the temperature in degree Celsius? Use global averages.
4. If the atmosphere warmed up by 5°C, would the atmospheric pressure at 5 km above sea level increase or decrease, and by approximately how much? (Use eq. (1.6).)
5. Compute the difference of saturation vapor pressure between 0°C and 30°C. Compare the results you get with eqs (1.10) and (1.11).
6. Explain why the North Polar temperature inversion is present in winter but not in summer?
7. Why do you think the salinity at 45°N–180°E is so much less than the salinity at 25°N–180°E (Fig. 1.11)?

The Global Energy Balance

2.1 WARMTH AND ENERGY

Temperature, a key climate variable, is a measure of the energy contained in the movement of molecules. Therefore, to understand how the temperature is maintained, one must consider the *energy balance* that is formally stated in the first law of thermodynamics. The basic global energy balance of Earth is between energy coming from the Sun and energy returned to space by Earth's radiative emission. The generation of energy in the interior of Earth has a negligible influence on its energy budget. The absorption of solar radiation takes place mostly at the surface of Earth, whereas most of the emission to space originates in its atmosphere. Because the atmosphere is mostly transparent to solar radiation and mostly opaque to terrestrial emission of radiation, the surface of Earth is much warmer than it would be in the absence of its atmosphere. When averaged over a year, more solar energy is absorbed near the equator than near the poles. The atmosphere and the ocean transport energy poleward to reduce the effect of this heating gradient on Earth's surface temperature. Much of the character of Earth's evolution and climate has been determined by its position within the solar system.

2.2 THE SOLAR SYSTEM

The source for the energy to sustain life on Earth comes from the Sun. Earth orbits the Sun once a year while keeping a relatively constant distance from it, so that our Sun provides a stable, comfortable source of heat and light. Our Sun is one of about 10^{11} stars in our galaxy, the Milky Way (Table 2.1). It is a single star, whereas about two-thirds of the stars we can see are in multiple star systems.

The luminosity is the total rate at which energy is released by the Sun. We know of stars that are 10^{-4} times less luminous and 10^5 times brighter than the Sun. Their emission temperatures range from 2000 K to 30,000 K,

Global Physical Climatology. http://dx.doi.org/10.1016/B978-0-12-328531-7.00002-5

TABLE 2.1   Characteristics of the Sun

Mass	1.99×10^{30} kg
Radius	6.96×10^{8} m
Luminosity	3.9×10^{26} W
Mean distance from Earth	1.496×10^{11} m

whereas the temperature of the photosphere of the Sun is about 6000 K. The photosphere is the region of the Sun from which most of its energy emission is released to space. Stellar radii range from 0.1 to 200 solar radii. Energy is produced in the core of the Sun by nuclear fusion, whereby lighter elements are made into heavier ones, releasing energy in the process. For a smallish star such as the Sun, the projected lifetime on the main sequence is about 11 billion years, of which about half has passed. The Sun is thus a solitary, middle-aged, medium-bright star. Theories of stellar evolution predict that the luminosity of the Sun has increased by about 30% during the lifetime of Earth, about 4.5 billion years. The cause of this increased energy output is the gradual increase in the density of the Sun, as lighter elements convert to heavier ones.

The solar system includes eight planets. They may be divided into the *terrestrial*, or inner, planets and the *Jovian*, or outer, planets (Table 2.2). The terrestrial planets include Mercury, Venus, Mars, and Earth. The Jovian planets include Jupiter, Saturn, Uranus, and Neptune.

2.2.1 Planetary Motion

The planets orbit about the Sun in ellipses, which have three characteristics: the mean planet–Sun distance, the eccentricity, and the orientation of the orbital plane. The mean distance from the Sun controls the amount of solar *irradiance* (energy delivered per unit time per unit area) arriving

TABLE 2.2   Characteristics of Inner and Outer Planets

Characteristic	Outer (Jovian)	Inner (terrestrial)
Density	Small	Large
Mass	Large	Small
Sun distance	Large	Small
Atmosphere	Extensive	Thin or none
Satellites	Many	Few or none
Composition	H, He, CH_4, NH_3	Mostly silicates, rocks

at the planet. The mean distance from the Sun also controls the length of the planetary year, the time it takes the planet to complete one orbit. The planetary year increases with increasing distance from the Sun.

The *eccentricity* of an orbit is a measure of how much the orbit deviates from being perfectly circular. It controls the amount of variation of the solar irradiance at the planet as it moves through its orbit during the planetary year. If the eccentricity is not zero, so that the orbit is not circular, then the distance of the planet from the Sun varies through the year. The orientation of the orbital plane does not have too much direct bearing on the climate. Because the solar system formed from a spinning cloud of gas and rock, most planets are in more or less the same orbital plane, which is perpendicular to the axis of the Sun's rotation.

In addition to the orbit parameters, the parameters of the planets' rotation and their relationship to the orbit are very important. The rotation rate controls the daily variation of the insolation at a point (diurnal cycle) and is also an important control on the response of the atmosphere and ocean to solar heating, and thereby on the patterns of winds and currents that develop.

The *obliquity* or tilt is the angle between the axis of rotation and the normal to the plane of the orbit. It influences the seasonal variation of insolation, particularly in high latitudes. It also strongly affects the annual mean insolation that reaches the polar regions. Currently the obliquity of Earth's axis of rotation is about $23.45°$.

The *longitude of perihelion* measures the phase of the seasons relative to the planet's position in the orbit. For example, at present Earth passes closest to the Sun (*perihelion*) during Southern Hemisphere summer, on about January 5. As a result the Southern Hemisphere receives more top-of-atmosphere insolation during summer than the Northern Hemisphere, although the annual mean insolation is the same in both hemispheres.

The effect of these orbital parameters on climate will be discussed in more detail in Chapter 12, where the orbital parameter theory of climate change is described. For the time being, we consider only the distance from the Sun and the obliquity or declination angle.

2.3 ENERGY BALANCE OF EARTH

2.3.1 First Law of Thermodynamics

The first law of thermodynamics states that energy is conserved. The first law for a closed system may be stated as: "The heat added to a system is equal to the change in internal energy plus the work extracted." This law may be expressed symbolically as

$$\delta Q = dU + dW \tag{2.1}$$

Here δQ is the amount of heat added, dU is the change in the internal energy of the system, and dW is the work done by the system. Heat can be transported to and from a system in the following three ways:

1. *Radiation*: No mass is exchanged, and no medium is required. Pure radiant energy moves at the speed of light.
2. *Conduction*: No mass is exchanged, but a medium transfers heat by collisions between atoms or molecules.
3. *Convection*: Mass is exchanged. A net movement of mass may occur, but more commonly parcels with different energy amounts change places, so that energy is exchanged without a net movement of mass (Fig. 2.1).

The transmission of energy from the Sun to Earth is almost entirely radiative. Some mass flux is associated with the particles of the solar wind, but the amount of energy is small compared to the energy arriving as photons. Moreover, the work done by Earth on its environment can also be neglected. To calculate the approximate energy balance of Earth, we need only consider radiative energy exchanges. The amount of matter

FIGURE 2.1 **Cumulonimbus cloud over Africa photographed from the international space station expedition 16, NASA, February 5, 2008.** Note the three-dimensional structure of the clouds, the upper-level spreading anvil, its shadow, and the various shapes and sizes of the smaller clouds. Air movements associated with clouds move heat and moisture vertically by convection and also influence radiative transfer in the atmosphere. The distance across the photo is approximately 70 km and the top of the cloud is approximately 15 km high.

in space that could affect the flow of energy between the Sun and Earth is small, and we may thus consider the space between the photosphere of the Sun and the top of Earth's atmosphere to be a vacuum. In a vacuum, only radiation can transport energy.

2.3.2 Energy Flux, Irradiance, and Solar Constant

The Sun puts out a nearly constant flux of energy that we call the solar *luminosity*, $L_0 = 3.9 \times 10^{26}$ W. We can calculate the average irradiance at the photosphere by dividing this energy flux by the area of the photosphere.

$$\text{Irradiance}_{\text{photo}} = \frac{\text{Flux}}{\text{Area}_{\text{photo}}} = \frac{L_0}{4\pi r_{\text{photo}}^2}$$
$$= \frac{3.9 \times 10^{26} \text{ W}}{4\pi [6.96 \times 10^8 \text{ m}]^2} = 6.4 \times 10^7 \text{ Wm}^{-2} \tag{2.2}$$

Since space is effectively a vacuum and energy is conserved, the amount of energy passing outward through any sphere with the Sun at its center should be equal to the luminosity, or total energy flux from the Sun. If we assume that the irradiance is uniform over the sphere, and write the irradiance at any distance d from the Sun as S_d, then conservation of energy requires

$$\text{Flux} = L_0 = S_d 4\pi d^2 \tag{2.3}$$

From this we deduce that the *total solar irradiance* (TSI) S_d is inversely proportional to the square of the distance to the Sun. We define the TSI as the irradiance of the solar emission at the mean distance of the Earth from the Sun.

$$\text{TSI} = \text{irradiance at distance } \bar{d} = S_d = \frac{L_0}{4\pi \bar{d}^2} \tag{2.4}$$

The measurement of the TSI has a long and interesting history, including trying to estimate it from the reflection from the Moon. The most reliable estimates come from special instruments placed on Earth-orbiting satellites, so that a direct measurement of the irradiance can be made without an intervening atmosphere. These instruments have achieved great precision, so that the variability of the TSI on periods of hours to decades has been measured with great reproducibility by different instruments, but the mean absolute value has varied from instrument to instrument. The best and most recent estimate of the TSI is 1360.8 ± 0.5 Wm^{-2} (Kopp and Lean, 2011). For the purposes of doing exercises in this book, we will assume that the TSI is 1360 Wm^{-2} at the mean distance of Earth from the Sun (1.5×10^{11} m). We will give this the symbol S_0.

2.3.3 Cavity Radiation

The radiation field within a closed cavity in thermodynamic equilibrium has a value that is uniquely related to the temperature of the cavity walls, regardless of the material of which the cavity is made. This cavity radiant intensity, which is uniquely related to the wall temperature, is also called the *blackbody radiation*, since it corresponds to the emission from a surface with unit emissivity and absorptivity. Perfect blackbodies may not be easily found, but the radiation inside a cavity in equilibrium will always equal the blackbody radiation. The dependence of the irradiance of blackbody emission on temperature follows the Stefan–Boltzmann law.

$$E_{BB} = \sigma T^4; \quad \sigma = 5.67 \times 10^{-8} \, \text{Wm}^{-2}\text{K}^{-4} \tag{2.5}$$

2.3.4 Example: Emission Temperature of the Sun

We calculated earlier that the solar irradiance at the photosphere is about $6.4 \times 10^7 \, \text{Wm}^{-2}$.

We can equate this to the Stefan–Boltzmann formula (2.5) and derive an effective emission temperature for the photosphere.

$$\sigma T_{photo}^4 = 6.4 \times 10^7 \, \text{Wm}^2$$

$$T_{photo} = \sqrt[4]{\frac{6.4 \times 10^7 \, \text{Wm}^2}{\sigma}} = 5796 \, \text{K} \sim 6000 \, \text{K}$$

2.3.5 Emissivity

In equilibrium, the irradiance inside a cavity at temperature T is $E_{BB} = \sigma T^4$. We may define the emissivity, ε, as the ratio of the actual emission of a body or volume of gas to the blackbody emission at the same temperature.

$$E_R = \varepsilon \sigma T^4 \rightarrow \varepsilon = \frac{E_R}{\sigma T^4} \tag{2.6}$$

2.4 EMISSION TEMPERATURE OF A PLANET

The emission temperature of a planet is the blackbody temperature at which it needs to emit in order to achieve energy balance. The basic idea is to equate the solar energy absorbed by a planet with the energy emitted by a blackbody. This defines the emission temperature of the planet.

Solar radiation absorbed = planetary radiation emitted

To calculate the solar radiation absorbed, we begin with the TSI, which measures the energy flux of solar radiation in Watts per square meter

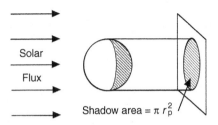

FIGURE 2.2   **The shadow area of a spherical planet.**

(Wm^{-2}) arriving at the mean distance of the planet from the Sun. The energy flux is defined relative to a flat surface perpendicular to the direction of radiation. Solar radiation is essentially a parallel and uniform beam for a planetary body in the solar system, because all the planets have diameters that are small compared to their distance from the Sun. The amount of energy incident on a planet is equal to the TSI times the area that the planet sweeps out of the beam of parallel energy flux. We call this the *shadow area* (Fig. 2.2). Since the atmosphere of Earth is very thin, we can ignore its effect on the shadow area and use the radius of the solid planet, r_p, to calculate the shadow area.

We must also take account of the fact that not all the solar energy incident on a planet is absorbed. Some fraction is reflected back to space without being absorbed and so does not enter into the planetary energy balance. We call this planetary reflectivity the planetary albedo[1] and give it the symbol α_p. If S_0 is the TSI at the average position of the planet, we have

$$\text{Absorbed solar radiation} = S_0(1 - \alpha_p)\pi r_p^2 \qquad (2.7)$$

Since the *planetary albedo* for Earth is about 29%, only 71% of the TSI is absorbed by the climate system. This amount of energy must be returned to space by terrestrial emission. We assume that the terrestrial emission is like that of a blackbody. The area from which the emission occurs is the surface area of a sphere, rather than the area of a circle. The terrestrial emission flux is thus written as

$$\text{Emitted terrestrial radiation} = \sigma T_e^4 4\pi r_p^2 \qquad (2.8)$$

If we equate the absorbed solar flux with the emitted terrestrial flux, we obtain the planetary energy balance, which will define the emission temperature.

$$\frac{S_0}{4}(1 - \alpha_p) = \sigma T_e^4 \qquad (2.9)$$

or

[1]Albedo comes from a Latin word for 'whiteness.'

$$T_e = \sqrt[4]{\frac{(S_0/4)(1-\alpha_p)}{\sigma}} \qquad (2.10)$$

The factor of 4 dividing the TSI in (2.9) is the ratio of the global surface area of a sphere to its shadow area, which is the area of a circle with the same radius. The emission temperature may not be the actual surface or atmospheric temperature of the planet; it is merely the blackbody emission temperature a planet requires to balance the solar energy it absorbs. For a planet such as Earth, with a strong greenhouse effect, most of the emission comes from the atmosphere, not the surface.

2.4.1 Example: Emission Temperature of Earth

Earth has an albedo of about 0.29. The emission temperature of Earth from (2.10) is therefore

$$T_e = \sqrt[4]{\frac{(1360 \text{ Wm}^{-2}/4)(1-0.29)}{5.67\times10^{-8} \text{ Wm}^{-2}\text{K}^{-4}}} \cong 255 \text{ K} \cong -18°C \cong 0°F$$

The emission temperature of 255 K is much less than the observed global mean surface temperature of 288 K $\cong$ +15°C. To understand the difference, we need to consider the greenhouse effect.

2.5 GREENHOUSE EFFECT

One may illustrate the greenhouse effect with a very simple elaboration of the energy balance model used to define the emission temperature. An atmosphere that is assumed to be a blackbody for terrestrial radiation, but is transparent to solar radiation, is incorporated into the global energy balance (Fig. 2.3). Since solar radiation is mostly visible and near infrared, and Earth emits primarily thermal infrared radiation, the atmosphere may affect solar and terrestrial radiation very differently. The energy balance at the top of the atmosphere in this model is the same as in the basic energy balance model that defined the emission temperature (2.9). Since the

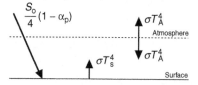

FIGURE 2.3 **Energy fluxes for a planet with an atmosphere that is transparent for solar radiation but opaque to terrestrial radiation.**

atmospheric layer absorbs all of the energy emitted by the surface below it and emits like a blackbody, the only radiation emitted to space is from the atmosphere in this model. The energy balance at the top of the atmosphere is thus

$$\frac{S_0}{4}(1-\alpha_p) = \sigma T_A^4 = \sigma T_e^4 \tag{2.11}$$

Therefore, we see that the temperature of the atmosphere in equilibrium must be the emission temperature in order to achieve energy balance. However, the surface temperature is much warmer as we can see by deriving the energy balance for the atmosphere and the surface. The atmospheric energy balance gives,

$$\sigma T_s^4 = 2\sigma T_A^4 \quad \Rightarrow \quad \sigma T_s^4 = 2\sigma T_e^4 \tag{2.12}$$

and the surface energy balance is consistent:

$$\frac{S_0}{4}(1-\alpha_p) + \sigma T_A^4 = \sigma T_s^4 \quad \Rightarrow \quad \sigma T_s^4 = 2\sigma T_e^4 \tag{2.13}$$

From (2.13) we obtain $T_s = \sqrt[4]{2}T_e = 303\,\text{K}$, which is warmer than the observed global mean surface temperature. This is because we have not taken into account the vertical transport of energy by atmospheric motions.

We can see from the diagram in Fig. 2.3 and the surface energy balance (2.13), that the surface temperature is increased because the atmosphere does not inhibit the flow of solar energy to the surface, but augments the solar heating of the surface with its own downward emission of longwave radiation. In this model, the surface receives the same amount of energy as downward emission of terrestrial radiation from the atmosphere as it receives from the Sun. The atmospheric greenhouse effect warms the surface because the atmosphere is relatively transparent to solar radiation and yet absorbs and emits terrestrial radiation very effectively.

2.6 GLOBAL RADIATIVE FLUX ENERGY BALANCE

The vertical flux of energy in the atmosphere is one of the most important climate processes. The radiative and nonradiative fluxes between the surface, the atmosphere, and space are key determinants of climate. The ease with which solar radiation penetrates the atmosphere and the difficulty with which terrestrial radiation is transmitted through the atmosphere determine the strength of the greenhouse effect. The decrease of temperature with altitude (lapse rate) is also a key part of the greenhouse effect.

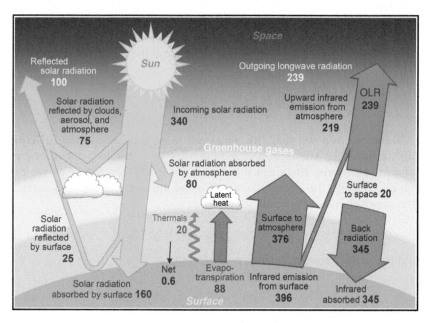

FIGURE 2.4 **Global and annual average radiative and nonradiative energy-flow diagram for Earth and its atmosphere.** Units are Wm^{-2}.

The fundamental energy exchanges in the climate system at the top of the atmosphere, within the atmosphere and at the surface are shown in Fig. 2.4. At the top of the atmosphere, absorbed solar radiation is about 240 Wm^{-2} and emitted terrestrial radiation is about 239 Wm^{-2}. The difference of about 0.6 Wm^{-2} is being stored in the ocean, which is heating up at the present time as a result of human production of greenhouse gases. More than 90% of the terrestrial emission to space originates in the atmosphere and only about 20 Wm^{-2} comes from the surface through the atmospheric infrared window. Of the 240 Wm^{-2} of solar radiation absorbed in the climate system, only about a third is absorbed in the atmosphere and two-thirds is absorbed at the surface. Although the surface receives 160 Wm^{-2} of solar radiation, this is less than half of the 345 Wm^{-2} that the surface receives as downward thermal emission from the atmosphere. The surface emits 396 Wm^{-2} of radiation upward, so that net sum of upward and downward terrestrial radiation cools the surface at a net rate of 51 Wm^{-2}, which is smaller than the sum of evaporation (88 Wm^{-2}) and convection of warmth away from the surface (20 Wm^{-2}).

One way to measure the strength of the greenhouse effect is to compare the emission from the surface with the emission from the top of the atmosphere. These differ by 157 Wm^{-2} = 396 Wm^{-2} – 239 Wm^{-2}, which is how much the atmosphere decreases the emission of energy to space for the current climate. This depends not only on the opacity of the atmosphere for longwave radiation but also on the transparency of the atmosphere to solar radiation and the fact that the temperature of the atmosphere decreases with altitude. The decrease of temperature with altitude is possible because the atmosphere is a compressible gas.

Using the numbers in Fig. 2.4, one can construct the energy budgets at the top of atmosphere, the atmosphere and the surface.

Top of atmosphere:

$$\text{Incoming solar} - \text{reflected solar} - \begin{array}{c}\text{emitted}\\\text{terrestrial}\\\text{radiation}\end{array} = \text{Storage}$$

$$340 \text{ Wm}^{-2} \quad - \quad 100 \text{ Wm}^{-2} \quad - \quad 239 \text{ Wm}^{-2} = 0.6 \text{ Wm}^{-2}$$

Atmosphere:

$$\begin{array}{c}\text{Absorbed}\\\text{solar}\end{array} + \text{thermals} + \begin{array}{c}\text{latent}\\\text{heating}\end{array} + \begin{array}{c}\text{surface}\\\text{longwave}\end{array} - \begin{array}{c}\text{outgoing}\\\text{longwave}\end{array} = 0$$

$$80 \text{ Wm}^{-2} \; + \; 20 \text{ Wm}^{-2} \; + \; 88 \text{ Wm}^{-2} \; + \; 51 \text{ Wm}^{-2} \; - \; 239 \text{ Wm}^{-2} = 0$$

The atmospheric heat budget can be rearranged in the following way,

$$\begin{array}{c}\text{Absorbed}\\\text{solar}\end{array} + \begin{array}{c}\text{surface}\\\text{longwave}\end{array} - \begin{array}{c}\text{outgoing}\\\text{longwave}\end{array} = \begin{array}{c}\text{Net}\\\text{radiative}\\\text{cooling}\end{array} = \text{Thermal} + \begin{array}{c}\text{latent}\\\text{heating}\end{array}$$

$$80 \text{ Wm}^{-2} \; + \; 51 \text{ Wm}^{-2} \; - \; 239 \text{ Wm}^{-2} = 108 \text{ Wm}^{-2} = 20 \text{ Wm}^{-2} + 88 \text{ Wm}^{-2}$$

So that 81% of the radiative cooling of the atmosphere is balanced by latent heating. This also marks a very significant constraint on the precipitation rate, since the heating of the atmosphere by the condensation of water vapor must be approximately balanced by the radiative cooling of the atmosphere. This is discussed more in Chapter 13.

Surface:

$$\begin{array}{c}\text{Absorbed}\\\text{solar}\end{array} - \text{thermals} - \begin{array}{c}\text{latent}\\\text{heating}\end{array} - \begin{array}{c}\text{surface}\\\text{longwave}\end{array} = \text{Storage}$$

$$160 \text{ Wm}^{-2} \; - \; 20 \text{ Wm}^{-2} \; - \; 88 \text{ Wm}^{-2} \; - \; 51 \text{ Wm}^{-2} = 0.6 \text{ Wm}^{-2}$$

The very strong downward emission of terrestrial radiation from the atmosphere is essential for maintaining the relatively small diurnal variations in surface temperature over land. If the downward longwave were not larger than the solar heating of the surface, then the land surface temperature would cool more rapidly at night, yielding a large diurnal variation of surface temperature.

2.7 DISTRIBUTION OF INSOLATION

Insolation is the amount of downward solar radiation energy incident on a plane surface. Seasonal and latitudinal variations in temperature are driven primarily by variations of insolation and average solar zenith angle. The amount of solar radiation incident on the top of the atmosphere depends on the latitude, season, and time of day. The amount of solar energy that is reflected to space without absorption depends on the solar zenith angle and the properties of the local surface and atmosphere. The climate depends on the insolation and zenith angle averaged over a 24-h period, over a season, and over a year. In this section, the geometric factors that determine insolation and solar zenith angle will be described. The TSI of 1360 Wm^{-2} is the mean solar irradiance per unit area at the mean position of Earth and is measured for a surface that is perpendicular to the solar beam. Because Earth is approximately spherical, virtually all the planet's sunlit surface is inclined at an oblique angle to the solar beam. Unless the Sun is directly overhead, the solar irradiance is spread over a surface area that is larger than the perpendicular area, so that the irradiance per unit surface area is smaller than the TSI. We define the *solar zenith angle*, θ_s, as the angle between the local normal to Earth's surface and a line between a point on Earth's surface and the Sun. Figure 2.5 indicates that the ratio of the shadow area to the surface area is equal to the cosine of the solar zenith angle. We may write the solar flux per unit surface area as

$$Q = S_0 \left(\frac{\overline{d}}{d} \right)^2 \cos\theta_s \qquad (2.14)$$

where $\overline{d}$ is the mean distance for which the TSI, S_0, is measured, and d is the actual distance from the Sun.

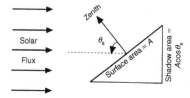

FIGURE 2.5   The relationship of solar zenith angle to insolation on a plane parallel to the surface of a planet.

The solar zenith angle depends on the latitude, season, and time of day. The season can be expressed in terms of the *declination angle* of the Sun, which is the latitude of the point on the surface of Earth directly under the Sun at noon. The declination angle (δ) currently varies between +23.45° at northern summer solstice (June 21) to –23.45° at northern winter solstice (December 21). The *hour angle, h*, is defined as the longitude of the subsolar point relative to its position at noon. If these definitions are made, then the cosine of the solar zenith angle can be derived for any latitude (ϕ), season, and time of day from spherical trigonometry formulas (see Appendix A).

$$\cos\theta_s = \sin\phi\sin\delta + \cos\phi\cos\delta\cos h \tag{2.15}$$

If the cosine of the solar zenith angle is negative, then the Sun is below the horizon and the surface is in darkness. Sunrise and sunset occur when the solar zenith angle is 90°, in which case (2.15) gives

$$\cos h_0 = -\tan\phi\tan\delta \tag{2.16}$$

where h_0 is the hour angle at sunrise and sunset. At noon the solar zenith angle depends only on the latitude and declination angle, $\theta_s = \phi - \delta$.

Near the poles, special conditions prevail. When the latitude and the declination angle are of the same sign (summer), latitudes poleward of ±90 – δ are constantly illuminated. At the pole the Sun moves around the compass at a constant angle of δ above the horizon. In the winter hemisphere, where ϕ and δ are of the opposite sign, latitudes poleward of ±90 + δ are in polar darkness. At the poles, 6 months of darkness alternate with 6 months of sunlight. At the equator, day and night are both 12 h long throughout the year. The average daily insolation on a level surface at the top of the atmosphere is obtained by substituting (2.15) into (2.14), integrating the result between sunrise and sunset, and then dividing by 24 h. The result is

$$\overline{Q}^{\text{day}} = \frac{S_0}{\pi}\left(\frac{\overline{d}}{d}\right)^2 [h_0\sin\phi\sin\delta + \cos\phi\cos\delta\sin h_0] \tag{2.17}$$

where the hour angle at sunrise and sunset, h_0, must be given in radians. The daily average insolation is plotted in Fig. 2.6 as a function of latitude and season. Earth's orbit is not exactly circular, and currently Earth is somewhat closer to the Sun during Southern Hemisphere summer than during Northern Hemisphere summer. As a result, the maximum insolation in the Southern Hemisphere is about 6.9% higher than that in the Northern Hemisphere. Note that at the summer solstice the insolation in high latitudes is actually greater than that near the equator. This results from the very long days during summer and in spite of the relatively large solar zenith angles at high latitudes.

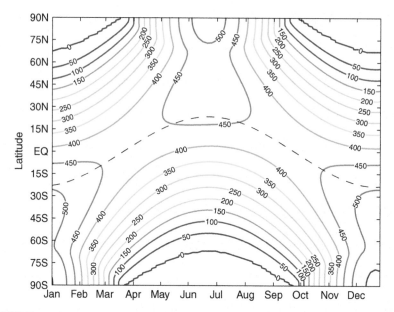

FIGURE 2.6 **Contour graph of the daily average insolation at the top of the atmosphere as a function of season and latitude.** The contour interval is 50 Wm^{-2}. The heavy dashed line indicates the latitude of the subsolar point at noon.

If the daily insolation is averaged over the entire year, the distribution given in Fig. 2.7 is obtained. The annual average insolation at the top of the atmosphere at the poles is less than half its value at the equator, where it reaches a maximum. By comparing the annual mean insolation at the equator with the insolation at the solstices, one can see that the insolation

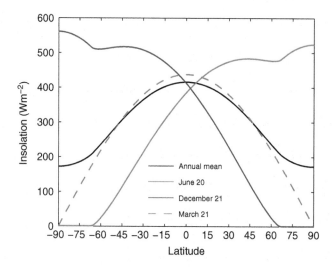

FIGURE 2.7 **Annual-mean, solstice and equinox insolation as functions of latitude.**

at the equator goes through a semiannual variation with maxima at the equinoxes and minima at the solstices. This is because the Sun crosses the equator twice a year at the equinoxes.

Because the local albedo of Earth depends on the solar zenith angle, the zenith angle enters in determining both the available energy per unit of surface area and the albedo. Therefore, it is of interest to consider the average solar zenith angle during the daylight hours as a function of latitude and season. In calculating a daily average zenith angle, it is appropriate to weight the average with respect to the insolation, rather than time. A properly weighted average zenith angle is thus computed with the following formula,

$$\overline{\cos\theta_s}^{\text{day}} = \frac{\int_{-h_0}^{h_0} Q\cos\theta_s\,dh}{\int_{-h_0}^{h_0} Q\,dh} \tag{2.18}$$

where Q is the instantaneous insolation given by (2.14).

The daily average solar zenith angle calculated according to (2.18) varies from a minimum of 38.3° at the subsolar latitude and increases to 90° at the edge of the polar darkness (Fig. 2.8). At the pole, the minimum value of the daily average solar zenith angle is achieved at the summer solstice, when it equals $\phi - \delta = 66.55°$. Because of the much higher average solar zenith angles in high latitudes, more solar radiation is reflected than would be from a similar scene in tropical latitudes.

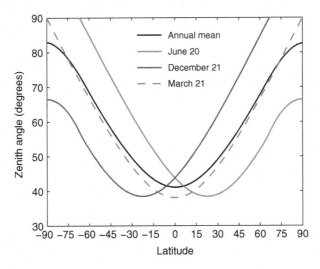

FIGURE 2.8 **Insolation-weighted daily average solar zenith angle as a function of latitude for the annual mean, solstices and equinoxes.**

2.8 THE ENERGY BALANCE AT THE TOP OF THE ATMOSPHERE

The amount of energy absorbed and emitted by Earth varies geographically and seasonally, depending on atmospheric and surface conditions as well as the distribution of insolation. The energy balance at the top of the atmosphere is purely radiative and can be measured accurately from Earth-orbiting satellites. The net radiation at the top of the atmosphere (R_{TOA}) is equal to the absorbed solar radiation (Q_{abs}) minus the *outgoing longwave radiation* (OLR).

$$R_{TOA} = Q_{abs} - OLR \qquad (2.19)$$

The absorbed solar radiation is the incident solar radiation at the top of the atmosphere S_{TOA} minus the fraction reflected, which is the albedo, α.

$$Q_{abs} = S_{TOA}(1-\alpha) \qquad (2.20)$$

The albedo is estimated by measuring the solar radiation reflected from a region of Earth and comparing that with the insolation. The global mean, or planetary, albedo of Earth is about 29%. The albedo shows interesting geographic structure (Fig. 2.9).[2] It is highest in the polar regions where cloud and snow cover are plentiful and where average solar zenith angles are large. Secondary maxima of albedo occur in tropical and subtropical regions where thick clouds are prevalent, and over bright surfaces such as the Sahara Desert. The smallest albedos occur over those tropical ocean regions where clouds are sparsely distributed. The ocean surface has an intrinsically low albedo so that when clouds and sea ice are absent, the Earth's albedo over ocean areas is only 8–10%.

Outgoing longwave radiation (OLR) is greatest over warm deserts and over tropical ocean areas where clouds occur infrequently (Fig. 2.10). It is lowest in polar regions and in regions of persistent high cloudiness in the tropics. The temperature of the emitting substance controls the OLR, so that the cold poles and the cold cloud tops produce the lowest values. A warm surface that is overlain by a relatively dry, cloudless atmosphere produces the highest OLR values.

The *net radiation* is negative near the poles and positive in the tropics (Fig. 2.11). The highest positive values of about 140 Wm^{-2} occur over the subtropical oceans in the summer hemisphere, where large insolation and relatively low albedos both contribute to large absorbed solar radiation. The greatest losses of energy occur in the polar darkness of the winter hemisphere, where the OLR is uncompensated by solar absorption. Dry desert areas such as the Sahara in northern Africa are interesting because,

---

[2]We use a Hammer equal-area projection at many points in this book. In displaying irradiance (Wm^{-2}), it is especially important that the fraction of the area of the projection occupied by a particular feature be proportional to the fraction of the area of Earth's surface that the feature occupies.

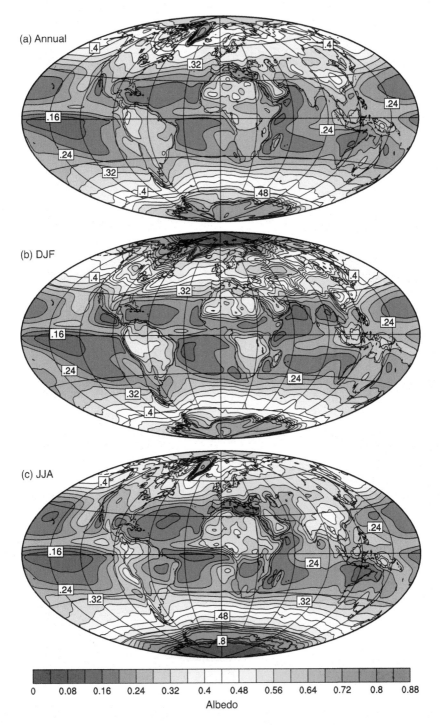

FIGURE 2.9 Global maps of planetary albedo measured from satellites for annual mean, DJF and JJA seasons. *Data from CERES 2000–2013.*

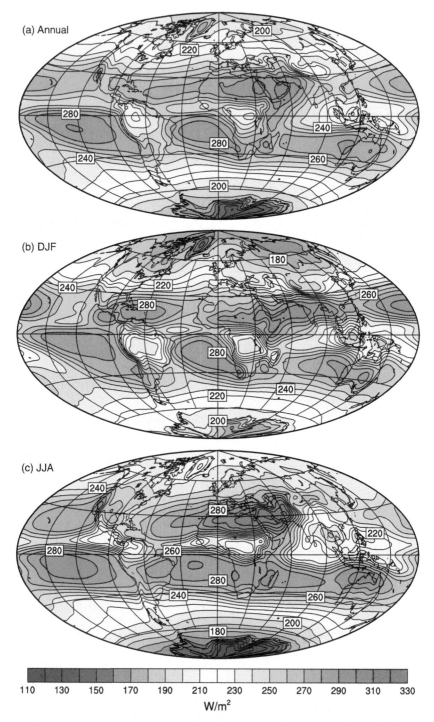

FIGURE 2.10   **Global maps of outgoing longwave radiation.** *Data from CERES 2000–2013.*

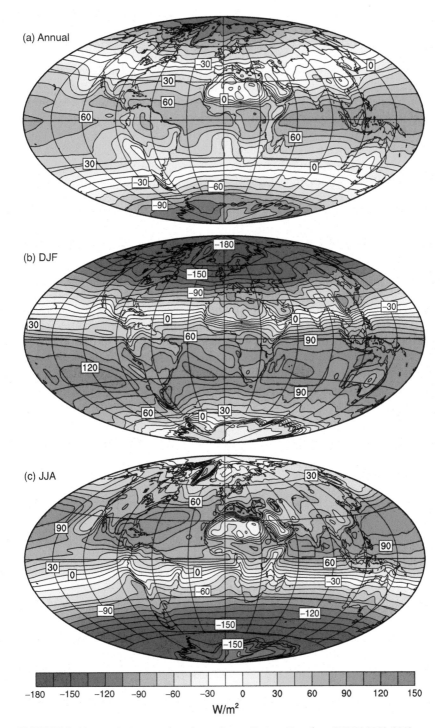

FIGURE 2.11   **Global maps of net incoming radiation.** *Data from CERES 2000–2013.*

despite their subtropical latitude, they lose energy in the annual average. Their energy loss is related to the relatively high surface albedo of dry deserts, combined with the large OLR loss associated with a dry, cloudless atmosphere above a warm surface. Since deserts are land, which cannot transport heat laterally, the radiative loss at the top of the atmosphere must be balanced by horizontal heat transport by the atmosphere into the desert region from elsewhere. Air with high energy at high altitudes sinks over the desert and warms adiabatically as it subsides. This air tends to have a low relative humidity and enhances the dryness of the desert.

When averaged around latitude circles, the components of the energy balance at the top of the atmosphere clearly show the influence of the latitudinal gradient of insolation (Fig. 2.12). The poleward decrease in absorbed solar energy is even greater than that of insolation because the albedo increases with latitude. Thus a smaller fraction of the available insolation is absorbed at higher latitudes than at the equator. The albedo increases with latitude because solar zenith angle, cloud coverage, and snow cover all increase with latitude. The energy emitted to space by the atmosphere does not decrease with latitude as rapidly as the absorbed solar radiation, because the atmosphere and ocean transport heat poleward and support a net energy loss to space in polar regions. The absorbed solar radiation exceeds the OLR in the tropics, so that the net radiation is positive there. Poleward of about 40°, the absorbed solar radiation falls below the OLR and the net radiation balance is negative, so that the climate system loses energy to space. The latitudinal gradient in annual mean net radiation must be balanced by a poleward flux of energy in the atmosphere and ocean.

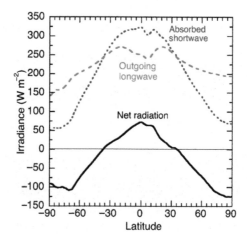

FIGURE 2.12 **Graphs of annual mean absorbed solar radiation, OLR, and net radiation averaged around latitude circles.** *Data from CERES 2000–2013.*

2.9 POLEWARD ENERGY FLUX

In the lowest curve of Fig. 2.12, we see that the annual mean net radiation is positive equatorward of about 40° of latitude and negative poleward of that latitude. As illustrated in Fig. 2.13, the energy balance for the climate system involves only the exchange at the top of the atmosphere, the transport through the lateral boundaries of the region in question by the atmosphere and ocean, and the time rate of change of energy within the region. Energy exchange with the solid earth can be neglected. We can write the energy balance for the climate system as

$$\frac{\partial E_{ao}}{\partial t} = R_{TOA} - \Delta F_{ao} \tag{2.21}$$

where $\partial E_{ao}/\partial t$ is the time rate of change of the energy content of the climate system, R_{TOA} is the net incoming radiation at the top of the atmosphere, and ΔF_{ao} is the divergence of the horizontal flux in the atmosphere and ocean. If we average over a year, then the storage term becomes small, and we have an approximate balance between net flux at the top of the atmosphere and horizontal transport.

$$R_{TOA} = \Delta F_{ao} \tag{2.22}$$

We can then use the observation of net radiation in Fig. 2.12 to derive the required annual mean energy transport in the northward direction. We assume that the energy lost to space poleward of any latitude must be supplied by poleward transport across that latitude by the atmosphere and the ocean. So if we integrate the net radiation over a polar cap area, we can calculate the total energy flux across each latitude belt in the following way:

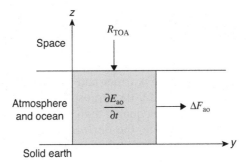

FIGURE 2.13 Diagram of the energy balance for a vertical column of the climate system in some latitude interval.

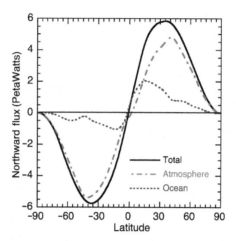

FIGURE 2.14 **Northward transport of energy for annual mean conditions.** Net radiation and atmospheric transport are estimated from observations, ocean transport is calculated as a residual in the energy balance. Units are petaWatts (10^{15} W). *Data from CERES EBAF-2.7 and ERA-Interim (Mayer and Haimberger, 2012).*

$$F(\phi) = \int_{-\frac{\pi}{2}}^{\phi} \int_{0}^{2\pi} R_{TOA} a^2 \cos\phi \, d\lambda \, d\phi \qquad (2.23)$$

The total northward energy flux inferred from the radiation imbalance peaks in mid-latitudes at about 5.8×10^{15} W, or 5.8 petaWatts (PW) (Fig. 2.14). This flux includes the contributions from both the atmosphere and the ocean. The flux of energy in the atmosphere can be estimated from analyzed fields of wind, temperature, and humidity. If this flux is subtracted from the total flux, it gives an estimate of the oceanic energy flux, which is more difficult to estimate from direct measurements. At 20°N, the contributions from the atmospheric and oceanic poleward fluxes are each about 2.0 PW. The ocean flux peaks at about 20°N in the subtropics, whereas the atmospheric flux has a broad maximum peaking near 40°N and 40°S. Although the oceanic heat flux is somewhat weaker in the Southern Hemisphere, it peaks in the subtropics. In Chapters 6 and 7 we investigate in more detail how the atmosphere and the oceans accomplish this poleward flux of energy. If the fluid envelope of Earth did not transport heat poleward, the tropics would be warmer and the polar regions would be much colder. Transport of heat by the atmosphere and the oceans makes the climate of Earth much more equable than it would otherwise be.

If we integrate over the globe, the horizontal transport is zero, since energy cannot be removed from spherical Earth through horizontal transport. Then in the annual global average, the net radiation is very close to zero. Any global mean net radiative imbalance will lead to a heating or cooling of Earth.

TABLE 2.3 Physical Data for the Planets

Planet*	Mass (10^{26} g)	Mean radius (km)	Mean density (g cm^{-3})	Average distance from Sun (10^6 km)	Length of year (days)	Obliquity (degrees)	Orbital eccentricity	Period of rotation (days)	Albedo
Mercury	3.3	2,440	5.42	58	88	~0	0.206	58.7	0.1
Venus	48.7	6,052	5.25	108	224.7	177.4	0.007	−243**	0.65
Earth	59.8	6,378	5.52	150	365.25	23.45	0.017	1.00	0.29
Mars	6.42	3,396	3.94	228	687	25.2	0.093	1.03	0.15
Jupiter	19,100	71,490	1.34	778	4,340	3.1	0.048	0.41	0.52
Saturn	5,690	60,260	0.69	1,430	10,800	26.7	0.054	0.45	0.47
Uranus	870	25,560	1.32	2,878	30,700	97.7	0.047	−0.72**	0.50
Neptune	1,030	24,765	1.64	4,510	60,200	28.3	0.009	0.67	0.40

* The first four planets are similar in size, mass, density, and probably chemical composition. They are the inner planets. The remaining four are very different from Earth, but are similar to one another. They are the outer planets.
** Venus and Uranus rotate in the opposite sense to the other planets.

EXERCISES

1. Use the data in Table 2.3 to calculate the emission temperatures for all of the planets. The actual energy emission from Jupiter corresponds to an emission temperature of about 124 K. How must you explain the difference between the number you obtain for Jupiter and 124 K?

2. Calculate the emission temperature of Earth if the solar luminosity is 30% less, as it is hypothesized to have been early in solar system history. Use today's albedo and Earth–Sun distance.

3. Calculate the emission temperature of Earth, if the planetary albedo is changed to that of ocean areas without clouds, about 10%.

4. Using the model illustrated in Fig. 2.3, calculate the surface temperature if the insolation is absorbed in the atmosphere, rather than at the surface.

5. Using the data in Fig. 2.4, estimate the blackbody temperature of Earth's surface and the blackbody temperature of the atmosphere as viewed from Earth's surface. How well do these temperatures agree with the temperatures derived from the model shown in Fig. 2.3?

6. Calculate the solar zenith angle at 9 AM local sun time at Seattle (47°N) at summer solstice and at winter solstice. Do the same for Miami (26°N).

7. A mountain range is oriented East–West and has a slope on its north and south faces of 15°. It is located at 45°N. Calculate the insolation per unit surface area on its south and north faces at noon at the summer and winter solstices. Compare the difference between the insolation on the north and south faces in each season with the seasonal variation on each face. Ignore eccentricity and atmospheric absorption.

8. In his science fiction novel, *Ringworld*, Larry Niven imagines a manufactured world consisting of a giant circular ribbon of superstrong material that circles a star. The width of the ribbon normal to a radius vector originating at the star's center is much greater than the thickness of the ribbon in the radial direction. However, the width of the ribbon is much less than the distance of the ribbon from the star. If the diameter of the ribbon is the same as the diameter of Earth's orbit about the Sun, the luminosity is the same as Earth's Sun, and the albedo of the ribbon is 0.3, what is the emission temperature of the sunlit side of the ribbon? Consider both the cases in which heat is conducted very efficiently from the sunlit to the dark side of the ribbon, and the case in which no heat is conducted from the sunlit to the dark side of the ribbon. Explain why your emission temperature is different from Earth's. At what distance from the center of the star would the ribbon need to be to reproduce the emission temperature of Earth?

Atmospheric Radiative Transfer and Climate

3.1 PHOTONS AND MINORITY CONSTITUENTS

The energy that sustains warmth and life at the surface of Earth travels from the Sun in the form of radiation. The interaction of incoming solar radiation with the atmosphere and the surface determines the total amount of solar energy absorbed and the distribution of solar heating among atmospheric layers and the surface. Because the atmosphere is relatively transparent to solar radiation, about half of the incoming solar radiation is absorbed at the ocean or land surface (Fig. 2.4). To achieve energy balance, heat provided by the absorption of solar radiation must be returned to space by emission from Earth. Transmission of thermal infrared radiation through the atmosphere and the upward transport of heat by atmospheric motions are the critical factors in returning energy to space. The transmission properties of the atmosphere are determined by its gaseous composition and the nature of aerosols and water clouds present in it. The composition of the atmosphere is such that it efficiently absorbs and emits thermal infrared radiation. The efficient absorption and emission of thermal infrared radiation by the atmosphere, combined with its relative transparency to solar radiation, causes the surface to be much warmer than it would be in the absence of an atmosphere.

Molecules that comprise a tiny fraction of the atmosphere's mass do nearly all the absorption and emission by the air. The dependence of the climate on the abundance of these minority constituents makes the climate sensitive to natural and human-induced changes in atmospheric composition. Relatively small changes in composition can affect the flow of energy through the climate system and thereby produce surprisingly large climate changes.

To understand how climate depends on atmospheric composition, it is necessary to understand the nature of electromagnetic radiation and the physical processes through which it interacts with gases and particles.

Global Physical Climatology. http://dx.doi.org/10.1016/B978-0-12-328531-7.00003-7

The equation of radiative transfer forms the mathematical basis for keeping track of these physical processes in an aggregate sense, so that radiative energy fluxes and heating rates can be computed. One-dimensional models, in which the vertical fluxes of energy by radiation are calculated from the radiative transfer equation, can be used to estimate the effect of trace-gas concentrations, clouds, and aerosols on the global mean surface temperature.

3.2 THE NATURE OF ELECTROMAGNETIC RADIATION

Electromagnetic radiation can be thought of either as a wave or as a particle that represents the movement of energy through space. For scattering of light by particles and surfaces, electromagnetic wave theory is most helpful. When considering absorption and emission of radiation, it is useful to think of radiant energy as discrete parcels of energy that we call photons. The speed of electromagnetic radiation in a vacuum is a constant $c^* = 3 \times 10^8 \, \text{ms}^{-1}$. This means that frequency v and wavelength λ are inversely related in a one-to-one correspondence.

$$v = \frac{c^*}{\lambda}; \quad \lambda = \frac{c^*}{v} \tag{3.1}$$

High frequencies correspond to short wavelengths, whereas low frequencies correspond to long wavelengths. Most of the time we will describe radiation in terms of its wavelength, which we will give in millimeters (mm = 10^{-3} m), micrometers (μm = 10^{-6} m), or nanometers (nm = 10^{-9} m). In explaining the photoelectric effect, Einstein postulated that radiant energy exists and propagates in quantum bits called photons. If we think of light as photons, then a photon has an energy, E_v, that is proportional to its frequency.

$$E_v = hv \tag{3.2}$$

where $h = 6.625 \times 10^{-34}$ Js is Planck's constant. Therefore, high-frequency, short-wavelength radiation has more energy per photon than low-frequency radiation.

Most of the Sun's radiant energy output is contained between wavelengths of 100 nm and 4 μm, and consists of ultraviolet, visible, and near infrared radiation. Of the Sun's emission, 99% comes from the sum of the visible (0.4–0.75 μm) and near infrared (0.75–5 μm) portions of the spectrum. Ultraviolet radiation makes up less than 1% of the total, but is nonetheless important because of its influence in the upper atmosphere,

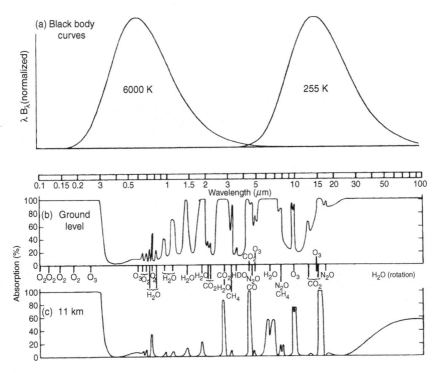

FIGURE 3.1 The normalized blackbody emission spectra for the Sun (6000 K) and Earth (255 K) as a function of wavelength (a). The fraction of radiation absorbed while passing from the surface to the top of the atmosphere as a function of wavelength (b). The fraction of radiation absorbed from the tropopause to the top of the atmosphere as a function of wavelength (c). The atmospheric molecules contributing the important absorption features at each frequency are indicated. *From Goody and Yung (1989). Reprinted with permission from Oxford University Press.*

and because it is harmful to life if it reaches the surface. Earth's energy emission is almost all contained between about 4 μm and 200 μm, and is therefore entirely thermal infrared (Fig. 3.1).

3.3 DESCRIPTION OF RADIATIVE ENERGY

The energy of radiation is measured by its intensity or radiance. The monochromatic intensity describes the amount of radiant energy (dF_v) within a frequency interval (v to $v + dv$) that will flow through a given increment of area (dA) within a solid angle ($d\omega$) of a particular direction in a time interval (dt).

$$dF_v = I_v \cos\theta \, d\omega \, dA \, dv \, dt \tag{3.3}$$

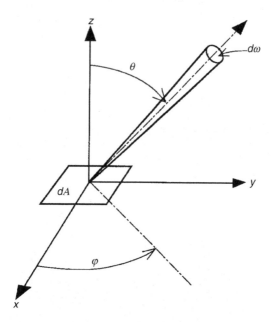

FIGURE 3.2   **The angles that define the radiance flowing through a unit area *dA* in the *x–y* plane.** These angles are in the direction defined by the zenith angle θ, and the azimuth angle φ, and within the increment of solid angle $d\omega$.

The direction is defined by the zenith angle, θ, and the azimuth angle, φ, as shown in Fig. 3.2. The magnitude of the radiant intensity, I_v, is given in energy per unit time, per unit area, per unit of frequency interval, per unit of solid angle, or watts per meter squared per hertz per steradian. In this book, we will be concerned mostly with the total energy per unit frequency passing across a unit area of a plane surface from one side to the other. To obtain this quantity, we integrate the radiant intensity over all solid angles in a hemisphere. To do this we need to make use of the definition of an increment of solid angle, which is equivalent to the area on the surface of a sphere with unit radius.

$$d\omega = \sin\theta d\theta d\varphi \tag{3.4}$$

Inserting (3.4) into (3.3) and integrating over the upper hemisphere we obtain

$$F_v = \int_0^{2\pi} \int_0^{\pi/2} I_v(\theta, \psi) \cos\theta \sin\theta \, d\theta \, d\varphi \tag{3.5}$$

This quantity is called the *spectral irradiance*. If F_v is integrated over all frequencies, we obtain the *irradiance*, which has units of watts per meter squared.

$$F = \int_0^\infty F_v \, dv \tag{3.6}$$

When a beam of radiation encounters an object such as a molecule, an aerosol particle, or a solid surface, several possible interactions between the radiation and the object can take place. The radiation can pass the object unchanged, which is called *perfect transmission*. The radiation can change direction without a change in energy, which is *pure scattering*. The radiation can be *absorbed*, where it ceases to exist and its energy is transferred to the object. The probability that a photon will be scattered, absorbed, or transmitted depends on the frequency of the radiation and the physical properties of the object it encounters. Pure water droplets in clouds scatter visible radiation very effectively with relatively little absorption. Water vapor and carbon dioxide are very effective absorbers of thermal infrared radiation at certain frequencies. Matter can also add to the intensity of a beam of radiation by emitting radiation in the direction of the beam. Emission of radiation by matter depends on the substance's physical properties and temperature.

3.4 PLANCK'S LAW OF BLACKBODY EMISSION

The intensity of radiation in a cavity in thermodynamic equilibrium is given uniquely as a function of frequency and temperature by Planck's law. An object that absorbs all radiations incident on it is called a *blackbody*. A blackbody with temperature T emits radiation at frequency v with an intensity given by Planck's law:

$$B_v(T) = \frac{2hv^3}{c^{*2}} \frac{1}{(e^{hv/kT} - 1)} \qquad (3.7)$$

where $h = 6.625 \times 10^{-34}$ Js (Planck's constant), $k = 1.37 \times 10^{-23}$ JK^{-1} (Boltzmann's constant), $c^* = 3 \times 10^8$ ms^{-1} (speed of light), v is the frequency of radiation in s^{-1}, T is the temperature in Kelvins.

The Stefan–Boltzmann law is an integral of Planck's law over all frequencies and over all angles in a hemisphere, and expresses the strong dependence of energy emission on temperature.

$$\pi \int_0^\infty B_v(T) dv = \sigma T^4 \qquad (3.8)$$

The factor of π arises from the integration over one hemisphere, assuming that the emission is independent of angle (isotropic). The Stefan–Boltzmann constant defined in (2.5) can be expressed in terms of the more fundamental constants of Planck's law:

$$\sigma = \frac{2\pi^5 k^4}{15 c^{*2} h^3} \qquad (3.9)$$

Planck's law of blackbody radiation contains within it Wien's law of displacement, which states that the wavelength of maximum emission is inversely proportional to temperature. That is, the hotter the object, the higher the frequency and the shorter the wavelength of emitted radiation. Note that for terrestrial temperatures ($\sim$255 K) the emission peaks around 10 μm, whereas for the temperature of the Sun's photosphere ($\sim$6000 K) the emission peaks around 0.6 μm (Fig. 3.1). The energy emission from both the Sun and Earth become very small near 4 μm, so that the frequencies at which they emit are almost completely distinct for energetic purposes. We can thus speak of solar and terrestrial radiation as separate entities. In climatology, solar radiation is often called *shortwave* radiation and terrestrial radiation is called *longwave*.

3.5 SELECTIVE ABSORPTION AND EMISSION BY ATMOSPHERIC GASES

In considering the global energy balance of Earth, we found that the effective emission temperature is 255 K, which is much less than the observed global mean surface temperature of 288 K. The explanation for this disparity is found in the different transmission properties of the atmosphere for terrestrial and solar radiation. The atmosphere is relatively transparent to solar radiation, whereas it is nearly opaque to terrestrial radiation. To understand the fundamental reasons behind this "greenhouse effect," we must understand a little about the interaction of radiation with matter.

In deriving his law of blackbody radiation, Planck found it necessary to postulate that the energy levels of an atomic or molecular oscillator are limited to a discrete set of values that satisfy

$$E_v = nh\nu \quad (n = 0, 1, 2, \ldots) \tag{3.10}$$

Equation (3.10) describes a set of discrete energy levels that differ one from the next by an amount $h\nu$. The oscillator may represent the periodic motions of an atom or molecule. A transition from one energy level to another, called a quantum jump, corresponds to the release or capture of an amount of energy equal to $h\nu$. One way to accomplish this energy transition is for the molecule to emit or absorb a photon of energy $h\nu$. This quantization effect only becomes apparent for very small oscillators such as molecules or atoms. For macroscopic oscillators such as a pendulum or a spring and weight apparatus, the amount of energy represented by $h\nu$ is too small to be noticed in comparison to the total energy of the system.

A photon is emitted from a substance in some finite amount of time ($\leq 10^{-8}$ s) then travels through space until it is absorbed. If it approaches a mass such as an air molecule or a solid particle, the photon can change phase or direction, a process called *scattering*, or it can be absorbed. When a photon is absorbed, it ceases to exist and its energy is transferred to the substance that absorbed it. The energy of a molecule can be stored in vibrational, rotational, electronic, or translational forms.

$$E_{\text{Total}} = E_{\text{Translational}} + E_{\text{Rotational}} + E_{\text{Vibrational}} + E_{\text{Electronic}} \qquad (3.11)$$

A molecule in the atmosphere can absorb a photon only if the energy of the photon corresponds to the difference between the energy of two allowable states of the molecule. Each mode of energy storage in a molecule corresponds to a range of energies, with electronic transitions corresponding to the largest energy differences and rotational transitions corresponding to the smallest. Allowable transitions between energy levels of the molecules making up the atmosphere determine the frequencies of radiation that will be efficiently absorbed and emitted by the atmosphere. If no transitions correspond to the energy of a photon, then it will have a good chance to pass through the gaseous atmosphere without absorption.

3.5.1 Translational or Kinetic Energy (Temperature)

Translational energy corresponds to the gross movement of molecules or atoms through space and is not quantized. At terrestrial temperatures, the kinetic energy of a molecule is generally small as compared to the energy required for vibrational transition. Collisions between molecules can carry away or supply energy during interactions between photons and matter, and this plays an important role in broadening the range of frequencies of radiation that can be absorbed by a particular transition between molecular energy levels. The Doppler effect associated with the movement of molecules can also broaden the range of frequencies absorbed or emitted by a particular energy transition of the molecule.

3.5.2 Rotational Energy

A macroscopic object such as a child's top can store an amount of energy that, in effect, varies continuously with its rotation rate. For tiny objects such as molecules in the atmosphere, the energy of rotation is quantized and can take on only discrete values. *Rotational energy transitions* involve energy changes that correspond to the energy of photons with wavelengths shorter than about 1 cm.

3.5.3 Vibrational Energy

Atoms are bonded together into stable molecules when the forces of attraction and repulsion are in balance at the appropriate interatomic distance. Molecular energy can be stored in the vibrations about this stable point. The states of the molecule and the allowable energy levels are again

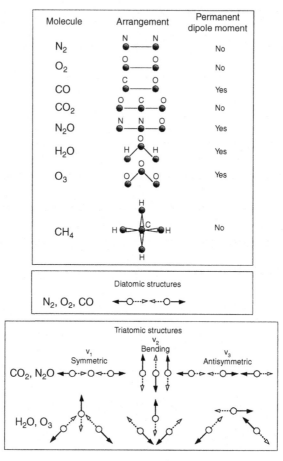

FIGURE 3.3 **Schematic diagrams showing the vibrational modes of diatomic and triatomic molecules.** *From McCartney (1983). Reprinted with permission from Wiley and Sons, Inc.*

quantized. Vibrational transitions require a photon with a wavelength of less than 20 μm. There are three independent modes of vibration for a triatomic molecule such as CO_2: two stretching modes and one bending mode (Fig. 3.3).

A symmetric, linear molecule such as CO_2 has no permanent dipole moment, since it looks the same from both ends. For this reason, it has no pure rotational transitions. During vibrational transitions the CO_2 molecule develops temporary dipole moments so that rotational transitions can accompany a vibrational transition. The combination of possible vibrational and rotational transitions allows the molecule to absorb and emit photons at a large number of closely spaced frequencies, composing an absorption band.

Water vapor is a good absorber of terrestrial radiation because it is a bent triatomic molecule. Because it is bent, it has a permanent dipole moment and therefore has pure rotation bands in addition to vibration–rotation bands.

Vibration–rotation bands and pure rotation bands of polyatomic molecules account for the longwave absorption of the clear atmosphere that is shown in Fig. 3.1. The bending mode of CO_2 produces a very strong vibration–rotation absorption band near 15 μm, which is very important for climate and critical for the stratosphere where it accounts for much of the longwave absorption and emission. This absorption feature is particularly important because it occurs near the peak of the terrestrial emission spectrum. Water vapor has an important vibration–rotation band near 6.3 μm, and a densely spaced band of pure rotational lines of water vapor strongly absorbs terrestrial emission at wavelengths greater than about 12 μm. The rotational lines of water vapor are strong ones, and do not require a lot of water mass to be important. For this reason they dominate the cooling of the upper troposphere, where the specific humidity is fairly low. Between these two water-vapor features absorption by water vapor is relatively weak, and so this wavelength region is called the *water-vapor window,* since only in this frequency range can longwave radiation pass relatively freely through the atmosphere (see middle panel of Fig. 3.1 and Fig. 3.4). In the middle of this atmospheric window sits the 9.6-μm band of ozone. Together, all these absorption features make the troposphere nearly opaque to longwave radiation. Complicating this picture a little bit is the continuum absorption by water vapor. When the water vapor concentration is high, as in the warm tropics near the surface, radiation is weakly absorbed and emitted at all thermal wavelengths by water molecules that stick together (dimers). Thus in warm humid regions, continuum absorption by water-vapor molecules causes some more significant absorption in the window region between 8 μm and 12 μm.

Significant absorption of solar radiation by vibration–rotation bands of atmospheric gases occurs at near-infrared wavelengths between about 1 μm and 4 μm, mostly by water vapor and carbon dioxide (Figs. 3.1 and 3.4). These absorption features account for most of the absorption of solar radiation by air molecules in the troposphere. Visible wavelengths (~0.3–0.8 μm) are virtually free of gaseous absorption features, giving the cloud-free atmosphere its transparency in these wavelengths. Because a large fraction of the Sun's radiant energy is contained in visible wavelengths (Fig. 3.1), solar radiation penetrates relatively freely to the surface, where it provides heat and light.

3.5.4 Photodissociation

If a photon is sufficiently energetic, it can break the bond that holds together the atoms in a molecule. To photodissociate the molecules in the atmosphere, photons with energy corresponding to wavelengths shorter

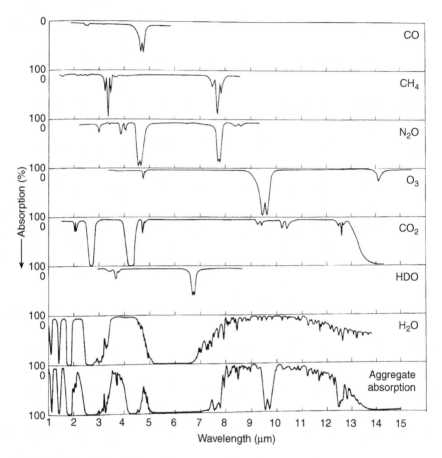

FIGURE 3.4 Infrared absorption spectra for various atmospheric gases. *From Valley (1965). Used with permission from McGraw-Hill, Inc.*

than ~1 μm are required. Oxygen is dissociated in the upper atmosphere by radiation with wavelengths shorter than ~200 nm. When a photon participates in the dissociation of a molecule, it ceases to exist and the atmosphere absorbs its energy.

Ozone is a bent molecule made up of three oxygen atoms and is more loosely bonded than molecular oxygen. It can be dissociated by radiation between 200 nm and 300 nm, and ozone absorbs most of the radiation in this band before it reaches the surface (Fig. 3.1), where it would otherwise do considerable damage to life.

3.5.5 Electronic Excitation

Photons with energies in excess of that corresponding to a wavelength of 1 μm can also excite the electrons in the outer shell of an atom.

Sometimes when oxygen or ozone is dissociated by solar radiation, one or both of the resulting oxygen atoms are in an electronically excited state.

3.5.6 Photoionization

If the wavelength of radiation is shorter than about 100 nm, then it can actually remove electrons from the outer shell surrounding the nucleus and produce ionized atoms. This process is responsible for the ionosphere. If the photon is even more energetic (shorter than ~10 nm), it can also photoionize the inner shell of electrons.

3.5.7 Absorption Lines and Line Broadening

The atmosphere is a very effective absorber at those discrete frequencies corresponding to an energy transition of an atmospheric gas. We may call each of these discrete absorption features an *absorption line*. The collection of such absorption lines in a particular frequency interval can be called an *absorption band*. Vibrational and rotational transitions are of primary interest for absorption and emission of terrestrial radiation in the atmosphere, since the energy levels associated with these transitions correspond to the energies of photons of thermal infrared radiation. Polyatomic molecules, such as H_2O, CO_2, O_3, CH_4, N_2O, and many others, have vibration rotation bands of importance in the thermal infrared portion of the electromagnetic spectrum (Fig. 3.4, Table 3.1). Since lower energies are required for

TABLE 3.1   Wavelengths of Vibrational Modes of Some Important Atmospheric Molecules

Species	Vibrational modes		
	$v1$	$v2$	$v3$
CO	4.67		
CO_2		15.0	4.26
N_2O	7.78	17.0	4.49
H_2O	2.73	6.27	2.65
O_3	9.01	14.2	9.59
NO	5.25		
NO_2	7.66	13.25	6.17
CH_4	3.43	6.52	3.31
CH_4	5.25		

Units are in microns (μm).
From Herzberg and Herzberg © 1957 from McGraw Hill, Inc. and Shimanouchi, 1967a,b, 1968.

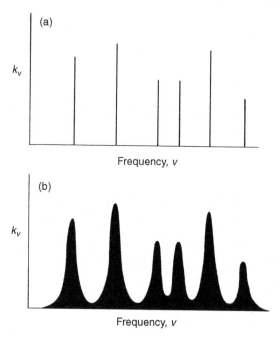

FIGURE 3.5   **Hypothetical line spectrum (a) before broadening, (b) after broadening.**

rotational transitions, molecules with pure rotational transitions can give a densely packed band of rotation lines. This is the case for water vapor, which has many rotational absorption lines at closely spaced frequencies, which form a rotation band that absorbs much of Earth's emission at wavelengths between 12 μm and 200 μm. Many of these lines are very strong, in the sense that photons with these wavelengths are very likely to be absorbed by water, and they are very important in the cold upper troposphere where water vapor is present, but not very abundant.

A portion of a line absorption spectrum might look like Fig. 3.5a. Lines are not always evenly spaced or equally strong, since some transitions are more probable than others. The line width depends on broadening processes, which include natural, pressure, and Doppler broadening. After broadening, the absorption spectrum may have substantial absorption between the line centers (Fig. 3.5b).

Natural broadening is associated with the finite time of photon emission or absorption and with the uncertainty principle. If we know the energy exactly, then we can only know the frequency to a finite precision. This mechanism is usually less important than pressure or Doppler broadening.

Pressure broadening (also called *collision broadening*) is brought about by collisions between molecules or atoms, which can supply or remove small amounts of energy during radiative transitions, thereby allowing photons

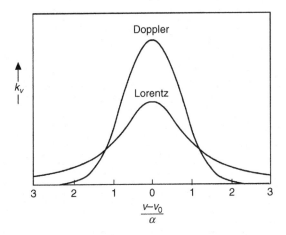

FIGURE 3.6 **Line shapes produced by pressure (Lorentz line shape) and Doppler broadening for the same line width (α) and intensity at some central frequency v_0.** *From Goody and Yung (1989). Reprinted with permission from Oxford University Press.*

with a broader range of frequencies to produce a particular transition of a molecule. This is the primary broadening mechanism in the troposphere.

Doppler broadening results from the movement of molecules relative to a photon, which can cause the frequency of radiation to be Doppler-shifted. This again allows a broader range of frequencies of radiation to effect a particular molecular transition. Doppler broadening becomes the dominant mechanism at low pressures (high altitudes), where collisions are less frequent.

Pressure broadening is dominant in the troposphere and gives the lines a characteristic shape and width (Fig. 3.6). The line shape is important. Absorption is most probable at the frequency corresponding to the change in the energy of the molecule between two states, and this frequency locates the line center. The probability of absorption decreases away from the line center but remains significant at some frequency interval away from it. These weak absorption features away from the line center are called the *wings* of the absorption line. For strong absorption lines, the radiation at frequencies near the center of the line is completely absorbed after traveling a relatively small distance through the atmosphere. Under these conditions, most of the transmission of energy is carried by frequencies in the wings of the absorption lines, where absorption is weaker. Pressure broadening, which produces the Lorentz line shape and has broad wings, is particularly effective at producing absorption and emission far from the line centers.

Most of the atmosphere is made up of molecular nitrogen and oxygen. These are diatomic molecules, which have no dipole moment even when vibrating. Therefore, they have no vibration–rotation transitions at the small energies corresponding to terrestrial radiation. Thus, it is the minor trace concentrations of polyatomic molecules that determine the infrared transmissivity of the atmosphere. The most important gases are water vapor, carbon dioxide, and ozone (in that order), but many other gases contribute significantly (Fig. 3.2). Except for the region between 8 μm and 12 μm, the atmosphere is nearly opaque to terrestrial radiation. The key absorption features for terrestrial radiation are a water-vapor vibration–rotation band near 6.3 μm, the 9.6-μm band of ozone, the 15-μm band of carbon dioxide, and the dense rotational bands of water vapor that become increasingly important at wavelengths longer than 12 μm.

Visible radiation is too energetic to be absorbed by most of the gases in the atmosphere and not energetic enough to photodissociate them, so that the atmosphere is almost transparent to it. Solar radiation with wavelengths between about 0.75 μm and 5 μm, which we will call near-infrared radiation, is absorbed weakly by water, carbon dioxide, ozone, and oxygen (Fig. 3.1). Most of the ultraviolet radiation from the Sun with wavelengths shorter than 0.2 μm is absorbed in the upper atmosphere through the photodissociation and ionization of nitrogen and oxygen. Radiation at frequencies between 0.2 μm and 0.3 μm is absorbed by ozone in the stratosphere.

3.6 THE LAMBERT–BOUGUER–BEER LAW: FORMULATION OF FLUX ABSORPTION

In Section 3.5, the physical processes whereby molecules absorb and emit radiation were discussed. This section and the following one illustrate how knowledge of the absorption properties of the atmosphere can be incorporated into a formula to calculate the flux of radiation. To simplify the initial discussion, we will not consider emission by the atmosphere. Suppose, for example, that we are interested in the absorption of solar radiation in the atmosphere. Atmospheric gases can absorb solar radiation, but because the atmosphere is much colder than the Sun, the energy reemitted from the atmospheric gases is at longer wavelengths. For the moment, we will ignore scattering and consider that the atmosphere can only transmit or absorb solar radiation.

For many applications, the plane–parallel approximation is accurate and greatly simplifies radiative transfer calculations. Under this approximation the sphericity of Earth is ignored and atmospheric properties are assumed to be functions only of the vertical coordinate. This situation is illustrated in Fig. 3.7. Since incoming solar radiation can be considered to be a parallel beam of radiation, we need consider only one direction of radiation, which is characterized fully in this case by the zenith angle θ.

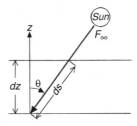

FIGURE 3.7   **The extinction path of a solar beam through a plane–parallel atmosphere.**

The *Lambert–Bouguer–Beer law of extinction* states that absorption is linear in the intensity of radiation and the absorber amount. The absorption by a layer of depth dz is proportional to the irradiance (F) times the mass of absorber along the path the radiation follows. The proportionality constant derives from the quantum-mechanical considerations outlined in Section 3.5, which determine the probability that a photon with a particular energy will be absorbed by a particular molecule. We call this constant the absorption coefficient and give it the symbol k_{abs}. In general, it depends on the pressure and temperature, since these affect the strength and shape of the absorption lines of the absorber in question.

The change in the irradiance (dF) along a path of length ds, where the density of the absorber is ρ_a and the absorption coefficient is k_{abs}, may be written

$$dF = -k_{abs}\rho_a F ds \qquad (3.12)$$

In (3.12), F and ds are both measured positive downward. F and k_{abs} depend on frequency, but we have dropped the frequency subscript for economy. The units of k_{abs} in (3.12) must be m² kg⁻¹. Because its units are area per unit mass, k_{abs} is sometimes also called the absorption cross-section of the gas in question. From Fig. 3.7, the path length is related to altitude according to

$$dz = -\cos\theta \, ds \qquad (3.13)$$

Therefore, (3.12) becomes

$$\cos\theta \frac{dF}{dz} = k_{abs}\rho_a F \qquad (3.14)$$

We can define the optical depth (τ) along a vertical path.

$$\tau = \int_z^\infty k_{abs}\rho_a \, dz \qquad (3.15)$$

Note that (3.15) implies that $d\tau = -k_{abs}\,\rho_a\,dz$, so that we can write (3.14) as

$$\cos\theta\frac{dF}{d\tau} = -F \qquad (3.16)$$

This equation has a very simple solution,

$$F = F_\infty e^{-\tau/\cos\theta} \qquad (3.17)$$

where F_∞ is, in this case, the downward irradiance at the top of the atmosphere. Thus, the incident radiation decays exponentially along the slant path ds where the optical depth is given by $\tau/\cos\theta$.

3.6.1 Absorption Rate

In an isothermal atmosphere in hydrostatic balance, the density (ρ_a) of an absorber with constant mass-mixing ratio is given by (Section 1.4):

$$\rho_a = \rho_{as}e^{(-z/H)} \qquad (3.18)$$

Here $H = RT/g$ is the scale height, $R = 287$ J K^{-1} kg^{-1}, $g = 9.81$ ms^{-2}, and ρ_{as} is the density of the absorber at the surface. If we introduce this into the equation for optical depth (3.15), assume that k_{abs} is a constant, and perform the integration, we obtain

$$\tau = \frac{p_s}{g}M_a k_{abs}e^{-z/H} \qquad (3.19)$$

where $M_a = \rho_a/\rho$ is the mass mixing ratio of the absorber. From (3.19) we can see that the total optical depth is $(p_s/g)M_a k_{abs}$, which is the total mass of the absorber per square meter times the absorption cross-section. We can use (3.19) to show that

$$\frac{d\tau}{dz} = -\frac{\tau}{H} \qquad (3.20)$$

To calculate the energy absorption rate per unit volume, we multiply the irradiance times the density times the absorption coefficient. Using (3.14), (3.17), and (3.20), and defining $\mu = \cos\theta$, we obtain that

$$\text{Absorption rate} = \frac{dF}{dz} = \frac{k_{abs}\rho_a F}{\mu} = -\frac{d\tau}{dz}\frac{F}{\mu} = \frac{F_\infty}{\mu}e^{-\tau/\mu}\frac{\tau}{H} \qquad (3.21)$$

The absorption rate per unit volume peaks where the product of irradiance and absorbing mass cross-section reaches a maximum. We can find this point by differentiating the absorption rate in (3.21) with respect to optical depth and setting the derivative to zero to find the maximum.

$$\frac{d}{d\tau}\left[\frac{F_\infty}{\mu}e^{-\tau/\mu}\frac{\tau}{H}\right] = \frac{F_\infty}{\mu H}e^{-\tau/\mu}\left[1-\frac{\tau}{H}\right] = 0 \tag{3.22}$$

Thus, we derive that the absorption rate peaks at $\tau/\mu = 1$. One may show that the pressure level, where the absorption rate per unit volume is maximum, is given by

$$\frac{p_{\text{max abs}}}{p_s} = \frac{\cos\theta}{H\,k_{\text{abs}}\rho_{\text{as}}} \tag{3.23}$$

where p_s is the surface pressure. The pressure of maximum absorption is proportional to the cosine of the solar zenith angle, so that as the Sun moves toward the horizon the absorption occurs higher in the atmosphere. The pressure at the level of maximum absorption is inversely proportional to the mass of absorber per unit surface area, $H\rho_{\text{as}}$, and to the absorption coefficient k_{abs}.

The heating rate associated with absorption of a downward flux of radiant energy is given by

$$\left.\frac{\partial T}{\partial t}\right|_{\text{rad}} = \frac{1}{c_p\rho}\frac{\partial F}{\partial z} \tag{3.24}$$

where c_p is the specific heat at constant pressure and ρ is the air density. Utilizing (3.14), (3.24) can be written as

$$\left.\frac{\partial T}{\partial t}\right|_{\text{rad}} = \frac{k_{\text{abs}}M_a}{c_p\mu}F \tag{3.25}$$

If the mass mixing ratio of the absorber, M_a, is independent of altitude, then the heating rate will be proportional to the irradiance itself, which is maximum at the outer extremity of the atmosphere. This is the case for absorption of ultraviolet radiation by molecular oxygen and nitrogen in the upper atmosphere, which produces maximum heating rates at very high altitudes and accounts for the rapid increase of temperature with altitude in the thermosphere shown in Fig. 1.2. For an absorber such as ozone, whose mixing ratio peaks sharply in the stratosphere, the heating rate will also peak in the stratosphere. This local maximum in the ozone-heating rate produces the local maximum in the climatological temperature at the stratopause near 50 km.

3.7 INFRARED RADIATIVE TRANSFER EQUATION: ABSORPTION AND EMISSION

We will further develop the radiative transfer equation for a plane, parallel atmosphere in which both emission and absorption by gases occur. The goal of this section is to provide an intuitive understanding of the transfer of longwave radiation through the atmosphere and its importance for climate. Toward this end, some simplifications will be made that cannot be made in accurate calculations of radiative transfer. Readers not interested in the details of infrared radiative transfer in the atmosphere may skip ahead to Sections 3.8 and 3.9, where simple heuristic models are described. In Section 3.10 the results of more accurate calculations will be presented.

We again make the plane, parallel atmosphere approximation under which Earth is considered to be a flat plane, and the properties of the atmosphere depend only on altitude. These are good approximations if the properties of the atmosphere vary slowly in the horizontal compared to the vertical. This is generally true for temperature and humidity because the atmosphere is so thin compared to the radius of Earth. It may not be a good approximation when clouds are present whose horizontal dimension is comparable to their vertical dimension.

3.7.1 Schwarzschild's Equation

Consider the situation depicted in Fig. 3.8 in which a beam of intensity I_v passes upward through a layer of depth dz, making an angle of θ with the vertical direction. The change of intensity along the path through this layer will be equal to the emission from the gas along the path minus the absorption

$$dI_v = E_v - A_v \tag{3.26}$$

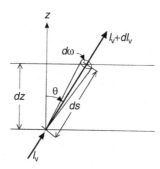

FIGURE 3.8 **The path of upward-directed terrestrial radiance through a plane-parallel atmosphere.**

The absorption can be assumed to follow the Lambert–Bouguer–Beer law.

$$dI_v = E_v - \rho_a\, ds\, k_v I_v \tag{3.27}$$

The emissivity ε_v is defined as the ratio of the emission of a substance to the intensity of radiation in a cavity at equilibrium. The latter is given by *Planck's function* (3.7), so that the emission may be written as follows:

$$E_v = \varepsilon_v B_v(T) \tag{3.28}$$

When collisions with other molecules are much more frequent than radiative transitions of a molecule, the assumption of local thermodynamic equilibrium may be made. This assumption is warranted for most absorption lines in the relatively high-pressure environment of the troposphere and stratosphere. In thermodynamic equilibrium, according to Kirchhoff's law, the emissivity of a substance must be equal to its absorptivity.

With the assumption of local thermodynamic equilibrium, we have $\varepsilon_v = \rho_a\, ds\, k_v$ so that (3.27) becomes

$$dI_v = \rho_a\, ds\, k_v (B_v(T) - I_v) \tag{3.29}$$

Utilizing $dz = \cos\theta\, ds$ and taking the limit of small dz, we obtain a radiative transfer equation

$$\cos\theta \frac{dI_v}{dz} = \rho_a k_v (B_v(T) - I_v) \tag{3.30}$$

In this equation, I_v is a function of both altitude z and zenith angle θ, and all other dependent variables are functions of only z. For infrared radiation, we may reasonably assume that the intensity is independent of azimuth angle, φ (Fig. 3.1), and that the blackbody emission, $B_v(T(z))$, is isotropic, being independent of both θ and φ, but still dependent on z.

To proceed toward a solution of (3.30) we again introduce the optical depth, defined this time from the surface ($z = z_s$) upward.

$$\tau_v(z) = \int_{z_s}^{z} \rho_a k_v\, dz \tag{3.31}$$

Using the definition (3.31) in (3.30) we obtain

$$\cos\theta \frac{dI_v(\tau_v(z), \theta)}{d\tau_v} = B_v(T(\tau_v(z))) - I_v(\tau_v(z), \theta) \tag{3.32}$$

Parentheses in (3.32) show the dependence of the radiance on zenith angle and altitude, the latter being represented parametrically by vertical optical depth.

Multiply (3.32) by the integrating factor $e^{\tau_v/\mu}$, where $\mu = \cos\theta$, to convert it to a form that can be integrated directly.

$$\mu\frac{d}{d\tau_v}\{I_v(\tau_v(z),\theta)e^{\{\tau_v(z)/\mu\}}\} = B_v(T(\tau_v(z)))e^{\{\tau_v(z)/\mu\}} \qquad (3.33)$$

We integrate (3.33) from the surface, where the optical depth is zero, to some arbitrary altitude where we wish to calculate the upward intensity and where the optical depth is $\tau_v(z)$. The integral is performed using τ'_v as the dummy variable of integration.

$$I_v(\tau_v(z),\mu) = I_v(0,\mu)e^{\{-\tau_v(z)/\mu\}} + \int_0^{\tau_v(z)} \mu^{-1}B_v(T(\tau'_v))e^{\{(\tau'_v-\tau_v(z))/\mu\}}d\tau'_v \qquad (3.34)$$

The first term on the right in (3.34) represents the emission from the surface, reduced by the extinction along the path from the surface to the altitude z. The second term represents the summation of the emissions from all of the layers of the atmosphere below the level z, each reduced by the absorption by the atmosphere between the level of emission and the level z.

3.7.2 Simple Flux Forms of the Radiative Transfer Equation Solution

Simplified flux forms of the solution (3.34) can be derived that yield easier understanding of how fluxes of thermal infrared radiation in the atmosphere depend on the vertical temperature profile and the infrared opacity of the atmosphere. A number of approximations are necessary to derive the following simplified equations for the upward $F^\uparrow(z)$ and downward $F^\downarrow(z)$ terrestrial irradiance at some height z in the atmosphere.

$$F^\uparrow(z) = \sigma T_s^4 \mathscr{T}\{z_s,z\} + \mathscr{T}\int_{\mathscr{T}\{z_s,z\}}^1 \sigma T(z')^4 d\mathscr{T}\{z',z\} \qquad (3.35)$$

$$F^\downarrow(z) = \int_{\mathscr{T}\{z,\infty\}}^1 \sigma T(z')^4 d\mathscr{T}\{z',z\}. \qquad (3.36)$$

These equations represent the fluxes of terrestrial radiant energy at all frequencies and integrated over all angles in the upward or downward hemispheres, and would be given in Wm^{-2} (Watts per square meter). The fluxes depend on the temperature and on the flux transmission, $\mathscr{T}\{z', z\}$, which is the fraction of the total terrestrial energy flux that can pass between altitudes z and z' without absorption. The transmission is always between zero, which indicates no transmission, and one, which indicates complete transmission. The transmission approaches one as the two altitudes get arbitrarily close together. As the two altitudes move apart, the transmission decreases at a rate that depends on the absorber amount between them.

The first term on the right in (3.35) is a boundary term and assumes that the ground emits like a blackbody at the temperature of the surface T_s. At the level z the flux density associated with the surface emission is reduced to the fraction $\mathcal{T}\{z_s, z\}$ of its surface value. The integral in (3.35) represents the contribution to the upward flux from the atmosphere below the level z, and in (3.36), the integral represents the contribution to the downward flux at z from the atmosphere above that level. The contributions to these integrals are greatest where the transmission function is changing most rapidly. This occurs when the difference in optical depth between the two locations is close to one.

The net flux of terrestrial radiation is given by the difference between the upward and downward flux:

$$F(z) = F^{\uparrow}(z) - F^{\downarrow}(z) \tag{3.37}$$

The heating rate associated with the divergence of the terrestrial flux density is then

$$\frac{\partial T}{\partial t} = -\frac{1}{\rho c_p} \frac{\partial F}{\partial z} \tag{3.38}$$

Of particular interest for climate are the upward flux of terrestrial radiation at the top of the atmosphere and the downward flux at the surface. From (3.35) and (3.36) these are

$$F^{\uparrow}(\infty) = \sigma T_s^4 \mathcal{T}\{z_s, \infty\} + \int_{\mathcal{T}\{z_s,\infty\}}^{1} \sigma T(z')^4 \, d\mathcal{T}\{z', \infty\} \tag{3.39}$$

$$F^{\downarrow}(z_s) = \int_{\mathcal{T}\{z_s,\infty\}}^{1} \sigma T(z')^4 \, d\mathcal{T}\{z', z_s\}. \tag{3.40}$$

Figure 3.9 illustrates what the two transmission functions in (3.39) and (3.40) might look like, together with a representation of the temperature profile in the atmosphere. The primary contributions to the integrals come from the levels where the transmission functions are changing most rapidly. In the case of outgoing longwave radiation (OLR), $F^{\uparrow}(\infty)$, only a small amount of the emission from the surface is able to escape to space under average terrestrial conditions. Under typical conditions, most of the OLR originates in the troposphere at levels where the temperature is significantly less than the surface value. From (3.39)

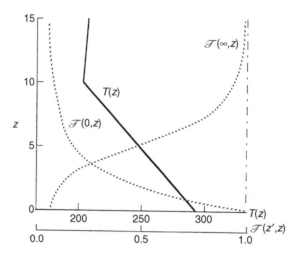

FIGURE 3.9 **Transmission functions and air temperature $T(z)$ as functions of altitude.**

it can be seen that the emission temperature defined in Chapter 2 is a weighted mean of the temperature profile from the surface to the top of the atmosphere, where the weight is the rate at which the transmission function changes with height.

When an absorbing atmosphere is present, the average emission temperature is less than the surface value, and the loss of energy by emission to space is much less than the infrared emission from the surface. This is one component of the greenhouse effect. Another aspect is the supply of energy to the surface by downward emission of terrestrial radiation from the atmosphere. The downward flux of terrestrial radiation at the surface originates in the lower troposphere, where most of the water vapor resides. The temperatures there are relatively warm. In the global mean, this downward flux is larger by nearly a factor of two than the input of energy to the surface from the Sun (Fig. 2.4).

It is interesting to consider the case of an isothermal atmosphere at temperature T_A, overlying a surface of temperature T_s. In this case, the temperature can be taken outside the integrals in (3.39) and (3.40).

$$F^{\uparrow}(\infty) = \sigma T_s^4 \mathcal{T}\{z_s, \infty\} + \sigma T_A^4 (1 - \mathcal{T}\{z_s, \infty\}) \qquad (3.41)$$

$$F^{\downarrow}(z_s) = \sigma T_A^4 (1 - \mathcal{T}\{z_s, \infty\}) \qquad (3.42)$$

From these expressions we obtain the limiting cases for an opaque and a transparent atmosphere:

$$\mathcal{T}\{z_s, \infty\} = 0 \quad \Rightarrow \quad F^{\downarrow}(z_s) = \sigma T_A^4, \quad F^{\uparrow}(\infty) = \sigma T_A^4 \qquad (3.43)$$

$$\mathscr{T}\{z_s,\infty\}=1 \quad \Rightarrow \quad F^{\downarrow}(z_s)=0, \quad F^{\uparrow}(\infty)=\sigma T_S^4 \tag{3.44}$$

In the first limit (3.43) the atmosphere is perfectly opaque to long-wave radiation and emits both upward and downward like a blackbody. In this case, the net longwave at the ground is simply related to the difference in blackbody emission by the atmosphere and surface, and the emission to space is entirely from the atmosphere. In the next section, we consider a model atmosphere made up of layers that are opaque to longwave radiation. In the case of perfect transmissivity equal to one (3.44) the atmosphere has no effect on terrestrial radiation, so that the downward longwave is zero and the surface emission escapes directly to space.

Clouds strongly affect the transmission of terrestrial radiation through the atmosphere. If a cloud is reasonably thick and has a sharp top and bottom, then it can be treated accurately as a perfect absorber of longwave radiation. If such a cloud is present with a bottom at z_{cb} and a top at z_{ct}, then (3.39) and (3.40) are changed.

$$F^{\uparrow}(\infty) = \sigma T_{z_{ct}}^4 \mathscr{T}\{z_{ct},\infty\} + \int_{\mathscr{T}\{z_{ct},\infty\}}^{1} \sigma T(z')^4\, d\mathscr{T}\{z',\infty\} \tag{3.45}$$

$$F^{\downarrow}(z_s) = \sigma T_{z_{cb}}^4 \mathscr{T}\{z_{cb},z_s\} + \int_{\mathscr{T}\{z_{cb},z_s\}}^{1} \sigma T(z')^4\, d\mathscr{T}\{z',z_s\} \tag{3.46}$$

In this case, the downward flux at the ground (3.46) also has a boundary term coming from the bottom of the cloud. It can be shown using the mean value theorem of calculus that if the temperature decreases with altitude, then a cloud will decrease the outgoing flux at the top of the atmosphere, and increase the downward flux at the ground. The amount of the change will depend on the lapse rate and the position of the cloud top and base relative to the transmission function for the clear atmosphere.

3.8 HEURISTIC MODEL OF RADIATIVE EQUILIBRIUM

A layer of atmosphere that is almost opaque for longwave radiation can be crudely approximated as a blackbody that absorbs all terrestrial radiation that is incident upon it and emits like a blackbody at its temperature. For an atmosphere with a large infrared optical depth, the radiative transfer process can be represented with a series of blackbodies arranged in vertical layers. Two layers centered at 0.5 km and 2.0 km altitudes provide a simple approximation for Earth's atmosphere. If we assume that the

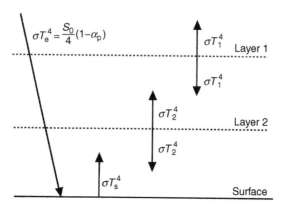

FIGURE 3.10 **Simple two-layer radiative equilibrium model for the atmosphere–earth system, showing the fluxes of radiant energy.**

atmospheric layers are transparent to solar radiation, we have the schematic energy flow diagram shown in Fig. 3.10.

We can solve for all of the unknown temperatures by using the energy balance at each of the layers. If no net energy gain or loss occurs at any of the levels, then the temperatures obtained are the radiative equilibrium values. At the top of the atmosphere we must have energy balance, so that

$$\frac{S_0}{4}(1-\alpha_p) = \sigma T_e^4 = \sigma T_1^4 \qquad (3.47)$$

We thus know immediately that the top layer temperature must equal the emission temperature of the planet, since, in this approximation, the only longwave emission that escapes to space comes from the upper layer. The energy balance at layer 1 is

$$\sigma T_2^4 = 2\sigma T_1^4 \qquad (3.48)$$

The balance at layer 2 yields

$$\sigma T_1^4 + \sigma T_s^4 = 2\sigma T_2^4 \qquad (3.49)$$

And the balance at the surface is

$$\frac{S_0}{4}(1-\alpha_p) + \sigma T_2^4 = \sigma T_s^4 \qquad (3.50)$$

The critical effect of an atmosphere that absorbs and emits longwave radiation appears in (3.50). The energy supplied to the surface by the Sun is augmented by a downward flux of longwave radiation from the atmosphere. This allows the surface temperature to rise significantly above the value it would have in the absence of an atmosphere.

We can use (3.47) through (3.50) to solve for the surface temperature.

$$T_s^4 = 3\frac{(S_0/4)(1-\alpha_p)}{\sigma} = 3T_e^4 \tag{3.51}$$

By extension, if such a model atmosphere has an arbitrary number of layers, n, the surface temperature in equilibrium will be

$$T_s = \sqrt[4]{n+1}\,T_e$$

The radiative equilibrium surface temperature for a two-layer atmosphere is 335 K, which is much hotter than Earth's surface temperature. Radiative equilibrium is not a good approximation for the surface temperature, since we know that latent and sensible heat fluxes remove substantial amounts of energy from the surface and drive a lapse rate that is less than the radiative equilibrium lapse rate.

If the atmosphere absorbs no solar radiation, for a thin layer of emissivity ε near the top of the atmosphere, the energy balance is between absorption of the flux of terrestrial radiation from below and the emission from the layer itself.

$$\varepsilon\sigma T_e^4 = 2\varepsilon\sigma T_{strat}^4 \tag{3.52}$$

where T_{strat} is the temperature at the outer edge of the atmosphere, which we may take to be the stratosphere.

A thin layer of atmosphere near the surface absorbs a fraction ε of the emission from above and below and emits in both directions. The temperature of the air adjacent to the surface, T_{SA}, may be derived from the energy balance there.

$$\varepsilon\sigma T_s^4 + \varepsilon\sigma T_2^4 = 2\varepsilon\sigma T_{SA}^4 \tag{3.53}$$

We can solve for all of the temperatures and obtain the following values.

$$T_1 = 255\text{ K} \quad T_2 = 303\text{ K} \quad T_s = 335\text{ K}$$

$$T_{strat} = 214\text{ K} \quad T_{SA} = 320\text{ K}$$

These temperatures are plotted in Fig. 3.11. In pure radiative equilibrium, the temperatures of the surface and the air in contact with the surface are different. This discontinuity is caused by the absorption of solar radiation at the surface. Such discontinuities are usually greatly suppressed in reality because of efficient heat transport by conduction and convection.

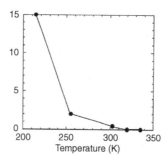

FIGURE 3.11   **Plot of the vertical temperature profile obtained from the simple two-level atmosphere radiative equilibrium model.**

Note that the lapse rate is much greater than the dry adiabatic lapse rate of 9.8 K km^{-1}, dropping by 80 K in the lowest 2 km, so that the radiative equilibrium temperature profile is unstable and any small motion will induce growth of convection.

3.9 CLOUDS AND RADIATION

Clouds consist of liquid water droplets or ice particles suspended in the atmosphere. They are formed by the condensation of atmospheric water vapor when the temperature falls below the saturation temperature. Water droplets and ice particles have substantial interactions with both solar and terrestrial radiation (Fig. 3.12). The nature of these interactions

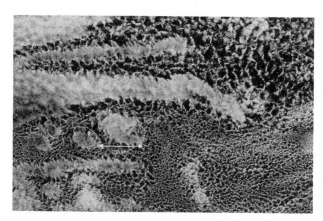

FIGURE 3.12   **Stratocumulus clouds in the southeast tropical Pacific Ocean.** Cells and structures of many scales are present. Some cells are closed and bright, while others are open and dark, revealing the lower albedo of the underlying ocean. *NASA MODIS Aqua image, April 17, 2010.*

depends on the total mass of water, the size and shape of the droplets or particles, and their distribution in space. The problem is often simplified by assuming that clouds are uniform and infinite in the horizontal, which is called the plane-parallel cloud assumption. If the droplet size distribution and the vertical distribution of humidity are assumed, then the cloud albedo and absorption depend on the total liquid water content of the cloud and the solar zenith angle. Cloud liquid water content is defined as the total mass of cloud water in a vertical column of atmosphere per unit of surface area.

For a particle of radius r the amount by which it reduces an incident beam of radiation I_0 is $\pi r^2 k_e I_0$, where k_e is the extinction efficiency. In general, this extinction will be composed of extinction by scattering and extinction by absorption, so that $k_e = k_s + k_a$. The optical depth for a layer of particles of depth h can be written

$$\tau = \pi h \int_0^\infty k_e(r) r^2 n(r) \, dr \tag{3.54}$$

where $n(r)$ is the number density of particles of radius r. For particles that are large compared to the wavelength of radiation the extinction efficiency $k_e \simeq 2$. If the distribution of particle radii is peaked near an average value $\bar{r}$, then one can write approximately that,

$$\tau = 2\pi h \bar{r}^2 N \tag{3.55}$$

where $N = \int_0^\infty n(r) \, dr$ is the total particle density and h is the depth of the cloud. For a water cloud, the total mass of liquid water per unit surface area of the cloud (LWC) is

$$\text{LWC} = \frac{4}{3} \pi \bar{r}^3 \rho_L h N \tag{3.56}$$

where ρ_L is the density of water and h is the depth of the cloud. Combining (3.55) and (3.56) the optical depth can be expressed as

$$\tau = \frac{3}{2} \frac{LWC}{\rho_L \bar{r}} \tag{3.57}$$

So for fixed liquid water content, the optical depth of the cloud varies inversely with the average particle radius. This is because refraction of light occurs at the interface between water and air, and smaller particles have a larger ratio of surface area to volume (and mass) than larger particles. Humans can potentially alter the clouds by releasing extra cloud condensation nuclei, the particles on which cloud droplets preferentially form. With more condensation nuclei, the water is spread over more droplets and the average radius decreases, brightening the clouds.

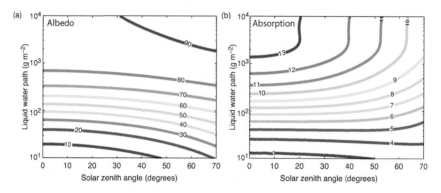

FIGURE 3.13   **The dependence of (a) cloud albedo and (b) cloud absorption on cloud liquid water path and solar zenith angle calculated for plane parallel clouds.** Values are given in percent. Calculation is for a cloud above a nonreflective surface and an effective radius of the cloud droplets of 14 μm was used.

Model calculations of the cloud albedo and absorption for plane-parallel water clouds are shown in Fig. 3.13. The albedo increases with the total water content or depth of the cloud and also with the solar zenith angle. The albedo increase with liquid water content is most rapid for smaller amounts of liquid water. As the cloud becomes very thick, the cloud albedo slowly approaches a limiting upper value and becomes insensitive to further increases in cloud mass. In very thick clouds, most of the solar radiation is scattered before it can reach the particles deep in the cloud, and radiation that is scattered from particles deep in the cloud is unlikely to find its way back out to space. For these reasons, increases in the optical thickness of a cloud eventually cease to make much difference in the reflectivity of the cloud. The variation of albedo with zenith angle is most rapid when the Sun is near the horizon, and least when the Sun is overhead. Absorption of solar radiation by plane-parallel clouds decreases with increasing zenith angle because radiation that is reflected to space at the higher zenith angles penetrates less deeply into the cloud and is therefore less likely to be absorbed. The absorption increases with liquid water content for an overhead sun, but the fraction absorbed is always much less than the fraction reflected.

The variations of the albedo of typical clouds in the atmosphere are dominated by the column amount of liquid water and ice in the cloud. Nonetheless, the albedo of clouds is sensitive to the droplet size as suggested by (3.57). Figure 3.14 shows how the calculated albedo of clouds changes as the radius of the droplets is varied, while keeping the liquid water content fixed. The albedo is greatest for smaller droplets, principally because these present a larger surface area for the same mass.

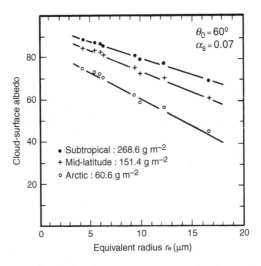

FIGURE 3.14 **The dependence of planetary albedo on the size of cloud droplets.** *From Slingo and Schrecker (1982). Reprinted with permission from the Royal Meteorological Society.*

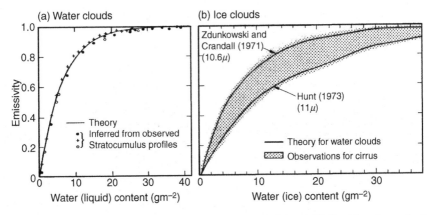

FIGURE 3.15 **The dependence of the longwave emissivity on (a) liquid water content and (b) ice content.** *(a) Slingo et al. (1982); reprinted with permission from the Royal Meteorological Society (b) Griffith et al. (1980); reprinted with permission from the American Meteorological Society.*

Clouds absorb terrestrial radiation very effectively. Figure 3.15 shows the emissivity of water and ice clouds as a function of liquid water content. Clouds become opaque to longwave radiation when the liquid water path exceeds about 20 gm^{-2}. If this liquid water path is achieved in an altitude range where the temperature is essentially uniform, then cloud surfaces can be assumed to absorb and emit terrestrial radiation essentially like black-bodies with the temperature at the edge of the cloud. This assumption is a good one except for thin clouds such as cirrus, which may be partially transparent to longwave radiation. Comparison of Figs. 3.13 and 3.15 shows that the albedo of clouds continues to increase with additional liquid water content long after the cloud has become opaque to longwave radiation.

3.10 RADIATIVE–CONVECTIVE EQUILIBRIUM TEMPERATURE PROFILES

As a reasonably straightforward method of attempting to understand the effects of radiative transfer on climate, one can solve the radiative transfer equation for global mean terrestrial conditions. This involves construction of appropriate models for the transmission of the various band systems of importance in the atmosphere, insertion of these into a computational analog of the radiative transfer equation, and iteration to obtain a steady balance solution. Such models are a much more sophisticated version of the simple radiative equilibrium model discussed in Section 3.8.

The variables that determine the fluxes of radiant energy in the atmosphere include the atmospheric gaseous composition, the aerosol and cloud characteristics, the surface albedo, and the insolation. Since horizontal transport of energy by atmospheric and oceanic motion affects the local climate, it is simplest to calculate the radiative equilibrium for conditions averaged over the globe. In a global-mean model the temperature and all other variables depend only on altitude, and the globally averaged insolation and solar zenith angle are appropriate. To understand the basic radiative energy balance of Earth, we need to specify the following:

1. H_2O: Water vapor is the most important gas for the transfer of radiation in the atmosphere. Its distribution is highly variable. The sources and sinks (evaporation and condensation) are determined by the climate itself, and they are fast compared to the rate at which the atmosphere's motion mixes moist and dry air together. Water vapor has a vibration–rotation band near 6.3 μm and a large number of closely spaced rotation lines at wavelengths longer than about 12 μm. It is also the principal absorber of solar radiation in the troposphere. Since the saturation vapor pressure of water is very sensitive to temperature, the relative humidity is sometimes specified in calculations, rather than the absolute humidity, when the humidity cannot be directly solved for. This allows the water-vapor feedback to be simulated in the model.

2. CO_2: The mixing ratio of carbon dioxide is increasing about 2 ppmv per year primarily because of fossil fuel combustion. In 2014 the value was about 400 ppmv, compared to a value in 1750 of about 280 ppmv. Because sources and sinks of CO_2 are slow compared to the time it takes the atmosphere to mix thoroughly, its mixing ratio is

approximately constant with latitude and altitude up to about 100 km. The strong vibration–rotation band of CO_2 at 15 μm is important for longwave radiative transfer. A significant amount of solar radiation is also absorbed by carbon dioxide.

3. O_3: Ozone has fast sources and sinks in the stratosphere where most of the atmospheric ozone resides. Near the surface, ozone is produced in association with photochemical smog. Its concentration in the middle and upper stratosphere is dependent on temperature, ultraviolet radiation, and a host of photochemically active trace species. Ozone has a vibration–rotation band near 9.6 μm that is important for longwave energy transfer, and also has a dissociation continuum that absorbs solar radiation between 200 nm and 300 nm, which protects life from harmful ultraviolet radiation. Absorption of solar radiation by ozone heats the middle atmosphere and causes the temperature increase with height that defines the tropopause and stratosphere. Stratospheric ozone has decreased because of the chlorine in manmade chemicals. Tropospheric ozone has increased because of human produced pollution, especially nitrous oxides from fuel combustion.

4. *Other greenhouse gases:* Many other atmospheric gases absorb and emit terrestrial radiation and contribute to the greenhouse effect. Many of these are changing in response to human activities. They will be discussed more fully in Chapter 13.

5. *Aerosols:* Atmospheric aerosols of various types affect the transmission of both solar and terrestrial radiation. A layer of sulfuric acid aerosols exists near 25 km in the mid-stratosphere, which increases and decreases in association with volcanic eruptions. Nearer the surface many different types of aerosols are produced by natural and human activities. Some of these absorb solar radiation (e.g., soot) whereas others mostly scatter radiation (e.g., sulfuric acid and organic aerosols).

6. *Surface albedo:* The surface albedo is highly variable from location to location in land areas, depending on the type and condition of the surface material and vegetation (see Chapter 4). Over open ocean the albedo is low but does vary with solar zenith angle and sea state. When the surface is snow covered, its albedo is generally much higher than when surface ice is not present.

7. *Clouds:* Clouds vary considerably in amount and type over the globe. They have very important effects on longwave and solar energy transfer in the atmosphere. The distribution in time and space and the optical properties of clouds are important for climate. For a global-mean radiative equilibrium calculation, cloud radiative properties must be specified. The simplest approach is to assume plane-parallel clouds and specify their distribution in the vertical. The optical properties of the clouds must also be specified. In global warming calculations, clouds must be determined as part of the climate model.

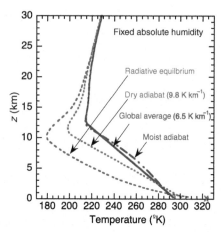

FIGURE 3.16 Calculated temperature profiles for radiative equilibrium, and thermal equilibrium with lapse rates of 9.8°C km^{-1}, 6.5°C km^{-1}, and the moist adiabatic lapse rate with clear skies. Absolute humidity, CO_2 (390 ppmv) and ozone have been specified from observations.

Figure 3.16 shows a calculated temperature profile that is in radiative equilibrium. Atmospheric temperatures in radiative equilibrium decrease rapidly with altitude near the surface and, as previously noted, the surface temperature is different from that of the air adjacent to the surface. In the troposphere, radiative equilibrium temperature profiles are hydrostatically unstable in the sense that parcels of air that are elevated slightly will become buoyant and continue to rise. In the real atmosphere, atmospheric motions move heat away from the surface and mix it through the troposphere. Figure 2.4 indicates that 60% of the energy removal from the surface is done by the transport of heat and water vapor by atmospheric motions and only 40% by net longwave radiation emission. The global mean temperature profile of Earth's atmosphere is not in radiative equilibrium, but rather in radiative–convective equilibrium. To obtain a realistic global-mean vertical energy balance, the vertical flux of energy by atmospheric motions must be included.

The simplest method by which the effect of vertical energy transports by motions can be included in a global-mean radiative transfer model is a procedure called *convective adjustment*. Under this constraint the lapse rate is not allowed to exceed a critical value, which must be specified. One can choose the observed global mean value of 6.5 K km^{-1}, or some other value as desired. Where radiative processes would make the lapse rate greater than the specified maximum value, a nonradiative upward heat transfer is assumed to occur that maintains the specified lapse rate while conserving energy. This artificial vertical redistribution of energy is intended to represent the effect of atmospheric motions on the vertical temperature profile without explicitly calculating nonradiative energy fluxes or atmospheric motions. In a global mean model, this "adjusted" layer extends from the surface to the tropopause.

A temperature profile that is in energy balance when radiative transfer and convective adjustment are taken into account may be called a *radiative–convective equilibrium* or *thermal equilibrium* profile. Thermal equilibrium profiles for assumed maximum lapse rates of 6.5°C km^{-1} and the dry adiabatic lapse rate of 9.8°C km^{-1} are also shown in Fig. 3.16. The thermal equilibrium profile obtained with a lapse rate of 6.5°C km^{-1} is close to the observed global mean temperature profile. No a priori reason exists for choosing a 6.5°C km^{-1} adjustment lapse rate other than that it corresponds to the observed global-mean value. However, the assumed lapse rate of 6.5°C km^{-1} is close to what is obtained if the moist adiabatic lapse rate is assumed. The maintenance of the lapse rate of the atmosphere is complex and involves many processes and scales of motion. Note that when the chosen lapse rate is larger, the surface temperature increases, when the specific humidity is fixed as a function of pressure. When the lapse rate is increased, the emission from the atmospheric water vapor occurs at a colder temperature relative to the surface, and the surface temperature must increase because of the enhanced greenhouse effect. Because the albedo has not been changed, the emission temperature of the Earth is about the same, roughly where the 6.5°C km^{-1} and 9.8°C km^{-1} profiles cross.

One use of a climate model is to understand what factors are most important and how changes in these factors will affect the climate. In particular, the one dimensional radiative–convective equilibrium model is useful for understanding the role of trace gases and clouds in determining the temperature profile. Figure 3.17 shows three equilibrium profiles obtained

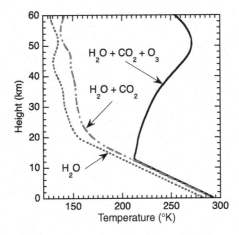

FIGURE 3.17 **Thermal equilibrium profiles for three cloudless atmospheres obtained with a critical lapse rate of 6.5 K km^{-1}.** One atmosphere has water vapor only; one includes water vapor and carbon dioxide; and the third contains water vapor, carbon dioxide, and ozone.

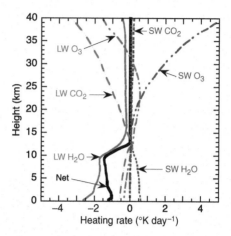

FIGURE 3.18 **Radiative heating rate profiles for a clear atmosphere.** LWH_2O, $LWCO_2$, and LWO_3 show the heating rates associated with longwave cooling by water vapor, carbon dioxide, and ozone, respectively. The SW prefix indicates the heating rate associated with solar absorption by each of these gases. Net is the sum of the solar and longwave radiative heating rates contributed by all gases[1].

with different gaseous compositions, but without clouds. With only water vapor present a reasonable approximation to the observed profile is obtained except that the stratosphere is absent. Carbon dioxide with a mixing ratio of 300 ppm raises the temperature about 10 K above the equilibrium obtained with only water vapor present. A sharp tropopause and the increase of temperature with height that characterizes the stratosphere appear only when solar absorption by ozone is included in the model.

The contributions of individual gases to the heating rate in radiative-convective equilibrium are shown in Fig. 3.18. In the stratosphere, the lapse rate for radiative equilibrium is never large and positive, so convective adjustment is not required. The first-order balance is between heating produced by solar absorption by ozone and cooling produced by longwave emission from carbon dioxide. The troposphere is not in radiative equilibrium, and convective heat transfer from the surface balances a net radiative cooling rate of about 1.5 K day^{-1}. A net positive radiative imbalance exists at the surface. In a clear atmosphere, cooling by water vapor alone provides a good approximation to the net

---

[1]Note that because of overlap between water vapor and carbon dioxide absorption lines, the heating rates in the troposphere do not add up. The contributions were calculated by taking the equilibrium temperature profile and calculating the heating rate for an atmosphere that only contained the gas in question.

longwave cooling. In the troposphere, longwave cooling from carbon dioxide is approximately balanced by solar absorption by water vapor. From the results of radiative–convective equilibrium calculations presented in Figs. 3.17 and 3.18, we conclude that water vapor is by far the preeminent greenhouse gas in the atmosphere for today's climate. Note that water vapor concentrations vary with temperature, so that processes internal to the climate system adjust water vapor concentrations on short time scales.

3.11 THE ROLE OF CLOUDS IN THE ENERGY BALANCE OF EARTH

It is possible to use Earth-orbiting satellites to accurately measure the radiative fluxes of energy entering and leaving Earth. If the spatial resolution of the measurements of the energy fluxes provided by the instrument on the satellite is great enough, then cloud-free scenes may be identified. These cloud-free scenes can be averaged together to estimate the clear-sky radiation budget. If these cloud-free scenes are taken to represent the atmosphere in the absence of clouds, then the difference between the cloud-free radiation budget and the average of all scenes represents the effect of clouds on the radiation budget. We can call the effect of clouds on the radiation budget the cloud radiative effect on the energy balance. The net cloud radiative effect is the difference between the net radiation at the top of the atmosphere and what the net radiation would be if clouds were removed from the atmosphere leaving all else unchanged.

$$\Delta R_{\text{TOA}} = R_{\text{Average}} - R_{\text{Clear}} \tag{3.58}$$

The net cloud radiative effect can be decomposed into its longwave and shortwave components. Since net radiation is absorbed solar minus outgoing longwave radiation,

$$R_{\text{TOA}} = Q_{\text{abs}} - \text{OLR} \tag{3.59}$$

then

$$\Delta R_{\text{TOA}} = \Delta Q_{\text{abs}} - \Delta \text{OLR} \tag{3.60}$$

The first term on the right is the shortwave cloud effect and the second is the longwave effect. Note that a reduction in OLR is an increase in net radiation.

Table 3.2 shows estimates of the globally and annually averaged radiation budget components for average conditions, cloud-free conditions, and the difference between them. In round numbers, the observations indicate that clouds increase the albedo from 15% to 29%, which results in

TABLE 3.2   Cloud Radiative Effect on the Top-of-Atmosphere Global Energy Balance as Estimated from Satellite Measurements

	Average	Cloud free	Cloud effect
OLR	240	266	+26
Absorbed solar radiation	240	288	−47
Net radiation	+0.56	+22	−21
Albedo	29%	15%	+14%

Irradiances are given in Wm^{-2} and albedo in percent.
From CERES data as described in Loeb et al. 2009.

a reduction of absorbed solar radiation of 47 Wm^{-2}. This cooling is offset somewhat by the greenhouse effect of clouds, which reduce the OLR by about 26 Wm^{-2}. The net cloud radiative effect is thus a loss of about 21 Wm^{-2}. The meaning of this number is that, if clouds could suddenly be removed without changing any other climate variable, then Earth would begin to gain 21 Wm^{-2} in net radiation and consequently begin to warm up. Of course clouds are not an external forcing to the climate system, but are rather part of it, and their radiative effects are part of the internal adjustment of the climate. It is important to consider how clouds of different types can affect the energy budget and thereby the temperature.

We can examine how clouds with different properties affect the equilibrium climate using the one-dimensional global thermal equilibrium model explored in the previous section. To do this we consider three simple types of clouds as defined in Table 3.3 and assume that they cover 100% of the Earth. The "average cloud" was chosen to produce the observed global mean effect of clouds on the Earth's energy balance shown in Table 3.2, and also give a surface temperature near the observed value of 288 K. In reality, the cloud effect on the energy balance is caused by clouds of varying types and characteristics that cover only a portion of the globe.

TABLE 3.3   Cloud Properties Used to Compute Fig. 3.19

Cloud	Base (km)	Top (km)	L/IWP (gm^{-2})	Particle radius (μm)
Average	2.0	2.5	36	15
Low	2.0	2.5	100	15
High	12.5	13	29	15

L/IWP means the liquid or ice (for high cloud) mass of the cloud.

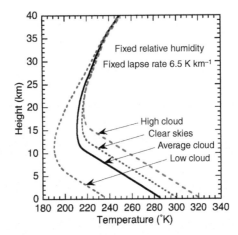

FIGURE 3.19 **Thermal equilibrium temperature profiles for atmospheres with various cloud distributions as given in Table 3.3.** The lapse rate is constrained to not exceed 6.5 K km^{-1} and the absolute humidity profile is set to approximately the currently observed value.

Here we represent that more complex effect with a single, relatively thin, relatively low cloud that covers 100% of the planet. To contrast with this average cloud we also consider a "high cloud", with the same albedo as the average cloud but at a higher altitude, and a "low cloud" with a higher albedo at the same altitude as the average cloud.

Figure 3.19 shows the effect on the thermal equilibrium temperature profile of inserting the clouds described in Table 3.3. In these calculations the relative humidity distribution with height is assumed fixed, so that as the climate warms or cools, the water vapor changes. In reality it is more likely that the relative humidity would remain nearly constant and the specific humidity would increase with surface temperature. We will discuss the effects of this later. The average cloud cools the surface compared to clear skies, as expected, since it introduces a negative net change to Earth's energy balance. Low clouds with high albedos greatly reduce the temperature at the surface and in the troposphere, because they increase the albedo of Earth, but their tops are at low altitudes and high temperatures, so they don't have a very strong greenhouse warming effect.

The addition of high clouds can cause the surface temperature to exceed the value obtained for both average cloud and cloud-free conditions, since they greatly reduce the outgoing longwave and, if they are relatively thin, do not reduce the absorbed solar radiation by enough to offset the reduced longwave emission. The high cloud specified in Table 3.3 was chosen to produce the same albedo as the average cloud, but because its

TABLE 3.4   Longwave and Shortwave Upward Top-of-Atmosphere Fluxes (Wm^{-2}) and Albedos for the Cases Shown in Fig. 3.19

	LW (up)	SW (up)	Albedo
Clear	291	51	0.15
Average	240	102	0.30
Low	193	149	0.44
High	240	102	0.30

Note how the average cloud and high cloud cases have the same top of atmosphere energy fluxes, but very different temperatures.

top is so high and its greenhouse effect so strong, it causes the surface temperature in equilibrium to be much warmer, even though the net absorbed and emitted radiation are the same as the average cloud case. The OLR, reflected shortwave and albedo for each of the cases shown in Fig. 3.19 are shown in Table 3.4. It is an intriguing aspect of high, thin clouds that they can greatly change the surface temperature in equilibrium without changing the energy balance at the top of the atmosphere. Thus we illustrate that the surface temperature can be very different for the same top-of-atmosphere radiation budget, depending on the types of clouds that the climate produces.

We can also calculate the cloud radiative effect of each of these cloud types. Remember that the cloud radiative effect is the effect on the energy balance of inserting the cloud without changing the temperature or humidity profiles. The numbers in Table 3.5 are obtained by using the temperature profile for the average cloud case shown in Fig. 3.19, and then calculating the top-of-atmosphere energy fluxes with and without each type of cloud. These numbers can be compared to the observed

TABLE 3.5   Cloud Forcing for the Cloud Types in Table 3.2 if inserted into an Atmosphere With the Average-Sky Humidity and Thermal Equilibrium Temperature Profile Shown in Fig. 3.19

	LW (up)	LWCF	SW (up)	SWCF	Net CF
Clear	265	0.0	51	0.0	0.0
Mean	240	24	102	−50	−26
Low	238	27	149	−98	−71
High	125	140	102	−50	+90

LWCF and SWCF are the changes in longwave and shortwave fluxes, respectively, as a result of inserting a cloud into the atmosphere. Net CF is their sum.

values shown in Table 3.2. Because the low cloud has a relatively warm top, its effect on escaping longwave radiation is modest and its net cloud radiative effect is -71 Wm^{-2}. On the other hand, the high cloud has a very cold top and greatly reduces the escaping longwave radiation, so that its net radiative effect is $+90$ Wm^{-2}. So the difference in net radiative effect between the high cloud and the low cloud is 161 Wm^{-2}, which is huge.

3.12 A SIMPLE MODEL FOR THE NET RADIATIVE EFFECT OF CLOUDINESS

We can illustrate the relative roles of the reflection of solar radiation and trapping of longwave radiation by clouds with a very simple model of their effect on the global energy balance at the top of the atmosphere. The energy balance at the top of the atmosphere is the difference between the absorbed solar radiation and the outgoing longwave radiation (3.59).

$$R_{\text{TOA}} = \frac{S_0}{4}(1-\alpha_{\text{p}}) - F^{\uparrow}(\infty) \tag{3.61}$$

where S_0 is the total solar irradiance and α_{p} is the albedo, so that $(S_0/4)$ $(1-\alpha_{\text{p}}) = Q_{\text{abs}}$ is the absorbed solar radiation.

We wish to calculate the difference in the net radiation that results from adding a cloud layer with specified properties to a clear atmosphere.

$$\Delta R_{\text{TOA}} = R_{\text{Cloudy}} - R_{\text{Clear}} = \Delta Q_{\text{abs}} - \Delta F^{\uparrow}(\infty) \tag{3.62}$$

Suppose that we can specify the albedo for both clear and cloudy conditions, so that the difference in absorbed solar radiation is

$$\begin{aligned} \Delta Q_{\text{abs}} &= \frac{S_0}{4}(1-\alpha_{\text{Cloudy}}) - \frac{S_0}{4}(1-\alpha_{\text{Clear}}) \\ &= -\frac{S_0}{4}(\alpha_{\text{Cloudy}} - \alpha_{\text{Clear}}) = -\frac{S_0}{4}\Delta\alpha_{\text{p}} \end{aligned} \tag{3.63}$$

To calculate the change in OLR, we subtract (3.39) from (3.45):

$$\Delta F^{\uparrow}(\infty) = F^{\uparrow}_{\text{Cloudy}}(\infty) - F^{\uparrow}_{\text{Clear}}(\infty) \tag{3.64}$$

$$\Delta F^{\uparrow}(\infty) = \sigma T^4_{z_{\text{ct}}} \mathcal{T}\{z_{\text{ct}}, \infty\} - \sigma T^4_s \mathcal{T}\{z_s, \infty\} - \int_{\mathcal{T}\{z_s, \infty\}}^{\mathcal{T}\{z_{\text{ct}}, \infty\}} \sigma T(z')^4 \, d\mathcal{T}\{z', \infty\} \tag{3.65}$$

If the top of the cloud is above most of the gaseous absorber of longwave radiation, which is water vapor, then we may make the approximation

$$\mathcal{T}\{z_{\text{ct}}, \infty\} \approx 1.0 \tag{3.66}$$

in which case (3.65) becomes

$$\Delta F^{\uparrow}(\infty) = \sigma T_{z_{ct}}^4 - \sigma T_s^4 \mathcal{T}\{z_s, \infty\} - \int_{\mathcal{T}\{z_s, \infty\}}^1 \sigma T(z')^4 \, d\mathcal{T}\{z', \infty\} \qquad (3.67)$$

or

$$\Delta F^{\uparrow}(\infty) = \sigma T_{z_{ct}}^4 - F_{\text{Clear}}^{\uparrow}(\infty) \qquad (3.68)$$

Inserting (3.63) and (3.68) into (3.62) gives an approximate formula for the change in net radiation at the top of the atmosphere that is produced by the addition of clouds to a clear atmosphere.

$$\Delta R_{\text{TOA}} = -\frac{S_0}{4}\Delta\alpha_p + F_{\text{Clear}}^{\uparrow}(\infty) - \sigma T_{z_{ct}}^4 \qquad (3.69)$$

If the cloud top is above most of the longwave absorber, then (3.69) indicates that the change in net radiation produced by the cloud depends on the albedo contrast between clear and cloudy conditions and on the temperature at the cloud top. Since most of the water vapor is in the first few kilometers of the atmosphere, the approximation (3.67) is qualitatively correct for cases with cloud tops above 4 or 5 km.

From (3.69), it is possible that the albedo contrast and the cloud top temperature can be such that the cloud produces no change in the net radiation. The condition for this is obtained by setting $\Delta R_{\text{TOA}} = 0$ in (3.69) and solving for the cloud top temperature.

$$T_{z_{ct}} = \left\{ \frac{-(S_0/4)\Delta\alpha_p + F_{\text{Clear}}^{\uparrow}(\infty)}{\sigma} \right\}^{1/4} \qquad (3.70)$$

If we assume that the temperature decreases with a lapse rate Γ from a surface value of T_s, then the temperature of the cloud top can be related to its altitude.

$$T_{z_{ct}} = T_s - \Gamma z_{ct} \qquad (3.71)$$

We can use (3.70) and (3.71) to solve for the cloud top altitude for which the reduction in OLR will just cancel the reduction in absorbed solar radiation associated with the presence of a cloud. For numeric values, we can use a solar irradiance of 1360 Wm^{-2}, a clear-sky OLR of 265 Wm^{-2}, a surface temperature of 288 K, and a lapse rate of 6.5 K km^{-1}. These are all reasonable global-mean values. The resulting curve of cloud top altitude versus albedo contrast is shown as the heavy curve in Fig. 3.20. Clouds with albedo contrasts and altitudes that fall along the heavy line have no

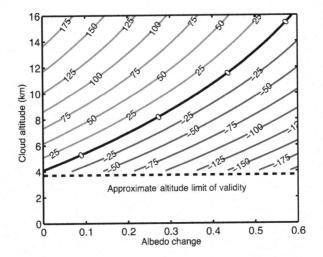

FIGURE 3.20 **Contours of change in net radiation at the top of the atmosphere caused by the insertion of a cloud into a clear atmosphere, plotted against cloud top altitude and the planetary albedo contrast between cloudy and clear conditions.** The net radiation changes are calculated with the approximate model described in Section 3.8 that is invalid for clouds with tops lower than about 4 km.

net effect on the energy balance at the top of the atmosphere. Those that fall below the line will produce a reduction in net radiation, or a cooling, and those above will produce warming. One can obtain the approximate cloud albedo enhancement for global average conditions by adding the clear-sky albedo, which is about 15%. As the cloud top rises, the albedo contrast between cloudy and clear conditions that will just balance the OLR change also increases. Clouds with high cold tops and low albedos can cause a significant positive change in net radiation, while low bright clouds can cause a large negative change in net radiation at the top of the atmosphere.

3.13 OBSERVATIONS OF REAL CLOUDS

The observed distribution of clouds has been estimated in two ways. Surface observers have recorded the type and fractional distribution of clouds and a long record of such observations has been compiled into a cloud climatology (Warren et al., 1986, 1988). Since about 1980, clouds have been systematically characterized from the observations of visible and infrared radiation taken from Earth orbiting satellites (Rossow and

Schiffer, 1991). Each of these data sets has its strengths and weaknesses related to the viewing geometry (up versus down) and the instrumentation (the human eye versus a radiometer). Surface observations have a much better view of cloud base, whereas satellite measurements see the tops of the highest clouds very well and provide a more direct means of estimating the visible optical depth of the clouds. Recently, radars and lidars have been flown in space that can actively scan for the vertical structure of clouds and aerosols.

Figure 3.21 shows global maps of the fractional area coverage of clouds with tops at pressures lower than 440 mb (high clouds), clouds with tops at pressures greater than 680 mb (low clouds) and clouds with tops at any pressure (total cloud amount). High clouds are concentrated in the convection zones of the tropics over equatorial South America and Africa, and a major concentration exists over Indonesia and the adjacent regions of the eastern Indian and western Pacific Oceans. Low clouds are most prevalent in the subtropical eastern ocean margins and in middle latitudes. The low cloud concentrations in the eastern subtropical oceans are associated with lower than average sea surface temperature (SST) (Fig. 7.14) and consist of stratocumulus clouds trapped below an inversion. Low clouds are heavily concentrated over the oceanic regions and are less commonly observed over land. The total cloud cover also shows a preference for oceanic regions, particularly in mid latitudes where the total cloud cover is greatest. Minima in total cloud cover occur in the subtropics in desert regions, but regions with low total cloud amounts also occur over the Caribbean Sea and over the southern subtropical zones of the Pacific, Atlantic, and Indian oceans.

Satellite measurements of the broadband energy flux also enable measurements of the spatial distribution of the cloud radiative effects whose global values are given in Table 3.2. The longwave cloud effect is the reduction of the OLR by the clouds, and so is a positive contribution to the radiation budget or warming influence on the surface climate. The largest contributions are made in the convective regions and the *intertropical convergence zone* (ITCZ) in the tropics where high clouds with cold tops are abundant (Fig. 3.22a). The reduction of absorbed solar radiation is also relatively large in these regions, since the deep convective clouds also have high albedos, but low clouds in middle latitudes are also very effective in reducing the absorbed solar radiation (Fig. 3.22b). The net effect of clouds on the energy budget at the top of the atmosphere is generally smaller than its longwave and shortwave components because in most cases they are of opposite sign. The largest net contributions are reductions in net radiation by low clouds in high latitudes and in the stratus cloud regions in the eastern subtropical oceans (Fig. 3.22c).

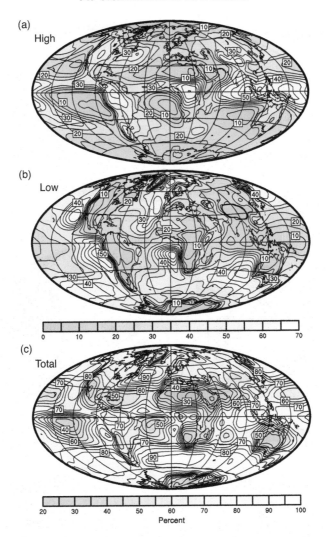

FIGURE 3.21 Annual average cloud fractional area coverage in percent estimated from satellite data under the international satellite cloud climatology project averaged for the period 1983–2009. Contour interval is 5%, lighter shade means more cloud. (a) Clouds with tops higher than 440 mb, (b) clouds with tops lower than 680 mb, and (c) all clouds. *ISCCP, Rossow and Schiffer (1991).*

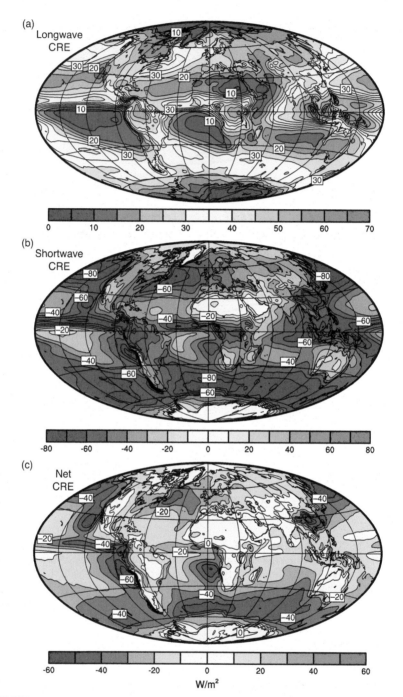

FIGURE 3.22 **Annual average cloud radiative effects in Wm^{-2} estimated from satellite data by the clouds and the Earth's radiant energy system (CERES) experiment.** (a) Reduction of OLR caused by clouds, (b) increase in absorbed solar radiation caused by clouds (note values are negative), and (c) increase in net radiation caused by clouds. *Wielicki et al. (1996).*

EXERCISES

1. Suppose a gas that absorbs solar radiation has a uniform mixing ratio of 1 g kg^{-1} and an absorption cross-section of $5 \text{ m}^2 \text{ kg}^{-1}$. At what altitude will the maximum rate of energy absorption per unit volume occur? Assume an isothermal atmosphere with $T = 260$ K, and a surface pressure of 1.025×10^5 Pa, and that the Sun is directly overhead.

2. In problem 1, if the frequency range at which the absorption is taking place contains 5% of the total solar energy flux, what is the heating rate in degrees per day at the level of maximum energy absorption? What is the heating rate one scale height above and below the level of maximum energy absorption? Use the globally averaged insolation.

3. Do problem 1 with a solar zenith angle of $45°$. Discuss the difference the angle makes.

4. Derive (3.23) using (3.22), (3.19) and the ideal gas law.

5. Use the model of Fig. 3.10, but distribute the solar heating such that $0.3\sigma T_e^4$ is absorbed in each of the two atmospheric layers and the remaining $0.4 \, \sigma T_e^4$ is absorbed at the surface. Calculate the new radiative equilibrium temperature profile. How does it differ from the case where all of the solar heating is applied at the surface?

6. Place the two layers in the model of Fig. 3.10 at 2.5 and 5.0 km. Assume a fixed lapse rate of 6.5 K km^{-1}. Derive energy balance equations that include an unknown convective energy flux from the surface to the lower layer and from the lower layer to the upper layer. Solve for the temperature profile in thermal equilibrium and find the required convective energy fluxes from the surface and the lower layer. (*Hint:* Start from the top and work down.) How do the radiative and convective fluxes compare with the proportions given in Fig. 2.4?

7. For the conditions of problem 5, calculate the pure radiative equilibrium temperature profile with no convective adjustment. Plot the thermal equilibrium and pure radiative equilibrium temperature profiles from this model with the levels at 2.5 and 5.0 km and compare them with the profiles shown in Fig. 3.16. What happens to this comparison if the top level is moved up or down by 2 km? Is the dependence on the height of the layer reasonable? How do the required convective fluxes change as you move the top level up and down?

8. Suppose that tropical convective clouds give an average planetary albedo of about 0.6 compared to the cloud-free albedo of about 0.1. The insolation is about 400 Wm^{-2}, and the cloud-free OLR is about 280 Wm^{-2}. Use the simplified model of Section 3.11 to find the cloud top temperature for which the net radiative effect of these clouds will be zero. Are such temperatures observed in the tropical troposphere, and, if so, where? (Refer to Fig. 1.3.) If the surface temperature is 300 K and

the average lapse rate is 5 K km^{-1}, at what altitude would the cloud top be to make the longwave and shortwave effects of the cloud just equal and opposite?

9. For the conditions of problem 7, what is the rate of net radiative energy loss if the cloud albedos are 0.7 rather than 0.6? By how much would you need to lower the cloud tops to produce an equal reduction in net radiation?

The Energy Balance of the Surface

4.1 CONTACT POINT

The surface of Earth is the boundary between the atmosphere and the land or ocean. Defining the location of this boundary can be difficult over a highly disturbed sea or over land surfaces with a variable plant canopy. We will assume that the location of the surface can be appropriately defined, and we will treat it as a simple interface between two media. However, in considering the important energy exchange processes we must include the atmosphere and oceanic boundary layers and the first few meters of soil. The energy fluxes across the surface are as important to the climate as the fluxes at the top of the atmosphere, especially because the climate at the surface is of most practical significance. The surface energy balance determines the amount of energy available to evaporate surface water and to raise or lower the temperature of the surface. Surface processes also play an important role in determining the overall energy balance of the planet.

The energy budget at the surface is more complex than the budget at the top of the atmosphere because it requires consideration of fluxes of energy by conduction and by convection of heat and moisture through fluid motion, as well as by radiation. The local surface energy budget depends on the insolation, the surface characteristics such as wetness, vegetative cover, albedo, and on the characteristics of the overlying atmosphere. The energy budget of the surface is intimately related to the hydrologic cycle, because evaporation from the surface is a key component in the budgets of both energy and water. Understanding the energy budget of the surface is a necessary part of understanding climate and its dependence on external constraints.

Global Physical Climatology. http://dx.doi.org/10.1016/B978-0-12-328531-7.00004-9

4.2 THE SURFACE ENERGY BUDGET

The energy budget can be written in terms of energy flux per unit area passing vertically through the air–surface interface and is measured in watts per square meter. The processes that determine energy transfer between the surface and atmosphere include solar and infrared radiative transfer, fluxes of energy associated with fluid motions of the atmosphere and ocean, and movement of energy through the soil. For the purposes of energy-budget computations, surface storage takes place in that volume between the boundary with the atmosphere and a depth below the surface where energy fluxes and the storage rate of energy are considered negligible. This depth can be as little as a few meters in dry land areas or as much as several kilometers in oceanic areas where deep water is formed. For water surfaces, the horizontal energy fluxes accomplished by the fluid motions under the surface can be very important. The surface energy balance can be written symbolically as (4.1),

$$\frac{\partial E_s}{\partial t} = G = R_s - \text{LE} - \text{SH} - \Delta F_{eo} \tag{4.1}$$

where $\partial E_s / \partial t = G$ is the storage of energy in the surface soil and water, R_s is the net radiative flux of energy into the surface, LE is the latent heat flux from the surface to the atmosphere, SH is the sensible heat flux from the surface to the atmosphere, and ΔF_{eo} is the horizontal flux out of the column of land–ocean below the surface.

Under steady-state conditions in which the storage of energy is small, such as one might assume to hold for annual averages or for daily averages over land, the energy balance is between radiative heating and the processes that remove energy from the surface.

$$R_s = \text{LE} + \text{SH} + \Delta F_{eo} \tag{4.2}$$

Under most conditions, radiation heats the surface and latent and sensible heat fluxes cool it, so that the radiative, latent, and sensible heat flux terms in (4.1) and (4.2) are most often positive (Fig. 4.1).

The physical meaning of (4.1) is that the storage of energy below the surface is equal to the net radiative input minus the heat lost from the surface by evaporation, sensible heat flux, and horizontal heat transport to other latitudes or longitudes. In constructing (4.1) we have left out a multitude of other terms that can be important locally or for brief periods. These terms include the following:

- The latent heat of fusion required for melting ice and snow in spring may require 10% of the radiative imbalance for limited periods.
- Conversion of the kinetic energy of winds and waves to thermal energy is generally small.

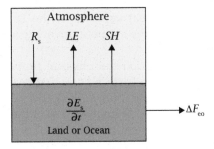

FIGURE 4.1 **The relationship of the various terms in the surface energy balance.** R_s, net radiation; LE, evaporative cooling; SH, sensible cooling; $\partial E_s / \partial t$, heat storage below the surface; ΔF_{eo}, divergence of horizontal energy flux below the surface.

- Heat transfer by precipitation can occur if the precipitation is at a different temperature than the surface. This mechanism is particularly important during summer showers, because the precipitation can be much cooler than the surface, and the thermal capacity of water is large.
- Some solar energy is not realized as heat, but is stored in the chemical bonds formed during photosynthesis. This is less than 1% globally, but can be ~5% locally for limited periods of time.
- Heat release by oxidation of biological substances, as in biological decay or forest fires, is the reverse of photosynthesis. Energy bound up in biological matter during photosynthesis is returned to the physical climate system through oxidation. This process takes place year round, but proceeds most rapidly when the surface is warm and moist.
- Geothermal energy release in hot springs, earthquakes, and volcanoes is small in a global sense.
- Heat released by fossil fuel burning or nuclear power generation can be important locally, but is not significant for the global energy balance.

4.3 STORAGE OF HEAT IN THE SURFACE

Energy storage in the surface is very important for the seasonal cycle of temperature over the oceans and the diurnal cycle over land and ocean. Using the simplest description, the amount of energy in the surface may be written as the product of an effective heat capacity for the earth–ocean system and a corresponding mean temperature,

$$E_s = \bar{C}_{co} T_{eo} \tag{4.3}$$

where $\bar{C}_{eo}$ is the effective heat capacity of the land or ocean system ($\mathrm{J\,m^{-2}\,K^{-1}}$) and T_{eo} is the effective temperature of the land or ocean energy-storing material (K).

The heat capacity depends on the physical properties of the surface materials and the depth of the surface layer that communicates with the atmosphere on the time scale of interest. It is generally only the first few meters of soil that respond to seasonal forcing of the surface energy balance, but the temperature of the top 50–100 m of ocean changes with the seasons. The depth of ocean that exchanges energy with the surface varies seasonally, so that possible time dependence of the effective heat capacity must be considered.

The heat capacity of the atmosphere is estimated by including only the energy associated with the motion of the molecules, which is related to the temperature. The heat capacity is the amount of energy that is required to raise the temperature by $1°$. We may approximate the specific heat of air by the specific heat at constant pressure for dry air. To obtain the heat capacity for the entire atmosphere, we integrate over the mass of the atmosphere to get

$$\bar{C}_a = c_p \frac{p_s}{g} = \frac{1004\,\mathrm{J\,K^{-1}\,kg^{-1}} \times 10^5\,\mathrm{Pa}}{9.81\,\mathrm{m\,s^{-2}}} = 1.02 \times 10^7\,\mathrm{J\,K^{-1}\,m^{-2}} \qquad (4.4)$$

One can estimate the thermal capacity of the ocean by using the thermal capacity of pure liquid water at $0°C$. The thermal capacity for an arbitrary depth of water, d_w, can be obtained from the density ρ_w, and the specific heat c_w,

$$\begin{aligned}\bar{C}_o &= \rho_w c_w d_w = 10^3\,\mathrm{kg\,m^{-3}} \times 4218\,\mathrm{J\,K^{-1}\,kg^{-1}} \times d_w \\ &= d_w \times 4.2 \times 10^6\,\mathrm{J\,K^{-1}\,m^{-2}\,m^{-1}}\end{aligned} \qquad (4.5)$$

Comparing (4.4) and (4.5) we see that the thermal capacity of the atmosphere is equal to that of a little over 2 m of water. As discussed in Chapter 7, about the top 70 m of ocean interact with the atmosphere on the time scale of a year, so that on the seasonal time scale the thermal capacity of the ocean is about 30 times that of the atmosphere.

4.3.1 Heat Storage in Soil

The land has a much smaller effective heat capacity than the ocean. Because the surface is solid, it does not have the efficient heat transport by fluid motions that occurs in the atmosphere and ocean. Heat is transferred through the soil mostly by the less efficient process of conduction. Only the top 1 or 2 m of the soil is affected by seasonal variations.

TABLE 4.1   Properties of Soil Components at 293 K

	Specific heat (c_p) (J kg^{-1} K^{-1})	Density (ρ) (kg m^{-3})	$\rho\, c_p$ (J m^{-3} K^{-1})
Soil inorganic material	733	2600	1.9×10^6
Soil organic material	1921	1300	2.5×10^6
Water	4182	1000	4.2×10^6
Air	1004	1.2	1.2×10^3

After Brutsaert, 1982. Reprinted with permission from Kluwer Academic Publishers.

The heat capacity of a land surface is typically slightly smaller than that of the atmosphere.

The vertical flux of energy by conduction in the soil is proportional to the vertical temperature gradient in the soil,

$$F_s = -K_T \frac{\partial T}{\partial z} \tag{4.6}$$

where K_T is the thermal conductivity. The heat balance in the soil is between storage in the soil and convergence of the diffusive heat flux.

$$C_s \frac{\partial T}{\partial t} = -\frac{\partial}{\partial z}(F_s) = \frac{\partial}{\partial z}\left(K_T \frac{\partial T}{\partial z} \right) \tag{4.7}$$

The volumetric heat capacity of the surface material C_s is the product of the specific heat of the soil c_s, and the soil density ρ_s. The heat capacity of the soil depends on the volume fractions of soil f_s, organic matter f_c, water f_w, air f_a, and the density and specific heat of each component of the surface material.

$$C_s = \rho_s c_s f_s + \rho_c c_c f_c + \rho_w c_w f_w + \rho c_p f_a \tag{4.8}$$

From Table 4.1, it can be seen that the heat capacity of the air in the soil is very small, so that when water replaces air in the open spaces in the soil the heat capacity greatly increases. Porosity is the volumetric fraction of the soil that can be occupied by air or water.

Thermal conductivity of soils depends on the material, the porosity, and the soil water content. The thermal conductivity increases with water content for soils with relatively high porosity. Values vary from 0.1 Wm^{-1} K^{-1} for dry peat to 2.5 Wm^{-1} K^{-1} for wet sand. Under the condition that the thermal conductivity, K_T, is independent of depth, (4.7) simplifies to the heat equation

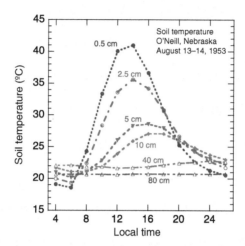

FIGURE 4.2　**Soil temperature at various depths under a grass field at O'Neill, Nebraska on August 13, 1953.** As a function of time of day. Measured thermal diffusivities on the day illustrated range from 2.5×10^{-7} m^2 s^{-1} at 1 cm to 6×10^{-7} m^2 s^{-1} at 5-cm depth in the soil. *Data from Lettau and Davidson (1957).*

$$\frac{\partial T}{\partial t} = D_T \frac{\partial^2 T}{\partial z^2} \qquad (4.9)$$

where $D_T = K_T / C_s$ is the thermal diffusivity of the surface material. A simple scale analysis of (4.9) can be used to determine the depth through which a temperature anomaly applied at the surface will penetrate in a given time. One can show that the penetration depth, h_T, of temperature anomalies associated with a periodic forcing of temperature at the surface is given by

$$h_T = \sqrt{D_T \tau} \qquad (4.10)$$

where τ is the time scale of the periodic forcing at the surface. Taking a typical value of soil diffusivity of $D_T = 5 \times 10^{-7}$ m^2 s^{-1}, we obtain a penetration depth of about 10 cm for diurnal forcing and about 1.5 m for annual forcing. A surface temperature variation with a time scale of 10,000 years would penetrate the surface material to a depth of about 150 m. Because of the slow rates at which heat can be transported through soil and rock by conduction, horizontal heat transport under the land surface can be ignored in climate modeling.

Figure 4.2 shows temperatures at various depths in the soil as a function of time during a clear day in summer. The near surface soil experiences a large diurnal variation in temperature with minimum temperatures just before sunrise and maximum temperatures shortly after noon. Deeper in the soil, the temperature variations are smaller and occur later in the day because of the time it takes the temperature pulse to diffuse into the soil. The amplitude of the temperature perturbation decreases to about e^{-1} of its surface value at about 10 cm below the surface and is very small by 40 cm below the surface. This observed decrease in the amplitude of the diurnal temperature perturbation is consistent with the simple dimensional arguments given previously.

The temperature data can be used to infer the thermal conductivity as a function of depth in the soil through the use of (4.7). Alternatively, if the vertical profiles of C_s and K_T are known and assumed constant with time, then measured temperature profiles in a deep layer of Earth can be used to estimate past variations in surface temperature on time scales of hundreds to thousands of years (Huang et al., 2000).

4.4 RADIATIVE HEATING OF THE SURFACE

The net input of radiative energy to the surface is the sum of the net solar and longwave flux densities at the surface,

$$R_s = S^{\downarrow}(0) - S^{\uparrow}(0) + F^{\downarrow}(0) - F^{\uparrow}(0) \tag{4.11}$$

where $S^{\downarrow}(0)$ and $S^{\uparrow}(0)$ are the downward and upward flux of solar radiation at the surface, respectively, and $F^{\downarrow}(0)$ and $F^{\uparrow}(0)$ represent similarly defined longwave fluxes.

4.4.1 Absorption of Solar Radiation at the Surface

The net downward solar energy flux can be written as the product of the downward solar flux at the surface multiplied by the absorptivity of the surface.

$$S^{\downarrow}(0) - S^{\uparrow}(0) = S^{\downarrow}(0)(1 - \alpha_s) \tag{4.12}$$

The surface albedo, α_s, is defined as the fraction of the downward solar flux density that is reflected by the surface.

The surface albedo varies widely depending on the surface type and condition, ranging from values as low as 5% for oceans under light winds to as much as 90% for fresh, dry snow. The numbers in Table 4.2 are characteristic, but each surface type can exhibit a range of albedos (Figs. 4.3 and 4.4). The most common surface is that of water, and its albedo depends on solar zenith angle, cloudiness, wind speed, and impurities in the water. The surface albedo of ocean under clear skies increases dramatically as the Sun approaches the horizon. Clouds scatter radiation very effectively, so that the solar radiation under a cloud is no longer a parallel beam but

TABLE 4.2   Albedos for Various Surfaces in Percent

Surface type	Range	Typical value
WATER		
Deep water: low wind, low altitude	5–10	7
Deep water: high wind, high altitude	10–20	12
BARE SURFACES		
Moist dark soil, high humus	5–15	10
Moist gray soil	10–20	15
Dry soil, desert	20–35	30
Wet sand	20–30	25
Dry light sand	30–40	35
Asphalt pavement	5–10	7
Concrete pavement	15–35	20
VEGETATION		
Short green vegetation	10–20	17
Dry vegetation	20–30	25
Coniferous forest	10–15	12
Deciduous forest	15–25	17
SNOW AND ICE		
Forest with surface snow cover	20–35	25
Sea ice, no snow cover	25–40	30
Old, melting snow	35–65	50
Dry, cold snow	60–75	70
Fresh, dry snow	70–90	80

is scattered in all directions. Under a cloud, the photons that reach the surface come from all possible directions with about equal probability, so that beneath a sufficiently thick cloud it is impossible to tell where in the sky the sun is located. Therefore, the surface albedo under overcast skies is insensitive to solar zenith angle. The amount of solar energy that reaches the surface under overcast skies is sensitive to solar zenith angle, however, since clouds are very effective reflectors of solar radiation and their albedo is somewhat sensitive to solar zenith angle (Fig. 3.13).

The reflectivities of various surfaces depend on the frequency of radiation (Fig. 4.5). Clouds and snow are most reflective for visible radiation,

FIGURE 4.3   NASA Natural Color Satellite Image of Southwestern Alaska on January 15, 2012. Fresh snow on land is very bright, while sea ice with tendrils in Bristol Bay is slightly darker. The ocean is very dark, except where clouds obscure the dark surface. *Image courtesy MODIS Rapid Response Team at NASA GSFC.*

and become less reflective at near-infrared wavelengths, where substantial absorption by water occurs. Green plants have a very low albedo for *photosynthetically active radiation*, where chlorophyll absorbs radiation efficiently. Radiation in the wavelength band from about 0.4–0.7 μm is effective for photosynthesis and growing plants absorb more than 90% of it. At about 0.7 μm the albedo of green plants increases sharply, so their albedo for near-infrared radiation can be as high as 50%. Since nearly half of the solar energy that reaches the surface is at wavelengths longer than 0.7 μm, this increase in albedo is significant for the energy budget of the surface. Plants need wavelengths shorter than 0.7 μm for photosynthesis, but the near-infrared energy absorption at wavelengths longer than 0.7 μm heats the leaves without any conversion of energy to plant tissue. The higher albedos at wavelengths longer than 0.7 μm thus help the leaves to stay cool. When green plants die and dry out, their chlorophyll content decreases and their albedo at visible wavelengths increases, as shown by the example of a field of straw.

The albedo of vegetated surfaces depends on the texture and physiological condition of the plant canopy. Leaf canopies with complex

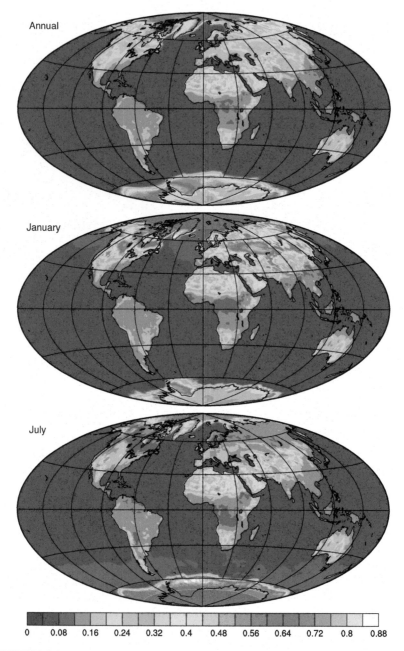

FIGURE 4.4 **Surface albedo of Earth for annual mean, January and July. Gray areas indicate missing data.** *Data from NASA CERES surface albedo product.*

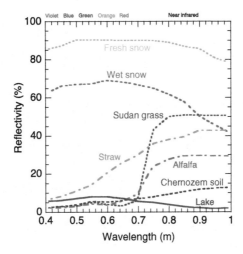

FIGURE 4.5 **Surface reflectivity as a function of wavelength of radiation for a variety of natural surfaces.** Human eyesight is sensitive to wavelengths from 0.4 μm (violet) to 0.7 μm (red). Alfalfa and sudan grass appear green because their albedo is higher for green light (∼0.55 μm) than for other visible wavelengths. *Data from Mirinova (1973).*

geometries and many cavities can have albedos that are lower than the albedo of an individual leaf. The ratio of near-infrared to visible radiation increases with depth below the top of the plant canopy, because the visible wavelengths are more effectively absorbed by leaves. The higher albedo of leaves for near-infrared radiation allows it to be scattered down through the plant canopy and heat the soil.

Pure water is most reflective for blue light. Natural water bodies contain many impurities and may be most reflective for green light, but their reflectance is generally higher at visible than at near-infrared wavelengths. Soils have higher reflectivities at near-infrared wavelengths than at visible wavelengths. Soils have significantly higher albedos when they are dry than when they are wet, and smooth soil surfaces have higher albedos than rough surfaces (Table 4.3).

Because surface albedo is highly variable and has a strong effect on absorbed solar radiation, it can have a large effect on surface temperature. Surface albedo can also have a strong effect on the sensitivity of climate, if it changes systematically with climatic conditions. Feedback processes involving surface albedo are discussed in Chapter 10.

4.4.2 Net Longwave Heating of the Surface

To calculate the net downward longwave radiation at the surface, one must know the downward longwave radiation coming from the atmosphere, the temperature of the surface, and the longwave emissivity of the surface, ε. If the frequencies of downward longwave radiation and

TABLE 4.3   Albedos for Dry and Moist Soil Surfaces

	Even surface		Tilled surface	
	Dry	Moist	Dry	Moist
Chernozem of dark gray color	13	8	8	4
Light chestnut soil of gray color	18	10	14	6
Chestnut soil of grayish red color	20	12	15	7
Gray sandy soil	25	18	20	11
White sand	40	20	–	–
Dark blue clay	23	16	–	–

From Mirinova, 1973.

radiation emitted from the surface are essentially the same, then the effective absorptivity of the surface is equal to its emissivity. Since this is approximately true, a fraction ε of the downward longwave radiation at the ground is absorbed, so that the upward longwave at the surface can be written

$$F^\uparrow(0) = (1 - \varepsilon)\, F^\downarrow(0) + \varepsilon \sigma T_s^4 \qquad (4.13)$$

We may thus write

$$F^\downarrow(0) - F^\uparrow(0) = \varepsilon(F^\downarrow(0) - \sigma T_s^4) \qquad (4.14)$$

Emissivities of some natural surfaces are given in Table 4.4. Most surfaces have very high emissivities. The great exceptions to this are polished metals, which have very low emissivities, so that metal films on plastic are very useful as emergency blankets or thermal shrouds for satellites, and

TABLE 4.4   Typical Thermal Infrared Emissivity for Selected Surfaces

Water	0.99	Glass, smooth uncoated	0.91
Ice and snow	0.98	Paper	0.87
Bare soil	0.96	Brick	0.90
Vegetation	0.96	Carbon	0.85
Stone pavement	0.96	Plastic	0.95
Wood	0.94	Mercury liquid	0.10
Asphalt, black	0.97	Silver, polished	0.02
Concrete	0.91	Gold, polished	0.02

Values vary based on the condition of the surface and wavelength.

metal films can also be used to make low emissivity glass for house windows that transmits visible light very well. Because of the strong greenhouse effect at work in Earth's atmosphere, the downward longwave from the atmosphere and the emission from the surface are both relatively large and tend to offset each other. The longwave emissivities of most natural surfaces are more than 0.95 and so variations of emissivity do not play a key role in the determination of surface climate. Inaccuracy in the estimation of surface emissivity can cause errors in the calculation of net longwave flux at the surface of about 5%. The errors in estimates of the surface temperature in equilibrium are much smaller, however, because in deriving the temperature from the energy flux balance, a one-fourth root is taken. This reduces the variation in surface temperature associated with emissivity variations to about 1%.

4.5 THE ATMOSPHERIC BOUNDARY LAYER

The *atmospheric boundary layer* is the lowest part of the troposphere, where the wind, temperatures, and humidity are strongly influenced by the surface. The wind speed decreases from its value in the free atmosphere to near zero at the surface. Fluxes of momentum, heat, and moisture by small-scale turbulent motions in the boundary layer communicate the presence of the lower boundary to the atmosphere and are critical to the climate. Aerosols and gaseous chemical constituents of the atmosphere are also exchanged with the surface through the atmospheric boundary layer. A characteristic of the atmospheric boundary layer is its quick response to changes in surface conditions. The response of the surface to the daily variation of insolation is felt strongly throughout the boundary layer, but in the free troposphere diurnal changes are usually small, unless thermal convection associated with daytime heating of the land surface penetrates deep into the atmosphere. Another important aspect of the boundary layer is the cloud that it contains. The most common type of cloud is marine boundary-layer cloud over the ocean, and this type of cloud appears to be both the most important in the energy balance of Earth and also probably the most important type of cloud in determining the sensitivity of climate change.

The depth of the atmospheric boundary layer can vary between about 20 m and several kilometers, depending on the conditions, but a typical boundary-layer depth is about 1 km. The boundary layer is generally deeper when the surface is being heated, when the winds are strong, when the surface is rough, and when the mean vertical motion in the free troposphere is upward.

Transports of mass, momentum, and energy through the boundary layer are accomplished by turbulent motions. If it were not for the

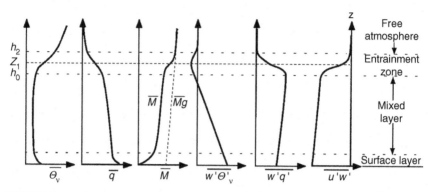

FIGURE 4.6 **Structure of a convective boundary layer.** It shows the distributions of mean virtual potential temperature $\overline{\Theta}_v$, water vapor mixing ratio $\overline{q}$, momentum $\overline{M}$, geostrophic momentum $\overline{M}_g$, and the vertical eddy fluxes of potential temperature, humidity, and momentum. *From Stull (1988) after Dreidonks and Tennekes (1984). Reprinted with permission from Kluwer Academic Publishers.*

chaotic swirls of turbulent motion in the boundary layer, the nonradiative exchange between the surface and the atmosphere would be extremely slow. The turbulent motions that carry the vertical fluxes in the boundary layer range in scale from the depth of the boundary layer to the smallest scales where molecular diffusion becomes an important transport mechanism. Turbulence can be generated thermally or mechanically. *Mechanical turbulence* is generated by the conversion of the mean winds to turbulent motions, and is strongest when the mean wind in the lower atmosphere is large. *Convective turbulence* is generated when warm air parcels near the surface are accelerated upward by their buoyancy. Convective turbulence is most easily observed over land surfaces during daylight hours, when strong solar heating of the surface provides a source of buoyant energy, but is also very common over the oceans. When the boundary layer is relatively unstable, so that buoyancy forces or shearing instabilities generate turbulence, the boundary layer may contain a well-mixed layer, where momentum, heat, and moisture are almost independent of height.

The structure of the planetary boundary layer varies widely, depending on the meteorological conditions and whether the surface is being heated or cooled. When a surface heat source is present, such as over land during the daytime, the boundary layer is often unstable and has a structure that is generally like that shown in Fig. 4.6. The lowest part of the boundary layer is called the *surface layer*, where the vertical fluxes of momentum, heat, and moisture are almost constant with height. In the mixed layer, buoyancy drives turbulent motions that maintain the potential temperature, Θ_v, the humidity, $\overline{q}$, and the momentum, $\overline{M}$, at values that are almost independent of height. Heat and moisture are transported upward in the mixed layer and momentum is transported downward toward the

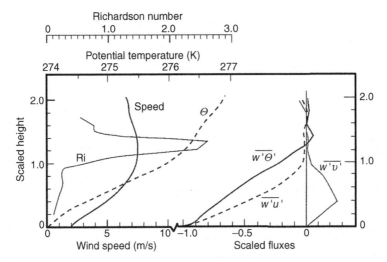

FIGURE 4.7 **Averaged profiles of wind speed, potential temperature, Richardson number and vertical fluxes of potential temperature ($\overline{w'\Theta'}$), and horizontal momentum ($\overline{w'u'}$) and ($\overline{w'v'}$) from nocturnal observations at Haswell, Colorado, on March 24, 1974.** The height is scaled by the depth in which turbulence is observed to occur, which on average is about 100 m in this case. Vertical eddy fluxes are scaled by their surface values. *From Mahrt et al. (1979). Reprinted with permission from Kluwer Academic Publishers.*

surface. The top of the boundary layer is a transition zone between the boundary layer and the free atmosphere, which is often called the *entrainment zone*. Across this transition, the air properties change rapidly from those of the mixed layer to those of the free atmosphere above, generally marked by a decrease in humidity, an increase in potential temperature, and a decrease in the magnitude of the vertical fluxes of heat, moisture, and momentum by turbulent motions. *Entrainment* is the process whereby air from the free atmosphere is incorporated into the boundary layer. Entrainment is measured by a small downward eddy potential temperature flux in the entrainment zone. Entrainment adds mass to the boundary layer and is necessary when the boundary layer is deepening, or when the large-scale flow is downward into the boundary layer.

At night, longwave emission cools the land surface more rapidly than the air above it and the boundary layer can become very stable, with cold, dense air trapped near the surface (Fig. 4.7). Under these conditions, turbulence and the vertical fluxes it produces can be greatly suppressed, and the surface becomes mechanically uncoupled from the free atmosphere, although radiative transports can still occur. The potential temperature increases rapidly with height near the surface, and the transport of potential temperature is downward, so that less dense air is being forced downward against buoyancy. The energy for mixing less dense air downward toward the surface is provided by the mean wind speed shear, which tends to be quite strong under these conditions, often with a low-level wind maximum near the top of the boundary layer and weak winds near the surface. The minimum surface-air temperature achieved on a clear night is thus generally lower when the wind speed in the free atmosphere is weak and provides little energy for mixing warm air downward to the radiatively cooled surface. On cool nights with high boundary layer stability, farmers may try to prevent crops from freezing by using propellers to generate mechanical turbulence that will mix warm air downward toward the surface.

The atmospheric boundary layer can contain clouds that play an important role in boundary-layer physics and vertical transports. The release of latent heat in clouds can provide buoyancy to drive vertical motions in the boundary layer. Boundary-layer clouds that are important for climate include the fair weather cumulus clouds and stratocumulus clouds. Though less widespread, fog is also an important boundary-layer cloud. The boundary layer also interacts in important ways with deep convective clouds, since the high potential temperature air that drives deep convection is produced in the boundary layer. Except when fog is present, the tops of boundary-layer clouds generally occur near the top of the boundary layer and their bases are some distance above the surface, so that a cloud and a subcloud layer exist within the boundary layer. Stratocumulus clouds modify the boundary-layer physics both through their convective heat and moisture fluxes and through their radiative effects. Because stratocumulus cloud tops are relatively warm and emit longwave radiation efficiently, longwave cooling from cloud tops can be an important mechanism for generating buoyancy within the boundary layer, since it cools the air at the top of the boundary layer, which then tends to sink and be replaced by warmer parcels of air rising from below.

Figure 4.8 shows the diurnal variation of temperature in the lowest 1500 m of the atmosphere over Nebraska during a relatively clear summer day. At sunrise, the surface is colder than the air 1 km above the surface. This temperature inversion quickly disappears after sunrise, as insolation warms the surface and this heat is transferred to a shallow layer of air near the ground. Near the middle of the day, the surface reaches its maximum temperature and a lapse rate near the dry adiabatic value of 9.8 K m^{-1} is observed near the surface. At this time, buoyancy raises warm parcels of air near the surface, and turbulent convection efficiently moves sensible heat upward in the boundary layer. Even before sunset, the surface begins to cool in response to efficient upward transport of energy by turbulent motions. After sunset the surface cools rapidly, so that by 10 PM the surface temperature has reached its nighttime value, leaving a very sharp inversion near the ground.

When mean wind speeds are light to moderate, diurnal variations in the temperature profile will affect the exchange of heat, moisture, and

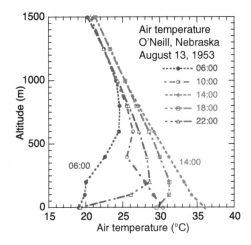

FIGURE 4.8 **Plot of air temperature at various local times in the lowest 1500 m of the atmosphere at O'Neill, Nebraska on August 13, 1953.** Times are given using a 24-h clock so that 1800 = 6 PM, etc. *Data from Lettau and Davidson (1957).*

momentum between the atmosphere and the surface. The strong density stratification associated with a nocturnal temperature inversion suppresses turbulent transport of momentum from the free atmosphere. Sensible and latent heat fluxes are also suppressed by the strong density stratification, so that the surface temperature responds primarily to the radiative forcing, which acts to cool the surface at night. Temperature inversions often develop when a high-pressure system dominates the local weather pattern. The weak surface winds, clear skies, and downward mean motion in the free atmosphere that are normally associated with high-pressure systems encourage the development of a strong inversion. Pollutants released under a temperature inversion can easily build up to unhealthy levels, because turbulent mixing into the free atmosphere is suppressed by the strong density gradient.

4.5.1 The Neutral Boundary Layer

When the static stability in the boundary layer is near neutral, buoyancy does not play an important role in the turbulent kinetic energy budget. Under neutral conditions, the source of energy for boundary-layer turbulence is the kinetic energy of the mean wind of the free atmosphere. The turbulence in the boundary layer produces a strong flux of momentum to the surface. The vertical flux of horizontal momentum at the surface, τ_0, constitutes a drag on the atmospheric flow. In the surface layer, the vertical gradient of wind speed (U) under conditions of neutral stability should depend only on the height (z), density (ρ), and surface drag (τ_0).

The characteristic wind speed for use in dimensional analysis is the friction velocity, u_*.

$$u_* = \left(\frac{\tau_0}{\rho}\right)^{1/2} \tag{4.15}$$

Using the friction velocity and the height to scale the wind shear, dimensional analysis suggests that the scaled wind shear should be a constant.

$$\left(\frac{z}{u_*}\right)\frac{dU}{dz} = \frac{1}{\kappa} \tag{4.16}$$

The von Karman constant, κ, is the same for all neutral boundary layers regardless of the surface characteristics, and has a measured value of approximately 0.4. Equation (4.16) can be integrated with respect to height to obtain the logarithmic velocity profile.

$$U(z) = \left(\frac{u_*}{\kappa}\right)\ln\left(\frac{z}{z_0}\right) \tag{4.17}$$

An additional constant, z_0, is introduced during the integration to obtain (4.17). It is called the *roughness height*, the height at which the wind speed reaches zero. For most natural surfaces the irregularities of the surface are larger than the 1 mm depth of the layer where molecular diffusion dominates, and this roughness can be characterized by the height z_0. Roughness heights for natural surfaces range from about 1 mm for average seas to more than 1 m for cities with tall buildings. Roughness heights are estimated by measuring the wind-speed profile under neutral conditions and then solving (4.17) for z_0. The logarithmic velocity profile has been shown to be a good approximation for many laboratory boundary layers and also for the planetary boundary layer under conditions of neutral stratification. It is valid for heights much greater than the roughness height, $z \gg z_0$, and so does not describe the mean wind-velocity profile within the plant canopy or very close to a rough surface.

The logarithmic velocity-profile law is useful for expressing the momentum flux at the surface in terms of the wind speed at some height in the surface layer. Substituting (4.15) into (4.17) yields an expression for the surface drag in terms of the wind speed, U_r, at some reference height z_r.

$$\tau_0 = \rho C_D U_r^2 \tag{4.18}$$

where

$$C_D = \kappa^2 \left\{ \ln \left(\frac{z_r}{z_0} \right) \right\}^{-2} \tag{4.19}$$

The drag coefficient, C_D, depends on the ratio of the reference height to the roughness height. The reference layer can be taken at any level within the surface layer where measurements can be conveniently acquired and where the logarithmic profile is a good approximation of the actual flow. The aerodynamic drag formula (4.18) and related formulas for the sensible and latent heat fluxes at the surface form the basis for empirical estimates of surface fluxes and the specification of surface fluxes in climate models. They allow the calculation of turbulent fluxes using only mean wind speed at a reference height and a few external parameters, and so are used in nearly all climate models and also for estimating surface turbulent fluxes from measurements of mean quantities.

4.5.2 Stratified Boundary Layers

The dimensional analysis for neutral boundary layers can be extended to stratified boundary layers (Monin and Obukhov, 1954; Arya, 1988). This theory adds heat flux and buoyancy variables to the dimensional analysis. Characteristic vertical profiles for both wind and temperature are derived in which the vertical coordinate is scaled by a dimensionless combination of the friction velocity, the heat flux, and the buoyancy. From these profiles, bulk aerodynamic formulas can be derived that describe the turbulent heat and momentum fluxes at the surface in terms of mean variables. The coefficients in these formulas now depend on the vertical stability of the atmosphere as well as the roughness height. This theory applies only to the surface layer.

The vertical stability can be characterized with the *Richardson number*. In differential form, the Richardson number depends on the vertical derivatives of potential temperature, Θ (see Chapter 1), and wind speed, U

$$\text{Ri} = \left(\frac{g}{T_0} \right) \frac{(\partial \Theta / \partial z)}{(\partial U / \partial z)^2} \tag{4.20}$$

where g is the gravitational acceleration and T_0 is the reference temperature. The bulk Richardson number for the boundary layer may be written as

$$\text{Ri}_B = \left(\frac{g}{T_0} \right) \left[\frac{z_r (\Theta(z_r) - \Theta(z_0))}{U(z_r)^2} \right] \tag{4.21}$$

The Richardson number indicates how likely the flow is to mix vertically in response to buoyancy and wind shear. The Richardson number is large when the potential temperature of the near surface air is high compared to the potential temperature at the surface. Under such conditions, the air is stably stratified, which inhibits vertical mixing. Parcels that are raised up from the surface become negatively buoyant and will be forced back downward by the gravity force. This stabilizing effect can be overcome by the kinetic energy available in the mean wind shear near the ground, which can generate turbulent velocities sufficient to mix stably stratified air. This influence of the kinetic energy is represented by the square of the wind speed in the denominator of (4.21). If the potential temperature decreases with height near the surface, then the boundary layer is unstable and the buoyancy force will accelerate small vertical displacements of parcels. The Richardson number for a buoyantly unstable boundary layer is negative. Under these conditions, vertical transfer of energy and moisture is relatively efficient because of the free exchange of parcels across level surfaces.

Over land areas, it is common for the boundary layer to become unstable during summer days as insolation heats the surface. At night, the surface cools faster than the overlying air so that a temperature inversion can develop (Fig. 4.8). The resulting strong-density stratification can suppress the nighttime exchanges of heat, momentum, and moisture between the surface and the free atmosphere. The effect of changes in the stability of the boundary layer on momentum fluxes can be seen in diurnal changes in the wind profile. Figure 4.9 shows the diurnal variations in wind speed

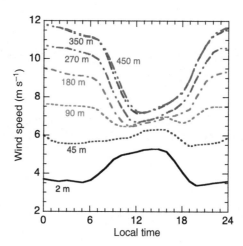

FIGURE 4.9 **Diurnal cycle of wind speed as a function of height measured from a tower in Oklahoma City and averaged over the period from June 1966 to May 1967.** *Adapted from Crawford and Hudson (1973). Reprinted with permission from the American Meteorological Society.*

measured at various heights on a TV tower in Oklahoma. At night, wind speeds measured very near the surface decrease, because the downward mixing of momentum from the free atmosphere is reduced by the greater static stability at night. The winds higher in the boundary layer increase at night because the drag from the surface is reduced. During the day, efficient mixing of momentum through a relatively unstable boundary layer causes the wind speed near the surface to increase at the expense of the wind speed higher in the boundary layer.

4.6 SENSIBLE AND LATENT HEAT FLUXES IN THE BOUNDARY LAYER

The turbulent fluid motions in the boundary layer produce sensible and latent heat fluxes from the surface. Transport by molecular diffusion is negligible compared to turbulent transport, except within about 1 mm of the surface. Turbulence is characterized by rapid chaotic fluctuations in wind velocity. Where mean vertical gradients in temperature or humidity exist, the turbulent fluctuations of wind velocity will be accompanied by fluctuations of scalar properties such as temperature and humidity. Vertical fluxes of mass, momentum, and energy are produced by turbulence when the parcels of air moving upward have different properties than parcels of air moving downward. Therefore, the flux can be measured by the spatial or temporal average of the product of vertical velocity and the property of interest.

For example, if we have measurements of the temperature, T, and vertical velocity, w, at a point near the surface, we may obtain the vertical flux of sensible heat from the time average of the product of vertical velocity and temperature, multiplied by the specific heat and average density of the air.

$$\text{Upward sensible heat flux} = c_p \rho \overline{wT} \tag{4.22}$$

For this estimate to be accurate, temperature and wind measurements must be taken at frequent enough intervals to define the turbulent fluctuations that produce the vertical transport. Since turbulent fluctuations are very rapid, measurements taken more frequently than every second can be required to directly measure turbulent fluxes. Inhomogeneities in surface conditions may also cause difficulty in obtaining representative fluxes from measurements taken at a single point.

By dividing the variables into a time mean and a deviation from the time mean, or eddy, we may write the upward sensible heat flux as a sum of time mean and eddy contributions.

$$w = \overline{w} + w', \ T = \overline{T} + T' \tag{4.23}$$

Here an overbar in (4.23) indicates a time average of the quantity under the overbar and a prime indicates a deviation from that time average. Substituting (4.23) into (4.22) and performing the averages, we obtain the mean and eddy contributions to the vertical flux of temperature.

$$\overline{wT} = \overline{w}\,\overline{T} + \overline{w'T'}$$
$$\text{Total} = \text{mean} + \text{eddy}$$
(4.24)

Near the surface, the mean vertical velocity is very small compared to the typical eddy or turbulent vertical velocities, and the eddy contribution to the vertical heat flux is dominant. We can then define the latent and sensible heat fluxes as the eddy fluxes of heat and moisture at some level in the atmospheric boundary layer.

$$\text{SH} = c_p \rho \overline{w'T'} \qquad \text{LE} = L\rho\overline{w'q'}$$
(4.25)

where ρ is air density, c_p is the specific heat of air at constant pressure, q is the specific humidity and L is the latent heat of vaporization. Measurements of the turbulent velocity, temperature, and moisture fluctuations necessary to calculate the sensible and latent energy fluxes are not routinely taken, and these rapid, small-scale fluctuations are not simulated in global climate models. In most cases, one must estimate the turbulent fluxes by using variables averaged over larger spatial and temporal scales than those of the turbulent motions in the boundary layer.

Several methods are available for estimating surface fluxes with observations of mean variables. The most common method is through the use of bulk aerodynamic formulas, which relate the turbulent fluxes to observable spatial or temporal averages. One might hypothesize that the sensible heat flux is proportional to the temperature difference between the surface and the air at some reference altitude, z_r, where mean variables are known. Since some of the kinetic energy of boundary layer turbulence comes from the mean winds blowing over the surface, we might assume that the turbulent fluxes are also proportional to the mean wind speed, U_r, at the standard height. These basic assumptions are consistent with the results of the similarity theory described previously, with which we obtain an expression that relates the sensible heat flux to the mean wind speed and temperatures.

$$\text{SH} = c_p \rho\, C_{\text{DH}}\, U_r\, (T_s - T_a(z_r))$$
(4.26)

The latent heat flux can be related to the difference of specific humidity, q, between the surface and the atmosphere at the reference height.

$$\text{LE} = L\, \rho\, C_{\text{DE}}\, U_r\, (q_s - q_a(z_r))$$
(4.27)

In the bulk aerodynamic formulas (4.26) and (4.27), ρ is the air density, c_p is the specific heat at constant pressure, L is the latent heat of vaporization and C_{DH} and C_{DE} are aerodynamic transfer coefficients for heat and moisture, respectively. Subscripts s and a indicate values for the surface and the air at the reference level, respectively.

The aerodynamic transfer coefficients depend on the surface roughness, the bulk Richardson number, and the reference height. Under ordinary circumstances, the values of the transfer coefficients for heat, moisture, and momentum would be nearly equal, and typical values for neutral stability and for a 10-m height above the surface would range from 1×10^{-3} over the ocean to 4×10^{-3} over moderately rough land. If the wind speed at 10 m is 5 m s^{-1} and $C_D = 3 \times 10^{-3}$, then from (4.26) the flux of sensible heat across the surface layer of atmosphere is about 15 Wm^{-2} for each degree of temperature difference between the surface and the air at 10 m.

4.6.1 Equilibrium Bowen Ratio for Saturated Conditions

The latent heat flux depends sensitively on the temperature through the dependence of saturation vapor pressure on temperature. Over water or wetland surfaces, we may assume that the mixing ratio of water vapor at the surface is equal to the saturation mixing ratio, q^*, at the temperature of the surface.

$$q_s = q^*(T_s) \tag{4.28}$$

The vapor mixing ratio of saturated air at the reference height can be approximated with a first-order Taylor series.

$$q_a^* = q^*(T_s) + \frac{\partial q^*}{\partial T}(T_a - T_s) + \dots \tag{4.29}$$

The actual vapor-mixing ratio of the air at the reference height can be expressed in terms of the relative humidity at that level.

$$RH = \frac{q}{q^*} \tag{4.30}$$

$$q_a \cong RH(q^*(T_s) + \frac{\partial q^*}{\partial T}(T_a - T_s)) \tag{4.31}$$

Substituting (4.31) into (4.27) yields an expression for the heat loss from the surface through evaporation in terms of the temperature difference and the relative humidity,

$$LE \cong \rho L C_{DE} U_r \left(q^*(T_s)(1 - RH) + RH \, B_e^{-1} \frac{c_p}{L}(T_s - T_a) \right) \tag{4.32}$$

where

$$B_e^{-1} \equiv \frac{L}{c_p} \frac{\partial q^*}{\partial T}\bigg|_{T=T_s} \tag{4.33}$$

The *Bowen ratio* is the ratio of the sensible cooling to the latent cooling of the surface. Comparing (4.32) and (4.26) we see that, when the surface is wet and the air is saturated, RH = 1, and assuming that $C_{DH} = C_{DE}$, the Bowen ratio takes a special value:

$$B_o \equiv \frac{SH}{LE} = B_e \tag{4.34}$$

When the surface and the air at the reference level are saturated, the Bowen ratio approaches the value B_e given by (4.33), which can be called the *equilibrium Bowen ratio*. We must assume that the flux of moisture from the boundary layer to the free atmosphere is sufficient to just balance the upward flux of moisture from the surface so that the humidity at the reference height is in equilibrium at the saturation value. The Bowen ratio in such an equilibrium is inversely proportional to the rate of change of the saturation mixing ratio of water vapor with temperature (4.33). The rate of change of the saturation-mixing ratio with temperature is very sensitive to the temperature itself. Using the approximate formula (1.9), it can be shown that

$$\frac{\partial q^*}{\partial T} \approx q^*(T)\left(\frac{L}{R_v T^2}\right) \tag{4.35}$$

The exponential dependence of the saturation-mixing ratio on temperature far outweighs the inverse square of temperature in (4.35), so that the equilibrium Bowen ratio decreases exponentially with temperature. The temperature dependencies of the saturation mixing ratio and the equilibrium Bowen ratio are shown graphically in a log–linear plot in Fig. 4.10. The equilibrium Bowen ratio is unity at about 0°C, and decreases to about 0.2 at 30°C. As the relative humidity in (4.32) is decreased from 1 to smaller values, the evaporative cooling increases, so that the equilibrium Bowen ratio is the maximum possible Bowen ratio for a wet surface. The actual Bowen ratio over a wet surface will generally be smaller than the equilibrium Bowen ratio, because the air at the reference height is usually not saturated. As a result of the strong temperature dependence of saturation vapor pressure, latent cooling of the surface dominates sensible cooling from a wet surface at temperatures

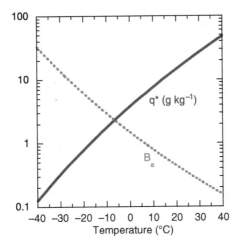

FIGURE 4.10   Saturation specific humidity q^*(g kg^{-1}) and equilibrium Bowen ratio B_e, as functions of temperature.

such as those in the tropics, but in high latitudes during winter sensible heat transport can be of greatest importance. To show this, assume a wet surface and saturated atmosphere, so that the actual Bowen ratio is the equilibrium Bowen ratio, we can write that

$$\frac{LE}{LE + SH} = \frac{1}{1 + B_e} \tag{4.36}$$

$B_e = 1$ at approximately 278K. At much warmer temperatures $B_e \ll 1$ and most of the surface cooling is by evaporation. At much colder temperatures $B_e \gg 1$ and most of the surface cooling is by sensible heat transfer.

The preceding discussion strictly applies only to conditions where the surface is wet, so that evaporative cooling is not constrained by lack of surface moisture. Over land areas, the evaporative cooling may be greatly reduced when moisture cannot be supplied from below the surface rapidly enough to keep the air in contact with the surface saturated. In desert areas, the surface is typically so dry that evaporative cooling is small regardless of the temperature, so that sensible cooling and longwave emission must balance solar heating. For vegetated terrain, cooling by evaporation and transpiration through leaves is controlled by the physical and biological condition of the plant canopy and the water content of the soil. The role of soil and vegetation in the surface water and energy balances is discussed further in Chapter 5.

4.7 DIURNAL VARIATION OF THE SURFACE ENERGY BALANCE

The surface energy balance is strongly influenced by the diurnal variation of insolation, except in polar regions. Figure 4.11 shows the diurnal variation of the surface radiation balance for grassland in Saskatchewan during a clear summer day with average winds. A dense mat of living and dead grass covers the surface and the surface albedo is about 16%. The average net radiation for the 24-h period shown was 155 Wm^{-2}, with 263 Wm^{-2} gained from the Sun and 108 Wm^{-2} net loss through infrared fluxes. The net downward solar radiation at the ground peaks near local solar noon at about 700 Wm^{-2}. The daytime solar heating is large because of the strong insolation, lack of cloudiness, and relatively low surface albedo. Downward longwave radiation is about 300 Wm^{-2} and does not change much during the day. The downward longwave radiation has almost no diurnal variation because of the small diurnal variation of air temperature in the free atmosphere. The surface upward emission is about 350 Wm^{-2} before sunrise and increases to about 500 Wm^{-2} at midday, in response to the warmer daytime surface temperatures. The surface temperature varies from about 10°C before sunrise to about 40°C at midday. The net longwave loss from the surface thus increases from about 50 Wm^{-2} at night to about 200 Wm^{-2} at midday. The longwave loss is relatively large because the skies are clear and the air humidity is low. The net radiation that results is an almost uniform 50 Wm^{-2} loss through longwave cooling during the night, and a strong, solar-driven gain peaking near 500 Wm^{-2} at midday.

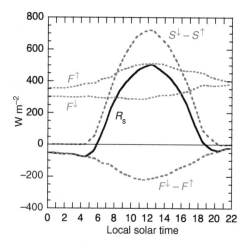

FIGURE 4.11 **Components of the radiative energy balance for a grass field in Matador, Saskatchewan on July 30, 1971.** $F^{\downarrow}$, downward longwave; $F^{\uparrow}$, upward longwave; $S^{\downarrow}-S^{\uparrow}$, net solar; $F^{\downarrow}-F^{\uparrow}$, net longwave; R_s, net radiation. After Ripley and Redmann (1976).

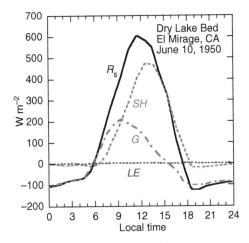

FIGURE 4.12 **Heat budget for a dry lake bed at El Mirage, California on June 10, 1950.** *Data from Vehrencamp (1953). © American Geophysical Union.*

The diurnal cycle in surface energy balance components for a dry, barren lake bed is shown in Fig. 4.12. The surface soil moisture was measured at about 2% at the time the measurements were taken, so evaporation plays virtually no role in the energy balance. At night the net radiation balance is negative as the surface emits radiation into the dry desert air. The radiative cooling of the surface is balanced by energy lost from the soil as it cools through the night. Sensible heat flux is nearly zero at night because the winds are weak and the cold air near the surface is denser than the warmer air above. The high vertical stability associated with the nighttime temperature inversion suppresses the turbulent exchange of energy. Sunrise occurs at about 5 AM and the surface cooling decreases rapidly. In the early morning, the net radiative heating is largely balanced by storage of heat in the ground. After mid-morning the strong daytime radiative heating is balanced primarily by upward sensible heat flux by atmospheric turbulence. The sensible heat flux peaks shortly after noon, and by mid-afternoon the surface has begun to cool because the sensible cooling exceeds the net radiation.

The heat budget for a mature cornfield during a clear day in late summer is shown in Fig. 4.13. Again the net radiation is weakly negative at night and goes through a strong positive maximum during the day. The nighttime radiation loss is balanced about equally by release of stored surface heat, downward sensible heat flux, and dewfall. Although the corn is nearly 3 m tall, a substantial amount of heat reaches the soil so that storage in the soil and corn stalks is an important part of the energy balance during the day. The surface soil is dry, but the corn roots have sufficient water. Sensible cooling is about half of the evaporative cooling, when averaged over the day. The peculiar change in the latent and sensible cooling near noon is thought to be related to the north–south orientation of the cornrows.

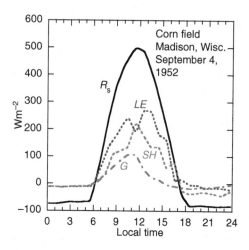

FIGURE 4.13 **Heat budget for a field of mature corn in Madison, Wisconsin, on September 4, 1952.** *Data from Tanner (1960). Reprinted with permission from the Soil Science Society of America.*

When a surface is wet, or when growing vegetation has ample soil water for evapotranspiration, the net radiation may be almost entirely used for evaporation, with the soil storage and sensible heat fluxes being of minor importance. When warm and dry air moves over a surface with ample water, the evaporative cooling may actually exceed the net radiation. Figure 4.14 shows the energy balance on a day when winds carry

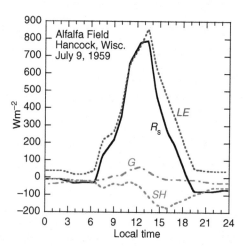

FIGURE 4.14 **Heat budget for a well-irrigated alfalfa field in Hancock, Wisconsin on July 9, 1956, when the air was advected to the field from a warm dry area.** *Data from Tanner (1960). Reprinted with permission from the Soil Science Society of America.*

warm dry air over a well-irrigated alfalfa field. Evapotranspiration equals or exceeds the net radiation at every hour of the day, and evaporation continues even through the night. The excess of the evaporative cooling over the net radiation is provided by downward sensible heat flux, with turbulent motions carrying heat downward to the evaporatively cooled surface. For small, irrigated plots in hot, arid regions, the sensible heating of evaporatively cooled surfaces can add to the water demand.

4.8 SEASONAL VARIATION OF THE ENERGY BALANCE OF LAND AREAS

The energy balance of the surface changes with season, especially in middle and high latitudes, where large seasonal variations of insolation and temperature occur. In tropical regions, the energy balance may change because of seasonal changes in precipitation, even when the temperature and insolation remain relatively constant. The seasonal variation of the components of the surface energy balance at several locations in middle latitudes is shown in Fig. 4.15. The annual variation of net radiation generally follows that of insolation, with peak values in summer approaching 200 Wm^{-2} in land areas, depending on the latitude, sky conditions, and surface albedo. Water surfaces in relatively cloudless areas can have summertime net radiation near 300 Wm^{-2}. The mechanism for balancing this net radiation depends on the local surface conditions. Over land areas, it is primarily a question of whether the mechanism is latent or sensible cooling, since storage is small and transport is zero. The apportionment between sensible and latent cooling depends on the availability of surface moisture, the temperature, and the humidity of the air.

In regions with significant precipitation during the summer season, where the surface remains relatively moist, the latent cooling is generally limited by and follows the annual cycle of radiative heating. Where the climate is exceptionally dry, such as in Yuma, Arizona, the latent cooling is negligible, except during months with precipitation. At Yuma, significant evaporation occurs in the springtime and during September. Flagstaff, Arizona, is at a higher elevation than Yuma and the mountains receive significant summertime precipitation. As a result, the evaporative cooling during summer is greater than at Yuma.

At West Palm Beach, Florida, the winters are relatively dry, so that the springtime insolation increase is initially balanced by equal contributions from latent and sensible cooling of the surface. As the summertime convective precipitation begins to wet the surface, the evaporative cooling takes over. At San Antonio, Texas, the surface dries out during spring and summer, so that a gradual increase in the importance of sensible heating occurs over the course of the summer months. At Astoria, Oregon, sensible

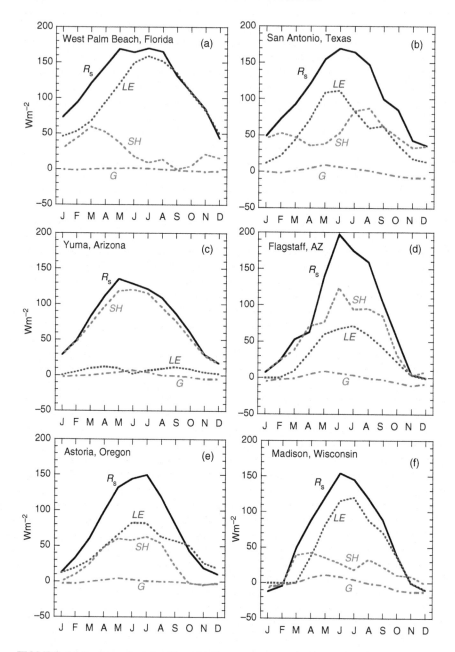

FIGURE 4.15 **Annual cycle of heat budget components for various mid-latitude land locations.** *Adapted from Sellers (1965). Reprinted with permission from the University of Chicago Press.*

cooling is important in the summer, even though Astoria gets more annual precipitation than Madison, Wisconsin, where evaporative cooling dominates during summer. This is because the precipitation at Astoria peaks in the winter, when little energy is available for evaporation, so that about 70% of the annual precipitation runs off before being evaporated. During the summer, relatively little precipitation falls at Astoria. At Madison, precipitation peaks in the summer when the large net radiation provides the energy to evaporate large amounts of water that later is released in convective weather systems such as thunderstorms.

In ocean areas, the heat capacity of the water is sufficient that the energy for evaporation may be derived from the energy of the water itself. The evaporative loss may be less correlated with net radiation than with factors such as the wind speed or the temperature and humidity contrast between the surface and the air above it. Much of the high-latitude evaporation over the oceans takes place in winter over the Kuroshio and Gulf Stream currents. These currents carry warm water poleward along the western margins on the Pacific and Atlantic oceans where it comes into contact with cold, dry air coming off the continents. The combination produces evaporation rates near 400 Wm^{-2} in these regions during winter. Figure 4.16 shows estimates of the annual cycle of surface energy fluxes over the Gulf Stream at 38°N. The annual variations in most terms are much larger than that of the net radiation. Enormous latent cooling rates in winter are balanced primarily by the release of energy stored in the water temperature and import of heat through horizontal transport.

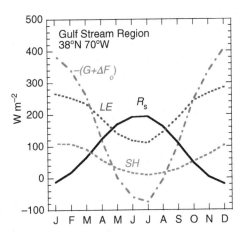

FIGURE 4.16 Annual cycle of the surface heat budget components for the region near 38°N, 70°W, which is affected by the Gulf Stream. Note that storage plus export has been plotted with reversed sign; heat converges and is lost to the atmosphere in winter. *Data from ERA-Interim Reanalysis.*

4.9 GEOGRAPHIC VARIATION OF THE SURFACE ENERGY BALANCE

Some intuition about the relative importance of the various components of the surface energy budget as a function of the climatic regime can be gained from their dependence on latitude. The net radiation at the surface peaks in the tropics, following the general pattern of insolation at the top of the atmosphere (Fig. 4.17). The upward and downward longwave fluxes are much larger than the net solar, but the downward longwave irradiance is almost as large as the upward irradiance, so that longwave radiation produces a weak cooling of the surface. The latent cooling term mirrors the insolation fairly closely (Fig. 4.18), since globally most of the radiative heating of the surface is used to evaporate water. The sensible cooling of the surface is much smaller and more uniform, with a slight increase over the Northern Hemisphere where more land is present. The ocean transports heat from near the equator toward the extratropics, but its influence is smaller than the other terms that represent the vertical movement of energy.

The geographic variation of the annual mean surface heat balance components are shown in Fig. 4.19. Net radiation is greatest over the tropical oceans, where the surface albedo is low and the surface temperature is moderate. In these regions, it often exceeds 180 Wm^{-2}. Most of the variation in net radiation comes from the latitudinal decrease of insolation and from cloudiness variations and their effect on surface solar heating. In high latitudes, the annual mean net radiative heating of the surface approaches zero.

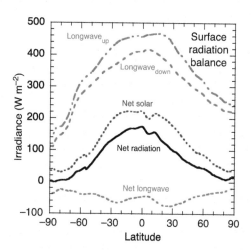

FIGURE 4.17 **Components of the annual average surface radiation budget plotted against latitude.** *Data from Lin et al., 2008.*

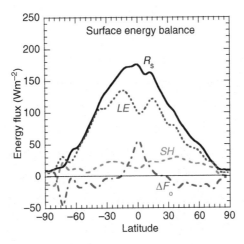

FIGURE 4.18 **Components of the annual average surface energy balance plotted against latitude.** *Data smoothed and adjusted from Lin et al., 2008.*

The evaporative heat loss from the surface has its greatest values over the mid-latitude, warm, western boundary currents, the Kuroshio and the Gulf Stream (Fig. 4.19b). In these regions at the western edges of the Pacific and Atlantic oceans, the evaporative heat loss may exceed 200 Wm^{-2} and is much greater than the local net radiative heating of the ocean surface. The evaporative cooling of the western boundary currents is greatest in the winter season. However, these areas are relatively small. The evaporative heat loss is also large over the expanses of the subtropical oceans, where evaporation consumes most of the energy provided to the surface by net radiation. Evaporation driven by insolation over the tropical and subtropical oceans is the boiler that drives the circulation of the atmosphere and the hydrologic cycle of Earth.

The sensible heat loss is largest from relatively dry land areas. The sensible heat loss from the ocean surface is small, except over the warm western boundary currents of the mid-latitude oceans (Fig. 4.19c). These large sensible heat fluxes occur when cold air from the continents flows over the warm ocean currents during winter. In these regions, sensible heat fluxes may exceed 50 Wm^{-2} in the annual mean, but they are still much less important than evaporative cooling.

The divergence of the heat flux in the ocean, or alternatively, the flux of heat from the atmosphere into the ocean, is large and negative over the western boundary currents (Fig. 4.19d). In these regions, the ocean is supplying heat by horizontal transport in the oceans, which is then lost to the atmosphere. The regions where the air is heating the water are found

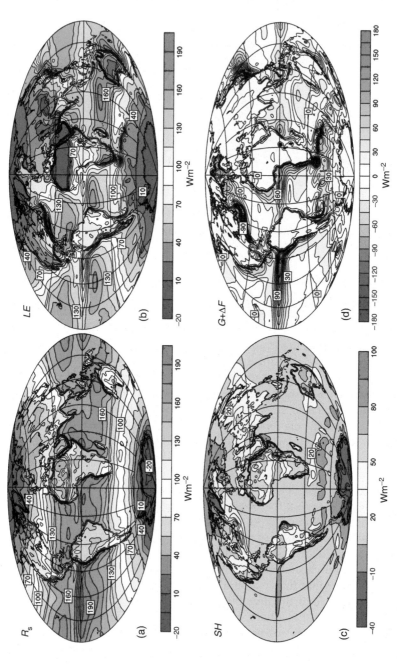

FIGURE 4.19 **Global maps of the annual average surface energy budget components.** (a) Net surface radiation; (b) latent heat flux; (c) sensible heat flux; (d) net downward heat flux into the ocean. *Based on ECMWF ERA-Interim reanalysis data products.*

along the equator and along the eastern margins of the oceans where upwelling brings cold water to the surface. In these regions of upwelling, latent heat loss is reduced and more of the net radiation is used to heat the ocean water. The large transfers of energy from the tropical and eastern oceans to the mid-latitude western oceans in the Northern Hemisphere play a critical role in determining the climate of maritime areas. The ocean currents that produce these important energy transports are described in Chapter 7.

EXERCISES

1. If the top 100 m of ocean warms by 5°C during a 3-month summer period, what is the average rate of net energy flow into the ocean during this period in units of Wm^{-2}? If the atmosphere warms by 20°C during the same period, what is the average rate of net energy flow into the atmosphere?

2. In (4.4) the heat content of the atmosphere is estimated as $c_p T(p_s / g)$, but in a compressible, hydrostatic atmosphere the potential energy of the atmosphere, gz, must also increase when it warms, since the atmosphere is raised as it warms. Use the hydrostatic relationship to estimate how much the required increase in potential energy adds to the heat capacity of the atmosphere, if the atmosphere warms by the same amount at all pressures.

3. Derive (4.18) from (4.15) and (4.17).

4. The blackbody emission from the surface can be linearized about some reference temperature T_0. $\sigma T_s^4 \approx \sigma T_0^4 + 4\sigma T_0^3 (T_s - T_0) + \cdots$. And the sensible cooling of the surface can be written as $SH \approx c_p \rho C_D U (T_s - T_a) + \cdots$. Calculate and compare the rates at which longwave emission and sensible heat flux vary with surface temperature, T_s. In other words, if the surface temperature rises by 1°C, by how much will the longwave and sensible cooling increase? Assume that $T_0 = 288$ K, T_a is fixed, $\rho = 1.2$ kg m^{-3}, $c_p = 1004$ J kg^{-1} K^{-1}, $C_D = 2 \times 10^{-3}$, and $U = 5$ m s^{-1}.

5. Air with a temperature of 27°C moves across a dry parking lot at a speed of 5 m s^{-1}. The insolation at the surface is 600 Wm^{-2} and the downward longwave radiation at the ground is 300 Wm^{-2}. The longwave emissivity of the surface is 0.85, and the albedo of the asphalt surface is 0.10. What is the surface temperature in equilibrium? What is the surface temperature if the asphalt is replaced with concrete with an albedo of 0.3 and the same emissivity? The air density and drag coefficient are as in problem 3. *Hint:* Linearize the blackbody emission around the air temperature and use the surface energy equation to show

that, $T_s - T_a = \dfrac{S^{\downarrow}(0)(1-\alpha_s) + \varepsilon(F^{\downarrow}(0) - \sigma T_a^4)}{c_p \rho C_D U + \varepsilon 4\sigma T_a^3}$.

6. Do problem 4 for the case in which the parking lot is wet and the air is maintained just at saturation, and include the effect of latent cooling of the surface. Ignore any effects of surface water on the albedo. Compare the surface temperature for wet and dry surfaces. How would the results differ if the air was not saturated? *Hint:* Make use of (4.34).

7. Give the reasons why the net radiation at the surface at Flagstaff is greater than the net radiation at Yuma during summer (Fig. 4.15c,d).

5

The Hydrologic Cycle

5.1 WATER, ESSENTIAL TO CLIMATE AND LIFE

Water continually moves between the oceans, the atmosphere, the cryosphere, and the land. The movement of water among the reservoirs of ocean, atmosphere, and land is called the hydrologic cycle. The total amount of water on Earth remains effectively constant on time scales of thousands of years, but it changes state between its liquid, solid, and gaseous forms as it moves through the hydrologic system. The amount of water moved through the hydrologic cycle every year is equivalent to about a 1-m depth of liquid water spread uniformly over the surface of Earth. This amount of water annually enters the atmosphere through evaporation and returns to the surface as precipitation in rain or snow. To evaporate 1 m of water in a year requires an average energy input of 80 Wm^{-2}. The sun provides the energy necessary to evaporate water from the surface. Once within the atmosphere, water vapor can be transported horizontally for great distances and moved upward. This horizontal and vertical movement of water vapor is critical to the water balance of land areas, since about one-third of the precipitation that falls on the land areas of Earth is water that was evaporated from ocean areas and then transported to the land in the atmosphere (Fig. 5.1). The excess of precipitation over evaporation in land areas supports the return of water from the land to the ocean in rivers.

The atmosphere contains a relatively small amount of water (Tables 5.1 and 1.2). If all the water vapor in the atmosphere were condensed to liquid and spread evenly over the surface of Earth, it would be only about 2.5 cm deep. Since 100 cm of water is evaporated and condensed per annum, the atmospheric water is removed by precipitation about 40 times a year, or every 9 days. Because net evaporation is a small residual of a more rapid two-way exchange of water molecules across the air–water interface, the actual residence time of water molecules in the atmosphere is about 3 days. Since nearly a 3-km depth of water is present near the surface of Earth, most of which is in the oceans, and only 2.5 cm can reside in the

Global Physical Climatology. http://dx.doi.org/10.1016/B978-0-12-328531-7.00005-0

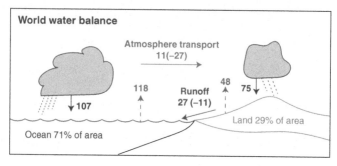

FIGURE 5.1 **Schematic diagram showing the basic fluxes of water in the global hydrologic cycle.** Units are centimeters per year spread over the area of the land or ocean. Since the areas of land and ocean are different, the land–ocean water exchanges by atmospheric transport and river runoff have different values depending on the reference area, as indicated by the parentheses. The smaller values are those referenced to the larger oceanic area.

TABLE 5.1   Water Volumes of Earth

Category	Volume (10^6 km^3)	Percent (%)
Oceans	1348.0	97.39
Polar ice caps, icebergs, glaciers	227.8	2.010
Ground water, soil moisture	8.062	0.580*
Lakes and rivers	0.225	0.020
Atmosphere	0.013	0.001
Total water amount	1384.0	100.0
Fresh water	36.00	2.60

Fresh water reservoirs as a percent of total fresh water

Polar ice caps, icebergs, glacier	77.2
Ground water to 800 m depth	9.8*
Ground water from 800 m to 4000 m	12.3*
Soil moisture	0.17*
Lakes (fresh water)	0.35
Rivers	0.003
Hydrated earth minerals	0.001
Plants, animals, humans	0.003
Atmosphere	0.040
Sum	100.000

Numbers uncertain.
(From Baumgartner and Reichel, 1975.)

atmosphere, an average water molecule must wait a very long time in the ocean, in an ice sheet, or in an aquifer, between brief excursions into the atmosphere.

In earlier chapters we saw the important role of water in many aspects of the climate system. Water is crucial to life, and the existence of oceans on Earth has dramatically influenced the character and evolution of Earth's atmosphere. Chemical and biological processes that take place in the oceans continue to regulate atmospheric composition. In the current atmosphere, water vapor is the most important gaseous absorber of solar and terrestrial radiation and accounts for about half of the atmosphere's natural greenhouse effect. Clouds of liquid water and ice contribute about 30% of the atmosphere's natural opacity to thermal radiation and contribute about half of Earth's reflectivity for solar radiation. The evaporation of water accounts for about half of the cooling of the surface that balances the net radiative heating of Earth's surface. The energy required to evaporate 1 kg of water is about 593 times the energy required to warm that water by 1°. As the water vapor rises into the atmosphere, it eventually condenses and precipitates, but the energy released during the condensation of atmospheric water vapor helps to drive the circulation systems of the atmosphere. Water can alter the surface albedo of Earth through the deposition of snow and ice and by fostering the development of vegetative cover on land surfaces.

5.2 THE WATER BALANCE

To understand how local climates are maintained, it is instructive to consider the water budget for the surface. In order to model the climate, the surface water balance must be accurately represented. The surface water balance may be written as

$$g_w = P + D - E - \Delta f \tag{5.1}$$

where g_w is the storage of water at and below the surface, P is the precipitation by rain and snow, D is the surface condensation (dewfall or frost), E is the evapotranspiration, and Δf is the runoff.

Averaged over a long period of time, the storage term is small. In addition, dewfall is usually small, or can be incorporated into a generalized precipitation. The resulting hydrologic balance for a long-term average is

$$\Delta f = P - E \tag{5.2}$$

A complementary balance for the atmosphere must also be satisfied. Precipitation minus evaporation is the net flux of water from the atmosphere to the surface and occurs with opposite sign in the atmospheric water balance.

$$g_{wa} = -(P + D - E) - \Delta f_a \qquad (5.3)$$

The terms have the same meaning as in (5.1), except that g_{wa} indicates storage of water in the atmosphere and Δf_a indicates horizontal export of water by atmospheric motions, primarily in the form of water vapor. Adding the budgets for the surface (5.1) and the atmosphere (5.3), we obtain a water balance for the surface–atmosphere system in which the exchange of water across the surface does not appear.

$$g_w + g_{wa} = -\Delta f - \Delta f_a \qquad (5.4)$$

When averaged over a year, the storage terms on the left of (5.4) are generally small, and the horizontal transport of water out of a region by the atmosphere must be equal and opposite to the net horizontal transport at or below the surface. This means that water carried to continents by atmospheric transport must equal the runoff from rivers, and atmospheric export of water from the subtropics must be balanced by ocean currents.

The distributions with latitude of the terms in the annually averaged surface water balance are shown in Fig. 5.2. Precipitation peaks near the equator, with secondary maxima in middle latitudes of each hemisphere. The equatorial maximum is associated with heavy precipitation in the *intertropical convergence zone*, where the trade winds converge from either hemisphere. Moisture-laden air near the surface flows toward the equator from both hemispheres and converges near the equator, where it is

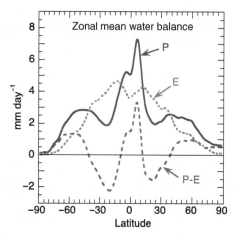

FIGURE 5.2 Latitudinal distribution of the surface hydrologic balance, showing evaporation E, precipitation P, and $P - E$ which equals runoff Δf. *Data from ERA Interim 1979–2009.*

released in thunderstorms, tropical cyclones, and other precipitation-producing weather systems (Fig. 5.3). The secondary maxima in mid-latitudes of each hemisphere are associated with the weather systems there. In middle latitudes, cyclonic disturbances with strong winds drive vertical motions that release water. Evaporation varies more smoothly

FIGURE 5.3   Satellite image of hurricane Katrina approaching the Gulf Coast of the United States on August 29, 2005. *NASA MODIS image.*

than precipitation, with a broad maximum in the tropics. Precipitation exceeds evaporation in the equatorial belt and again in middle to high latitudes. Evaporation exceeds precipitation in the belt from 15° to 40° of latitude, and these regions export water vapor to be condensed in the latitudes where the precipitation maxima occur. The runoff (also $P - E$) distribution shown in Fig. 5.2 implies transport of water vapor in the atmosphere from the subtropics to the equatorial and high latitude zones. A return flow in the oceans or rivers carries water back toward the subtropics.

The water balances of the continents and oceans are closely related to their climates and the processes that maintain climate (Table 5.2). The Atlantic and Indian oceans are net exporters of water vapor, whereas the Pacific and Arctic oceans receive more water in the form of precipitation than they give up to the atmosphere through evaporation. Comparison of the surface salinity of the Atlantic and Pacific oceans shows a much higher salinity in the north Atlantic than in the north Pacific (see Chapter 7). The surface hydrologic balance of the oceans plays an important role in determining their salinity and thereby the deep circulation of the oceans. The saline surface water of the Atlantic is a key factor for allowing surface

TABLE 5.2   Water Balance of the Continents and Oceans in mm/year

Region	E	P	Δf	$\Delta f/P$
LAND				
Europe	375	657	282	0.43
Asia	420	696	276	0.40
Africa	582	696	114	0.16
Australia	534	803	269	0.33
North America	403	645	242	0.37
South America	946	1564	618	0.39
Antarctica	28	169	141	0.83
All Land	480	746	266	0.36
OCEAN				
Arctic Ocean	53	97	44	0.45
Atlantic Ocean	1133	761	−372	−0.49
Indian Ocean	1294	1043	−251	−0.24
Pacific Ocean	1202	1292	90	0.07
All Ocean	1176	1066	−110	−0.10
Globe	973	973	0	

From Baumgartner and Reichel, 1975.

water to sink to the bottom of the ocean, since salinity is an important variable in determining seawater density, especially in high latitudes.

The *runoff ratio*, $\Delta f/P$, is a measure of the wetness of a continent. If it is large, then a significant fraction of the precipitation that falls on that continent flows into the ocean, rather than being evaporated over the land. The dry continents of Africa and Australia have relatively low runoff ratios. Typically, about 40% of the precipitation on a continent runs back to the global ocean in rivers. The evaporation from the surface of a continent typically makes up 60% of the precipitation that falls on that continent.

5.3 SURFACE WATER STORAGE AND RUNOFF

The storage term in (5.1) accounts for changes in the amount of water that is retained in the surface. Over land areas, this includes the water in the near surface soil and also water that flows deeper and becomes part of an underground water system. An additional important form of water storage is surface snow cover. Distinct seasons of precipitation and drying are a prominent feature of the climate in many regions. For such regions, storage of water in the soil and in snowpack is critical for determining the nature of the environment that develops during the dry season. In many mid-latitude regions, mountain snowpack is essential for spring and summer river flow, and at lower elevations spring snowmelt helps to replenish soil moisture and groundwater for the summer dry season. The combination of moist soil in springtime followed by summer warmth and sunshine makes many mid-latitude land areas agriculturally productive. Storage of precipitated water in snowpack depends only on the surface thermodynamics and physical structure. Storage of water that arrives at the surface as rainfall depends on the frequency and intensity of the precipitation and the characteristics of the soil, its vegetative cover, and the topography of the surface.

Climate interacts only with water that is on the surface or in the soil water zone. The soil water zone extends downward to the depth penetrated by the roots of the vegetation. Plants can draw water from this depth relatively quickly and release it to the atmosphere by *transpiration* through leaves. Because roots of plants can draw moisture from the soil more quickly than water is brought to the surface by nonbiological processes, vegetated surfaces normally release water more quickly to the atmosphere than bare soil does with the same water content. Depending on the conditions, one may need to consider a layer deeper than the root zone in order to predict surface moisture and evapotranspiration. Moisture stored deeper in the soil than the root zone must be brought upward by diffusion or capillary action. Transport through the soil in both liquid and vapor form is possible.

Water is suspended in the soil by adherence to soil particles in thin films. The amount of water that can be held in this manner is called the *field capacity*

of the soil. The moisture balance of the soil layer and the average soil moisture content are critical to the local climate of land areas. The water in this
zone is available for use by plants and can be transpired or evaporated.
The soil layer and associated vegetation determine the fate of precipitated
water, which may be quickly re-evaporated, absorbed by the soil, or run off
in stream flow. The transfer of surface water to the soil is called infiltration.
The fraction of precipitation that is retained by the soil is determined by soil
and vegetation properties and by the rate and frequency of precipitation.

If the soil water content increases above the field capacity, then gravitational forces carry the water downward to the water table, where it
becomes part of the groundwater. If the water encounters an impermeable
obstacle such as bedrock, then it may flow laterally, seeking lower pressure. Gradual collection of water in subsurface reservoirs results in the
formation of aquifers, from which freshwater can be extracted. In many
areas, subsurface fresh water reservoirs that have taken many centuries
to form, or that contain fossil water from previous wet epochs, are being
pumped out for human use.

If the surface soil is saturated and the precipitation or snowmelt is more
rapid than can be balanced by infiltration and evaporation, then surface
ponding will occur. Once the surface depressions in the soil are filled with
water, the surface water will begin to flow laterally toward streams and
drainage systems. Water runoff from land areas in streams and rivers
is important for navigation, fisheries, hydroelectric power generation,
irrigation of dry land areas, and municipal water supplies.

5.4 PRECIPITATION AND DEWFALL

Precipitation is produced when air parcels become supersaturated
with water vapor, condensation and droplet formation occur, the droplets
become large enough to fall, and the droplets or particles reach the surface
without re-evaporation. Supersaturation is most often caused by cooling
as air parcels expand adiabatically during ascent. Atmospheric motions
associated with midlatitude frontal and synoptic weather systems can
force ascent of air parcels. In the tropics and over continents during summer, ascent, condensation, and precipitation are often associated with convective instability, where parcels of air are forced upward by buoyancy
in cumulonimbus clouds. In stratiform cloud systems, light but steady
precipitation is generated through the radiative cooling of the tops of the
clouds and steady overturning of moist air from beneath. Heavy and persistent precipitation may result when moist air is forced over mountain
ranges by prevailing winds.

Global precipitation maps are produced by blending information from
surface gauges and remote sensing from satellites. The geographic distribution of precipitation is shown in Fig. 5.4. The general features of the

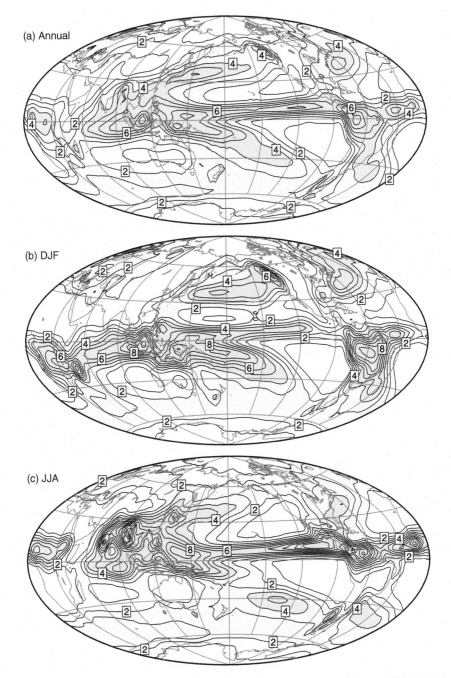

FIGURE 5.4 **Global maps of precipitation in millimeters per day for annual, DJF and JJA seasons.** Note that the Dateline is in the center of this plot. *Data from Global Precipitation Climatology Project.*

zonal-average precipitation in Fig. 5.2 are apparent, with the largest precipitation near the equator, where the average water content of the air is high and tropical convective systems are responsible for much of the rainfall. The tropical precipitation is concentrated near the Maritime Continent formed by the islands of the Western Pacific area and over the land areas of South America and Africa. In the eastern halves of the Pacific and Atlantic oceans, the precipitation forms a line north of the equator known as the Inter-Tropical Convergence Zone (ITCZ). Branching southeast from the ITCZ in the Pacific is the South Pacific Convergence Zone (SPCZ), which reaches from the Equator west of the Dateline toward mid-latitudes west of South America. In mid-latitudes of the Northern Hemisphere, the precipitation is concentrated near the western margins of the ocean, where storm tracks form, while in the Southern Hemisphere precipitation in mid-latitudes is strong at all longitudes. The tropical precipitation moves north and south following the Sun, particularly in the Asian sector, where heavy precipitation reaches nearly to 30°N in JJA. The forced ascent of moist surface air in mid-latitude weather systems and the westerly flow over obstacles such as coastal mountain ranges give rise to heavy precipitation, which can be seen in mid-latitudes west of the Rocky Mountains and the Andes. Precipitation declines toward polar regions. The entire hydrologic cycle is slowed down in polar regions because of the low temperatures and consequently low water-carrying capacity of the atmosphere.

When air comes into contact with a cold surface, usually on relatively clear nights, water vapor may condense directly onto the surface and form dew. Vapor flux from the soil may also be an important contributor to the accumulation of dew, especially at night when the underlying soil may be warmer than the surface. Dewfall is a significant contributor to the surface water balances in some arid climates, but is generally small and lumped together with the precipitation. Fog droplets that are too small to precipitate can be collected from the air by the leaves or needles of plants. In some climates such "combing" of liquid water from the air is an important mechanism whereby plants obtain moisture.

5.5 EVAPORATION AND TRANSPIRATION

Evapotranspiration is the removal of water from the surface to the air with an accompanying change in phase from the liquid to the vapor form. It is the sum of evaporation and transpiration. Evaporation refers to direct evaporation of water from the surface itself. *Transpiration* is the passage of water from plants to the atmosphere through leaf pores called *stomata*, which also serve as the point of entry for carbon dioxide required for photosynthesis. Water is absorbed from the soil and carried through the roots and stems of plants to the leaves, where it escapes as water vapor. Stomata normally close at night and open during the day, but they may also close at midday in response to high temperatures, temporary water deficit, or high carbon dioxide concentrations. The differences between evaporation and transpiration are important, but it is difficult to separate the effects of the two processes in practice, so they are generally added to form a single term in water budgets. Evapotranspiration may also include *sublimation*, which refers to the direct conversion of snow and ice to water vapor, without an intermediate liquid phase.

Evaporation from a wet surface is determined by the surface tension at the air–water interface and the rate of decrease of water-vapor concentration between the water surface and the adjacent air. The rate at which the water-vapor concentration changes with distance from a water surface depends on the molecular diffusivity and the ventilation of the air near the water surface by air motions. Normally, turbulent air motions are of primary importance for carrying water vapor away from a surface and dominate in determining gradients on scales larger than a few millimeters. The interaction of surface water waves with atmospheric turbulence can influence the rate of evaporation over the oceans. Over land, the structure of the surface and the vegetation covering it can have a substantial effect on the rate of evaporation. The collection of vegetable matter covering the land surface is called the *plant canopy*, which may be as thin as a layer of moss or as thick as a tall forest.

Plant canopies have important effects on the water and energy balances of the surface. Some of these effects are illustrated in Fig. 5.5. Precipitation that falls on a plant canopy can be intercepted by leaves and stems. Water that falls on the leaves can be evaporated from the leaves or drip to the surface. Interception of precipitation by leaves and evaporation from leaves can greatly decrease runoff if the rainfall rate is not too intense and the air is relatively dry. The leaf structure of a plant presents a much larger surface area on which evaporation can take place than the ground surface alone. The energy balance of the leaves is also of importance, since it determines how rapidly water can be evaporated. The structure and arrangement of leaves and branches affect the absorption of solar radiation, the emission of longwave radiation, and the ventilation of the surface by air motions. A parameter often used to characterize plant canopies is the *leaf area index* (LAI). It is defined as the ratio of the area of the tops of all the leaves in the canopy projected onto a flat surface to the area of the surface under the canopy. It is equal to the number of leaves that would be crossed by a vertical line passing through the canopy, on average.

5.5.1 Measurement of Evapotranspiration

Evapotranspiration can be estimated in a variety of ways. One of the most accurate methods is by weighing the moisture change in the soil

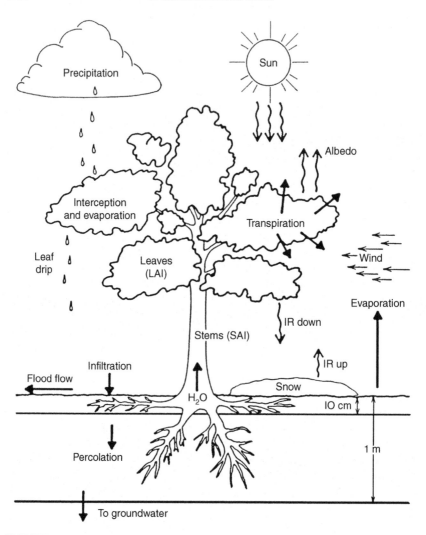

FIGURE 5.5 Diagram showing the effects of the vegetation canopy on the water and energy fluxes. *From Dickinson (1984). © American Geophysical Union.*

and its vegetative cover with a device called a *lysimeter*. A lysimeter is a container of soil set on a balance or provided with some other means of measuring water content. To obtain accurate results, the lysimeter must be large enough to contain the soil water zone and associated vegetation, and should be set in a larger environment where the surface conditions are similar to those under investigation, if results representative of the natural environment are desired. The lysimeter measures the weight of water in the soil. The net flux of moisture from the soil can be estimated from the change in weight.

Evapotranspiration can also be estimated by directly measuring the fluxes of moisture away from the surface by taking simultaneous measurements of vertical velocity and humidity. Because the moisture is carried upward by turbulent motions, the device used to measure wind and humidity fluctuations must respond on the time scale of seconds, but modern technology has made such measurements routine. It is also a challenge to obtain representative spatial and temporal means, particularly if the surface characteristics are spatially inhomogeneous.

An alternative to direct measurement of evapotranspiration is to infer it as a residual in the energy balance, if the other terms in the energy balance can be measured. Rearranging (4.1) to solve for the evaporation rate yields

$$E = \frac{1}{L}(R_s - SH - \Delta F_{eo} - G) \tag{5.5}$$

When the surface is moist, the net radiation and the evaporation are the largest terms in the surface energy balance (see Chapter 4), so that an accurate measurement of the net radiation and approximations to the other terms in the surface energy balance will provide a good estimate of evapotranspiration. Radiation can be measured very accurately and over long periods with relatively inexpensive instrumentation, and most weather stations have a device for measuring insolation. Sensible heat loss can be estimated from bulk aerodynamic formulas, if measurements of mean wind speed and temperature at two levels are available. Measurements of temperature profiles in the soil or water can be used to estimate energy storage below the surface.

5.5.2 Evaporation From a Wet Surface

Penman (1948) derived a method of calculating the evaporation from wet surfaces with minimal input data. The Bowen ratio is the ratio of sensible to latent surface energy flux. It may be estimated by using the bulk aerodynamic formulas (4.26) and (4.27).

$$B_o = \frac{SH}{LE} \cong \frac{c_p(T_s - T_a)}{L(q_s - q_a)} \tag{5.6}$$

Here we have assumed that the aerodynamic transfer coefficients for heat and moisture are equal. If the surface air is saturated, and the surface and reference-level air temperatures are not too different, we may make the following approximation:

$$\frac{(q_s^* - q_a^*)}{(T_s - T_a)} \cong \frac{dq^*}{dT} \tag{5.7}$$

where q^* is the saturation mixing ratio of water vapor. Using (5.7) in (5.6), and the assumption that the surface air is saturated, we obtain

$$B_o = B_e \left(1 - \frac{(q_a^* - q_a)}{(q_s^* - q_a)} \right) \tag{5.8}$$

where B_e is the equilibrium Bowen ratio defined in (4.33). It should be noted that the use of the Bowen ratio can be problematic in the presence of temperature inversions if the denominator in (5.7) is near zero.

The surface energy balance (5.5) may be rewritten as

$$E(1 + B_o) = E_{en} \tag{5.9}$$

where

$$E_{en} = \frac{1}{L}(R_s - \Delta F_{eo} - G) \tag{5.10}$$

E_{en} is the evaporation rate necessary to balance the energy supply to the surface by radiation, horizontal flux below the surface, and storage. Substituting (5.8) for the Bowen ratio in (5.9) yields

$$E(1 + B_e) = E_{en} + E\, B_e \frac{(q_a^* - q_a)}{(q_s^* - q_a)} \tag{5.11}$$

Using the bulk aerodynamic formulas, the evaporation may be eliminated from the second term on the right in (5.11) to yield an expression for the evaporation from a wet surface, which is often called Penman's equation.

$$E = \frac{1}{(1 + B_e)} E_{en} + \frac{B_e}{(1 + B_e)} E_{air} \tag{5.12}$$

The evaporating capacity of the air is defined by

$$E_{air} = \rho C_{DE} U(q_a^* - q_a) = \rho C_{DE} U q_a^* (1 - RH) \tag{5.13}$$

and depends on the relative humidity of the air, RH, as well as the air temperature and wind speed.

The advantage of (5.12) is that measurements of atmospheric variables at only one level are required. Over land surfaces the horizontal transport term is zero, and for time scales of a month or longer the storage term can also be ignored, so that only measurements of the net radiation, air temperature, specific humidity, and wind speed at one level are required to evaluate evaporation. The Penman equation (5.12) also shows the relative roles of air humidity and available radiation in driving evaporation over a wet surface. At high temperatures the equilibrium

Bowen ratio is small and evaporation is mostly dependent on available energy. As the equilibrium Bowen ratio becomes small, the evaporation rate approaches a value necessary to balance the energy input to the surface. This occurs at temperatures greater than about 25°C. At lower temperatures, and consequently higher equilibrium Bowen ratios, the evaporation rate is more dependent on the supply of unsaturated atmospheric air. At temperatures near or below freezing, the equilibrium Bowen ratio is large and the evaporation is dependent primarily on the drying capacity of the air.

5.5.3 Potential Evaporation

Evapotranspiration is constrained by the surface water supply, the energy available to provide the latent heat of vaporization, and the ability of the surface air to accommodate water vapor. The potential evaporation is defined as the rate of evaporation that would occur if the surface was wet, and is therefore the maximum possible evaporation for the prevailing atmospheric conditions. It measures the effect of energy supply and air humidity on the evaporation rate and avoids the more difficult issue of soil moisture availability and the physiological processes in plants that bring moisture from the soil to the atmosphere. If the potential evaporation exceeds the actual evapotranspiration, then a moisture deficit exists, and one may infer a dry surface. One method to calculate potential evaporation is from Penman's equation, which relates the evaporation from a wet surface to net radiative heating and mean air temperature, humidity, and wind speed at one level.

The potential evaporation can be used to understand how the hydrologic cycle at the surface might change with global mean temperature. The strongest variation in the potential evaporation is the saturation specific humidity, which increases at an exponential rate (1.11). We therefore expect that potential evaporation will increase in a warmer climate, meaning that water will be removed more efficiently from the surface in a warmed climate than in a cooler one. At the same time, if the atmospheric circulation does not change significantly, more moisture will be converged in regions of moisture convergence. We therefore expect that with warming will come greater contrast between areas in which precipitation exceeds evaporation and where evaporation exceeds precipitation. This is the "wet gets wetter, dry gets dryer" paradigm of global warming.

Since radiation dominates the energy supply for evaporation (5.10), we can write (5.12) as

$$PE = \frac{1}{(1+B_e)}\left(\frac{R_s}{L} + B_e E_{air}\right) \tag{5.14}$$

where we have introduced the acronym *PE* for potential evaporation. Also, we see that the second term in the parentheses can be written,

$$B_e E_{air} = \frac{\rho C_D U (1 - RH)}{\dfrac{L}{c_p} \dfrac{d \ln q^*}{dT}} = \frac{\rho C_D U (1 - RH)}{\dfrac{L}{c_p} \dfrac{L}{R_v T^2}} \tag{5.15}$$

Here we have used,

$$\frac{d \ln q^*}{dT} = \frac{L}{R_v T^2} \approx 6.5\% \, \text{K}^{-1} \tag{5.16}$$

Equation (5.16) states that the saturation specific humidity increases about 6.5% for each degree of warming for temperatures around 288 K. The exponential dependences of saturation vapor pressure in B_e and E_{air} cancel out in the second term of (5.14) so that neither term within the parentheses have the exponential dependence on temperature of saturation vapor pressure. If we assume, as is commonly done, that the relative humidity and the wind speed do not change very rapidly with warming, then the two terms inside the parentheses should vary only modestly with temperature, and the strongest temperature dependence resides in the $(1 + B_e)^{-1}$ in (5.14). This means that *PE* will increase with temperature, assuming that wind speed and relative humidity change slowly.

PE is, however, fairly sensitive to the relative humidity in the boundary layer. The relative humidity in the boundary layer is about 80%, so that if the relative humidity decreases to 79%, $1 - RH$ would change by 5% from 20% to 21%. So small changes in relative humidity in the boundary layer can cause big changes in the gradient of moisture in the boundary layer, which would strongly affect *PE*. The relative humidity in the boundary layer is maintained by a complex set of processes, and it is not a simple calculation to predict how it would change in response to climate warming.

Since we have established that the terms inside the parentheses in (5.14) do not change as rapidly as saturation vapor pressure, the $(1 + B_e)^{-1}$ term multiplying the parenthesis determines the fractional sensitivity of *PE* to temperature changes, assuming constant relative humidity and wind speed. We can take the derivative with respect to temperature and show that,

$$\frac{\partial}{\partial T} (1 + B_e)^{-1} = \left(\frac{B_e}{(1 + B_e)^2} \right) \left(C - \frac{2}{T} \right) \tag{5.17}$$

where $C = (L/R_v)(1/T^2)$. B_e decreases rapidly with temperature and is one at about 278 K (Fig. 4.10), so that the first part of (5.17) is about 0.5 at about 278 K and is smaller for both colder and warmer temperatures. At very cold temperatures, there is so little water vapor in the atmosphere that its changes have little influence. At warm temperatures, $B_e \ll 1$ and all of the available energy at the surface is already being used to evaporate water. The second part $(C - 2 / T)$ also decreases with increasing temperature, but the first part dominates the sensitivity of PE to temperature. The magnitude of the net effect is that the sensitivity of PE to temperature inferred from (5.17) decreases by about a factor of two between 0°C and 30°C, implying that PE is more sensitive to temperature in middle and high latitudes than in low latitudes. This is because at low temperatures the contribution to the total surface cooling from evaporation can increase at approximately the Clausius–Clapeyron rate, whereas at high temperatures virtually all the surface cooling is already being done by evaporation, which is constrained by the supply of energy to the surface (Scheff and Frierson, 2014). Since the supply of energy to the surface does not change rapidly with climate change, the potential evaporation does not change as rapidly in warmer latitudes as it does in colder latitudes where surface cooling can be shifted from sensible to latent cooling.

5.6 ANNUAL VARIATION OF THE TERRESTRIAL WATER BALANCE

The annual variation of the surface water balance at a location is intimately related to the local climate and its potential for human habitation and agriculture. The natural vegetation is adapted to the normal cycle of water surplus and water shortage that a region experiences. The annual variation of the water balance can be used as one means of classifying climates (Thornthwaite, 1948). The water balance depends on the annual variation of precipitation and evaporation, which together largely determine the soil moisture.

The accuracy with which the water balance can be determined depends on the amount and quality of the data available. Ideally, one would require good measurements of evaporation, precipitation, and soil moisture. However, evaporation and soil moisture are not routinely measured at most locations. Most climatological stations report surface air temperature and humidity, precipitation, and wind speed. With these variables, potential evaporation can be estimated from semi-empirical formulas such as (5.12), (5.20), or their simplified forms. If a soil moisture balance is calculated using the bucket model, then the actual evapotranspiration can be estimated from the potential evapotranspiration using (5.21). Figure 5.6 shows results of a simple water balance analysis for a variety of locations. These use daily ERA interim reanalysis data sets from 1979 to 2010, so that the surface radiation, precipitation, and evaporation are generated by the model, rather than by direct observation, but the model meteorology is very close to the observed weather, and the seasonal variations of

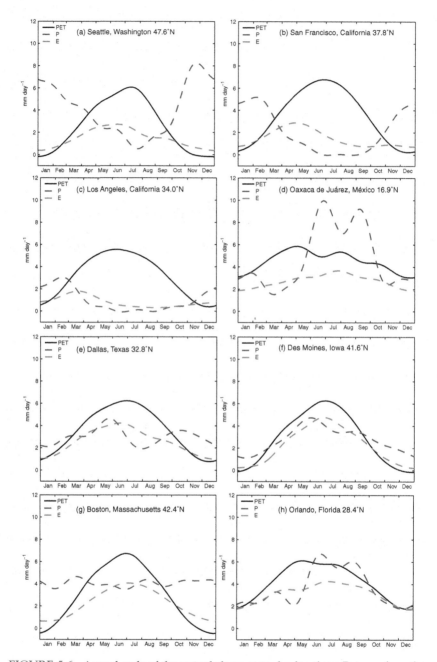

FIGURE 5.6   **Annual cycle of the water balance at twelve locations.** Data are from the ERA Interim Land Reanalysis data set for 1979–2010. Data are for the 0.75° box closest to the city indicated, and all data are from the model rather than direct observations. Potential evapotranspiration has been estimated simply as $R_s/\rho L$. The data have been smoothed.

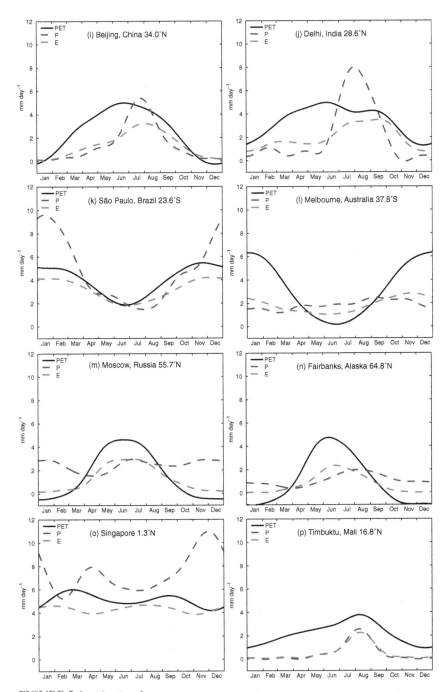

FIGURE 5.6   (*Continued*)

precipitation are very close to observations. Potential evapotranspiration (PET) is estimated from the net surface radiation alone as $PET = R_s/\rho_w L$.

The west coast of North America in middle latitudes experiences a wintertime maximum in precipitation associated with winter storms (Fig. 5.6a–c). The wintertime peak of precipitation is smaller and occurs later in southern California than on the Pacific Coast of Canada. In Juneau, Alaska the monthly precipitation peaks in September or October, whereas in Los Angeles it does not peak until January or February. The summertime minimum of precipitation becomes deeper and of longer duration toward the south. Because the summer minimum of precipitation corresponds with the season of strongest insolation and warmest temperatures, the soil moisture can become depleted in the summer months. At Seattle, evapotranspiration (E) exceeds precipitation (P) beginning in May, and during summer the soil is sufficiently dry that PET greatly exceeds the actual evapotranspiration. This continues until October, when the precipitation again exceeds the evapotranspiration. In San Francisco, the dry season begins in April and extends until November. In Los Angeles, the soil is nearly always dry, and only during the months of December through February does the actual evapotranspiration approach PET. Oaxaca, Mexico is in the tropics (~17°N), and its precipitation maximum comes in the summer months, when tropical convection reaches northward from the equatorial region (Fig. 5.6d). During the heavy precipitation months of June through September, the precipitation exceeds PET. The remainder of the year is drier.

In much of the interior of North America the precipitation peaks in the spring or early summer when the soil is moist and temperatures are high enough to allow the air to carry large amounts of water vapor. As the summer season progresses, the soil dries out and the precipitation amount decreases. During middle and late summer PET generally exceeds the actual evaporation, indicating relatively dry soil. At Des Moines, Iowa the precipitation peaks in late May and continues into the middle and late summer. The combination of warmth, insolation, and precipitation during the summer in this part of the American Midwest makes it well suited for agriculture, especially corn. Dallas, Texas has a similar May maximum in precipitation, but the dip in summertime precipitation is greater, and the average difference between PET and precipitation is greater, indicating a drier climate.

In the northeastern United States, the monthly precipitation amount is almost independent of season. The frontal precipitation of winter is replaced in summer by more convective precipitation, such that the total precipitation remains almost constant. PET follows the insolation and temperature and peaks in the summer. During the summer months the evaporation exceeds the precipitation, causing the soil to dry out somewhat. In Boston, Massachusetts the potential evaporation exceeds the precipitation from

May to September (Fig. 5.6g). Farther south on the Florida Peninsula, Orlando shows a summertime maximum in precipitation, not unlike Oaxaca, with very heavy precipitation from June to September. This precipitation is mostly associated with thunderstorms, which are driven by solar heating of the land and begin during the hottest part of the day.

In general, four basic types of water balance cycles can be identified for the USA, although many locations show a combination of several types. The west coast of North America in middle latitudes has a winter precipitation maximum and summer dry period. The interior of the continent experiences a spring or summer rainfall maximum, followed by a drying period of varying intensity in the late summer. On the east coast of North America in mid-latitudes, the precipitation amount is almost independent of season, but the potential evaporation peaks in the summer, producing some reduction in soil moisture. In tropical or subtropical latitudes, the precipitation is very small in winter, but thunderstorms yield large amounts of rain during the warmest part of the summer.

Beijing, China, and Delhi, India have a monsoonal climate, with very dry conditions in winter and spring, then heavy summer rains starting in June (Fig. 5.6i,j). São Paulo, Brazil has a summertime precipitation maximum that exceeds the potential evapotranspiration and during the rest of the year the precipitation about equals the PET, implying a wet climate. Melbourne, Australia has nearly constant precipitation through the year, which is less than PET, except in winter.

In high latitudes, the low saturation vapor pressure associated with the relatively cold temperature constrains the rates of evaporation and precipitation. At many locations, the precipitation is greater than PET during most of the year. Water evaporated at warmer latitudes is transported into high latitudes by atmospheric motions and is precipitated when large-scale motions drive saturated air upward. The energy available at the surface is insufficient to allow the evaporation of this water. The growing season is short, so that the vegetation is not especially effective in bringing water from the soil to the atmosphere. Therefore, soils in high latitudes typically have a high water content. At very high latitudes, this water is mostly frozen. In high latitudes, it is common for the precipitation to exceed PET except during the summer, as in Moscow, Russia and Fairbanks, Alaska (Fig. 5.6m,n).

Singapore is near the equator in the Maritime Continent region, where the precipitation exceeds PET for most of the year and the annual variation of PET is small (Fig. 5.6o). Timbuktu, Mali is in the Sahel region of Africa, which gets precipitation only in the summer for a brief period, but even then the precipitation does not match the PET (Fig. 5.6p). PET is relatively low for such a tropical location because of the high albedo of the desert, and because the surface gets hot and loses heat through longwave emission during the daytime.

5.7 MODELING THE LAND SURFACE WATER BALANCE

The water balance of the surface is intimately coupled with the surface energy balance. Over water surfaces, the joint energy–water balance problem is simplified because one can assume that the air at the surface is saturated, and the storage and retrieval of water from below the surface is not an issue. Over land surfaces, the heat and water balances are very sensitive to the amount of water below the surface and the rate at which it can be brought to the surface and evaporated or transpired through plants. Transpiration through plants is dependent not simply on the soil moisture and atmospheric conditions, but also on the physiological state of the plant cover. Modeling and understanding of the exchanges of energy, water, and carbon dioxide at the surface are critically important for understanding climate variability and change. Vegetation on the land (land cover) is an important mediator of these exchanges, is important in its own right for agricultural and other uses, and will evolve to adapt to changing climate (land cover change).

5.7.1 The Bucket Model of Land Hydrology

The simplest model for the soil water budget is the bucket model. The soil is assumed to have a fixed capacity to store water that is available for evapotranspiration. The rate of change of the mass of water in the soil per unit area w_w, is determined by the rainfall rate P_r, the evapotranspiration rate E, the melting of snow M_s, and the runoff rate Δf.

$$\frac{\partial w_w}{\partial t} = \rho_w \frac{\partial h_w}{\partial t} = P_r - E + M_s - \Delta f \tag{5.18}$$

The amount of available water in the soil can be expressed as an equivalent depth h_w, using a standard water density ρ_w. In the bucket model, the soil is assumed to have a fixed capacity to store moisture, corresponding to an equivalent water depth, h_c, which would typically be about 15 cm. If the soil moisture equals the capacity of the soil, then the soil is saturated. If the sum of rainfall plus snowmelt exceeds evaporation when the soil is saturated, then runoff at a rate just sufficient to keep the soil saturated is predicted.

To complete the soil moisture balance model for regions with snowfall, a separate budget for snowcover must be retained. If precipitation occurs when the surface temperature is below freezing P_s, it can be assumed to result in surface snow cover. The snow cover can be measured in terms of its water mass per unit area w_s, or an equivalent depth of water h_s.

$$\frac{\partial w_s}{\partial t} = \rho_w \frac{\partial h_s}{\partial t} = P_s - E_{sub} - M_s \tag{5.19}$$

The maximum carrying capacity of the surface for snow or ice is determined by the lateral flow of ice sheets and does not become a factor until the ice is several hundred meters thick. Snow cover is removed by sublimation, E_{sub}, or melting. The snow cover lies on top of the soil and does not enter into the soil moisture balance unless it melts. Melting occurs when the surface temperature rises to the freezing point of water. The latent heat of fusion must be supplied to the surface energy balance when melting occurs. Melting continues at the rate necessary to keep the surface temperature from rising above 0°C until the temperature falls below freezing or the snow cover is completely removed.

The rate of evaporation depends on the soil moisture. The soil moisture can be used to relate the actual evaporation to the potential evaporation – the evaporation that would occur if the surface were wet. If measurements of air humidity, air temperature, wind speed, and surface temperature are available, the bulk aerodynamic formula can be used to calculate PET.

$$PET = \rho_a\, C_{DE}\, U\, (q^*(T_s) - q_a) \tag{5.20}$$

If insufficient data to evaluate (5.20) are available, then another approximate formula can be used to estimate PET.

The actual evapotranspiration may be related to the potential evaporation and the soil moisture content.

$$E = \beta_E\, PET \tag{5.21}$$

Healthy vegetation may transpire at the rate of potential evaporation, even when the soil is not saturated. When the soil moisture falls below a certain level h_v, the vegetation will no longer transpire at the potential rate. For soil moisture availability less than h_v, it is simplest to assume that β_E varies linearly between zero and one.

$$\beta_E = \begin{cases} 1.0, & h_w \geq h_v \\ \left(\dfrac{h_w}{h_v}\right), & 0 < h_w < h_v \end{cases} \tag{5.22}$$

The simple bucket model can easily be elaborated by adding a deep layer that exchanges water with the upper layer at a slow rate depending on the relative saturation of the two layers. This allows the soil water zone to be replenished with moisture from below without the occurrence of precipitation. In this case, an additional budget equation for the deep layer is required, and a term describing the exchange with the deep layer must be added to the soil-moisture equation (5.19). A thin layer near the surface can also be added to allow better treatment of short time scales associated with rainstorms or diurnal variations.

5.7.2 More Elaborate Models of Land Surface Processes

To improve significantly on the bucket model of land-surface hydrology, much more complex models must be introduced that describe the interactions of the atmosphere with vegetation and soil. Such models must fully couple the momentum, heat, and moisture budgets near the surface and describe each with compatible levels of sophistication and detail. The processes that must be considered include those illustrated in Fig. 5.5. Plants play a central role in the momentum, energy, and moisture transfers at the surface, and must be included in a model. Plants have effects on boundary processes through their physical properties and biological processes, and they have the ability to move water through their leaves at a rapid rate to facilitate photosynthesis when water is available. However, in times of water stress, plants can reduce their transpiration rate by closing their stomata. More advanced models of land–atmosphere interaction also seek to predict the exchange of carbon between the land surface and the atmosphere and to predict land cover change as the vegetation and soil co-evolve with a changing climate.

The rate at which plants transpire water depends on the availability of photosynthetically active radiation, temperature, air humidity, the availability of water within the plant, and the physiological state of the plant. The rate of water movement through plants is limited by the vapor phase in the leaves, rather than by the uptake of liquid water in the roots. The vapor pressure gradients that drive transpiration are strongly related to leaf temperature. For this and other reasons, leaf temperature is important to model, which requires calculation of the energy transfer through the plant canopy and a heat budget for the leaf structure of the plant. The transfer of solar radiation through the plant canopy is important, since it determines the distribution of heat input throughout the plant canopy and at the soil surface. Much of the insolation will be absorbed by the vegetative cover, rather than by the underlying soil. Because plants and other natural surfaces have very different albedos for visible and near-infrared radiation, these two frequency bands of solar radiation must be treated separately in accurate calculations. The obstacle to free airflow presented by the physical structure of the plant canopy affects the ventilation within the canopy, which is important to the turbulent fluxes of momentum, heat, moisture and carbon dioxide. The similarity hypotheses and resulting velocity profiles that lead to the bulk aerodynamic formulas are not valid within the plant canopy. Mean wind speeds and turbulent kinetic energy are much smaller within the canopy than just above it. The air properties within the plant canopy and at the soil surface can be very different from those near the top of the canopy, which is in direct contact with the atmosphere.

In addition to drawing moisture from the soil, plants can also intercept a substantial amount of precipitated water and store it on leaves and stems. The effectiveness of a plant in intercepting rainfall is dependent on the leaf structure, leaf orientation, the leaf area per unit of surface area, and on the frequency and intensity of precipitation. Tall vegetation with a large leaf area index can greatly decrease the supply of moisture to the soil by intercepting precipitation and facilitating its re-evaporation before it reaches the ground. Interception losses of this nature can range from 15% to 40% of precipitation for coniferous forests and from 10% to 25% for deciduous forests in mid-latitudes. Interception loss is greater if the precipitation rate is low or is intermittent, and lesser if the precipitation rate is high or continuous. To model interception and storage on leaves, budgets must be calculated for the amount of water stored on the surfaces of leaves. The removal of this surface water by evaporation depends on the supply of energy and unsaturated air at each level within the plant canopy.

The soil is the main reservoir from which evapotranspiration is drawn. Three layers within the soil may be distinguished by their interaction with the atmosphere. A thin layer very near the surface determines the interaction of the atmosphere with the bare ground surface on the time scale of individual precipitation episodes. If this thin layer becomes saturated during a rain shower, then runoff may occur. If this layer becomes dry, then surface evaporation is very small and transpiration by plants becomes the only mechanism whereby water can be efficiently removed from the soil. Below the surface layer is a deeper layer in which the roots of the vegetation reside and draw moisture from the soil. Below the root zone is a still deeper layer to which moisture is carried by gravity if the soil is saturated, and from which moisture can be drawn by capillary action.

The ability of soil to hold moisture may be measured by its field capacity, which is defined as the maximum volume fraction of water that the soil can retain against gravity. Typical values for loam are about 30%, but field capacities range from 10% for sand to 55% for peat. If the volumetric water content of the soil falls below a certain level, plants are unable to draw moisture from the soil and will remain wilted at all times of day. This permanent wilting threshold is typically one-third to one-half of the field capacity. The soil moisture available to plants is the difference between the volumetric soil water fraction and the wilting threshold. The maximum available soil moisture is the difference between the field capacity and the wilting threshold, and is 15–20% for typical soils.

The hydraulic conductivities of most soils decrease rapidly as the soil dries out, so that conduction of water becomes very slow if the water content is much below field capacity. If the surface layer of the soil becomes very dry, then infiltration of subsequent rainfall may be inhibited. Similarly, the soil in the root zone may become very dry, whereas the soil moisture several centimeters below the deepest roots remains near field capacity. The amount of water that is available to the vegetation is thus approximately equal to the available volumetric soil water fraction times the rooting depth. If the roots extend down about 1 m, and the available field capacity is about 15%, then the total amount of water available to plants when the soil is saturated is equivalent to about 15-cm depth of water. This may be greater or less depending on the type of soil and vegetation. Plants that live in sandy soil tend to develop deep roots, which offsets the low volumetric fraction of available water in sand.

To generalize the bulk aerodynamic formulas (4.26 and 4.27) to systems with plant canopies, they can be written in the following way.

$$SH = c_p \, \rho \frac{(T_s - T_a)}{r_h} \quad LE = L\rho \frac{(q_s - q_a)}{r_v} \qquad (5.23)$$

In (5.23), r_h and r_v represent resistances to the transport of sensible and latent heat from the surface to the air above the plant canopy in units of s m^{-1}. Separate resistances can be specified for the plant canopy, the surrounding ground, and the air boundary above the plant canopy as in Fig. 5.7. Similar expressions could be written for carbon dioxide or other trace gases. One objective of land-surface modeling is to calculate values for these resistances that would cause a climate model to behave like nature.

Fig. 5.7 shows a schematic of the bucket model described previously along with a very generalized schematic of a more complex land-surface model including a plant canopy. The fluxes of sensible and latent heat that arise from the canopy model are a combination of what might escape directly from the soil and what is transmitted from the canopy. The great innovation of the more advanced land-surface models is to incorporate the biological effects of the plant canopy in controlling the exchanges of moisture and heat. The canopy has its own temperature and other variable

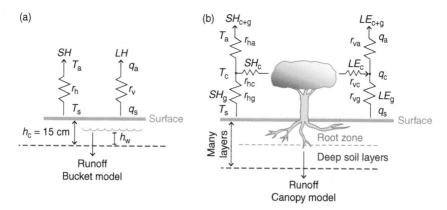

FIGURE 5.7 **Schematic diagrams of the bucket model and the canopy model of the land surface.** Subscripts g, c and a indicate values for the ground, canopy and air above the canopy. *Adapted from Sellers et al., 1997 and Pitman, 2003.*

properties that control the resistance to the transport of moisture and heat from the surface to the atmosphere. Variable stomatal-resistance models take into account the ability of plants to open or close their leaf pores depending on the condition of the plant and its environment. Further advances are the capability to model the response carbon uptake and release to climate conditions and CO_2 concentration. Even more advanced land-surface models can simulate change in the plant canopy and soil in response to changing climate conditions, such as the conversion of tundra to forests or the conversions of forest to grassland.

EXERCISES

1. The approximate volume of water retained in soil moisture and groundwater is given in Table 5.1. Use the data in Fig. 5.1 to calculate the time it would take for precipitation over land to deliver an amount of water equal to the soil water and groundwater. How long would it take to replace the groundwater and soil moisture if only 10% of the runoff could be redirected to replenishing the groundwater?

2. Use the bulk aerodynamic formula (4.32) to calculate the evaporation rate from the ocean, assuming that $C_{DE} = 10^{-3}$, $U = 5 \text{ m s}^{-1}$, and that the reference-level air temperature is always 2°C less than the sea surface temperature. Calculate the evaporation rate for the following:

   a. $T_s = 0°C$, $q_s^* = 3.75 \text{ g kg}^{-1}$, RH = 50%
   b. $T_s = 0°C$, $q_s^* = 3.75 \text{ g kg}^{-1}$, RH = 100%
   c. $T_s = 30°C$, $q_s^* = 27 \text{ g kg}^{-1}$, RH = 50%
   d. $T_s = 30°C$, $q_s^* = 27 \text{ g kg}^{-1}$, RH = 100%

   Assume a fixed air density of 1.2 kg m^{-3}. How would you evaluate the importance of relative humidity versus the importance of surface temperature for determining the evaporation rate?

3. Calculate the Bowen ratio using the bulk aerodynamic formulas for surface temperatures of 0, 15, and 30°C, if the relative humidity of the air at the reference level is 70% and the air–sea temperature difference is 2°C. Assume that the transfer coefficients for heat and moisture are equal.

4. Use the results of problem 3 to explain why high-latitude land areas often have high surface-moisture content.

5. Why is local winter and spring snow accumulation important for the summer soil moisture of mid-latitude continental land areas? How do you think the August climate would change if the winter and spring snowfall were replaced by rain showers?

6. What are some of the shortcomings of the bucket model of land hydrology? How are these limitations addressed by more sophisticated models for land-surface processes?

7. Derive (5.12) using the method outlined in the text.

6

Atmospheric General Circulation and Climate

6.1 THE GREAT COMMUNICATOR

The movement of air in the atmosphere is of critical importance for climate. Atmospheric motions carry heat from the tropics to the polar regions and thereby reduce the extremes of temperature that would otherwise result. Water from the oceans is evaporated and carried in the air to land, where rainfall supports plant and animal life. Winds supply momentum to ocean surface currents that transport heat and oceanic trace constituents such as salt and nutrients. The circulation of the atmosphere is a key component of the climate, since it both responds to temperature and humidity gradients and helps to determine them by transporting energy and moisture. The atmosphere provides the most rapid communication between geographic regions within the climate system.

The global system of atmospheric motions that is generated by the uneven heating of Earth's surface area by the Sun is called the *general circulation*. A complete description of the atmosphere's general circulation includes mean winds, temperature and humidity, the variability of these quantities, and the covariances between wind components and other variables that are associated with large-scale weather systems. A statistical description of the general circulation of the atmosphere has been constructed from the ensemble of daily global flow patterns estimated with data from balloons and satellites. The general circulation of the atmosphere can be simulated by solving the equations of motion on a computer, and such general circulation models form a part of global climate models. The best estimates of the observed circulation of the atmosphere are constructed by combining observations with model equations as in an analysis to initialize a weather prediction model. For climate purposes, past data can be assimilated into a modern weather or climate prediction model using the best current model in a process called reanalysis, which

Global Physical Climatology. http://dx.doi.org/10.1016/B978-0-12-328531-7.00006-2

produces the best possible estimate of the time history of the atmospheric general circulation.

6.2 ENERGY BALANCE OF THE ATMOSPHERE

Atmospheric motions are generated by geographic variations in heating of the surface caused by meridional gradients of insolation, albedo variations, and other factors. These gradients in energy input produce gradients in energy content that is available to generate atmospheric motions. By transporting energy, winds generally act to offset the effects of these heating variations on atmospheric temperature. The local energy balance of an atmospheric column of unit horizontal area includes the effects of radiation, sensible and latent heat exchange with the surface, and the horizontal flux of energy in the atmosphere.

$$\frac{\partial E_a}{\partial t} = R_a + LE + SH - \Delta F_a \qquad (6.1)$$

In (6.1), $\partial E_a / \partial t$ is the time rate of change of the energy content of an atmospheric column of unit horizontal area extending from the surface to the top of the atmosphere, R_a is the net radiative heating of the atmospheric column, LE is the supply of latent heat to the atmosphere from surface evaporation, SH is the sensible heat transfer from the surface to the atmosphere, and ΔF_a is the horizontal divergence of energy out of the column by transport in the atmosphere.

The net radiative heating of the atmosphere is the difference between the net downward irradiance at the top of the atmosphere and the net downward irradiance at the ground.

$$R_a = R_{TOA} - R_s \qquad (6.2)$$

The storage of energy in the atmosphere is negligible, particularly when averaged over a year, so that the atmospheric energy balance is the sum of radiative heating, sensible heating, and latent heating, balanced against the export of energy by atmospheric motions.

$$R_a + LE + SH = \Delta F_a \qquad (6.3)$$

The annually and zonally averaged net effect of radiative transfer on the atmosphere is a cooling of about $-90\ Wm^{-2}$, which is nearly independent of latitude (Fig. 6.1). To balance this, evaporation provides about $80\ Wm^{-2}$ and sensible heat transfer about $10\ Wm^{-2}$. The radiative cooling corresponds to an atmospheric temperature decrease of about 1.5°C per day (see Chapter 3). The energy lost from the atmosphere in 1 week

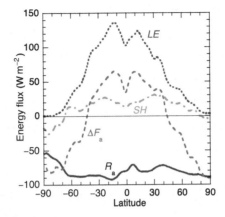

FIGURE 6.1   **Distribution with latitude of the components of the atmospheric energy balance averaged over longitude and over the annual cycle.** Units are Wm^{-2}. *Data from ERA-Interim.*

through radiative transfer equals about 2.5% of the atmosphere's total energy content. If only the atmosphere's thermal capacity were considered, this cooling rate, acting alone, would bring the global mean surface air temperature to below freezing in about 2 weeks. Under normal circumstances, the radiative cooling is balanced in the global mean by heating from latent heat of condensation and sensible heat transfer from the surface.

Heating of the atmosphere by the transfer of sensible heat from the surface is relatively small. The largest contribution to balancing the radiative loss from the atmosphere is the supply of latent heat of vaporization from the surface, which is later converted to sensible heat during precipitation. In contrast to the radiative cooling, the addition of latent heat has a very distinctive structure with latitude peaking near 140 Wm^{-2} in the tropics and diminishing to near zero in high latitudes. The latitudinal structure of the evaporation is reflected in the latitudinal structure of the atmospheric energy flux divergence. Atmospheric motions export about 50 Wm^{-2} from the equatorial region and import about 85 Wm^{-2} into the polar regions. This poleward transport of energy by the atmosphere is one of the important climatic effects of the general circulation of the atmosphere.

6.3 ATMOSPHERIC MOTIONS AND THE MERIDIONAL TRANSPORT OF ENERGY

Motions in the atmosphere can be associated with many physical phenomena, which have a wide variety of space and time scales (Fig. 6.2). Small-scale phenomena such as turbulence and organized mesoscale phenomena such as thunderstorms are effective primarily at transporting

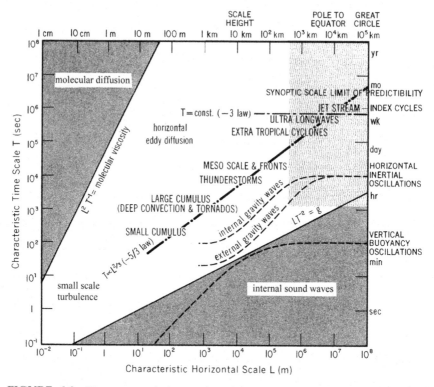

FIGURE 6.2 **The space and time scales of phenomena in the atmosphere.** Light shading represents approximately scales that can be resolved in climate models. *[From Smagorinsky (1974).]*

momentum, moisture, and energy vertically. Only very large-scale phenomena such as extratropical cyclones, planetary-scale waves, and slow meridional circulations that extend over thousands of kilometers are effective at transporting momentum, heat, and moisture horizontally between the tropics and the polar regions. The upward flux of energy and moisture in the boundary layer and the poleward flux of energy by planetary-scale circulations in the atmosphere have equal importance for climate. These phenomena have characteristic spatial scales that differ by nearly 10 orders of magnitude: from millimeters to 10 thousand kilometers.

6.3.1 Wind Components on a Spherical Earth

Wind velocities in the atmosphere are measured in terms of a local Cartesian coordinate system inscribed on a sphere. At each latitude (ϕ) and longitude λ on a sphere of radius a, the zonal and meridional components of horizontal velocity are defined in the following way (Fig. 6.3):

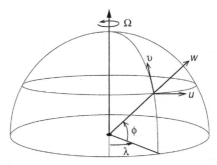

FIGURE 6.3   **Local Cartesian coordinates on a sphere and the zonal (u), meridional (v),**
and vertical (w) components of the local vector wind velocity.

$$u = a\cos\phi\,\frac{D\lambda}{Dt} = \text{zonal or eastward wind speed}$$

$$v = a\frac{D\phi}{Dt} = \text{meridional or northward wind speed} \tag{6.4}$$

Here D/Dt represents the material derivative – the temporal tendency
that is experienced by an air parcel moving with the flow. The vertical component of velocity can be measured in terms of the rate of change of altitude,
or the rate of change of pressure following the motion of air parcels.

$$w = \frac{Dz}{Dt} = \text{rate of change of altitude following an air parcel}$$

$$\omega = \frac{Dp}{Dt} = \text{rate of change of pressure following an air parcel} \tag{6.5}$$

The vertical velocity and the pressure velocity are related to each other
through an approximate equation, which is valid if a hydrostatic balance
is maintained.

$$\omega \cong -\rho g w \tag{6.6}$$

6.3.2 The Zonal Mean Circulation

In describing the circulations of the atmosphere, it is convenient to
consider the zonal average, which is the average over longitude, λ, at a
particular latitude and pressure, and is represented with square brackets.

$$[x] = \frac{1}{2\pi}\int_0^{2\pi} x\,d\lambda \tag{6.7}$$

Because of the relatively rapid rotation of Earth, and because diurnal-
ly averaged insolation is independent of longitude, averaging around a

latitude circle captures a physically meaningful subset of the climate. For climatological purposes, we are normally interested in averages over a period of time, Δt, that is long enough to average out most weather variations. This time interval may correspond to a particular month, season, or year, or it may be an average over an ensemble of many months, seasons, or years.

$$\bar{x} = \frac{1}{\Delta t} \int_0^{\Delta t} x \, dt \qquad (6.8)$$

Climatological zonal averages are usually obtained by averaging over both longitude and time.

The distribution of the zonal mean of the eastward component of wind, u, through latitude and height is one of the best-known characterizations of the global atmospheric circulation, and is often called the *zonal mean wind* (Fig. 6.4). In meteorology, winds are called westerly when they flow from west to east and easterly when they flow from east to west, following

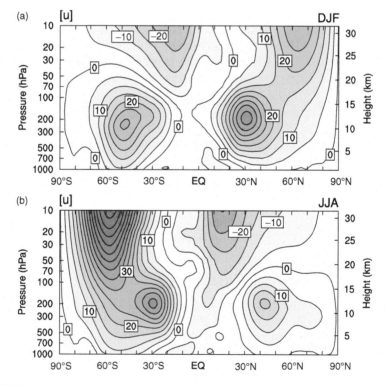

FIGURE 6.4 **Latitude–height cross-section of zonal-average wind speed for (a) DJF and (b) JJA.** Contour interval is 5 ms^{-1}; westerlies are red, easterlies are blue. *Data from ERA-Interim.*

nautical terminology. The zonal mean wind is westerly through most of the troposphere, and peaks at speeds in excess of 30 ms^{-1} in the subtropical jet stream, which is centered near 30° of latitude and at an altitude of about 12 km. The subtropical jet stream is strongest in the winter season. The zonal winds at the surface are westerly at most latitudes between 30° and 70°, but in the belt between 30°N and 30°S zonal mean easterly surface winds prevail. In the stratosphere, a strong westerly jet is present in the winter, stronger in the Southern Hemisphere, and shifts to easterlies in summer.

The zonal-average meridional and vertical components of wind are much weaker than the zonal wind. Maximum mean meridional winds are only about 1 ms^{-1}, and mean vertical wind speeds are typically a hundred times smaller than the mean meridional wind. The *mean meridional circulation* (MMC), which is composed of the zonal mean meridional and vertical velocities, can be described by a mass stream function, which is defined by calculating the northward mass flux above a particular pressure level, p.

$$\Psi_M = \frac{2\pi a \cos\phi}{g} \int_0^p [v]\,dp$$

The mass flow between any two streamlines of the mean meridional stream function is equal to the difference in the stream function values. The conservation of mass for the zonal mean flow implies a relationship between the mass stream function for the mean meridional circulation and the mean meridional velocity and pressure velocity.

$$[v] = \frac{g}{2\pi a \cos\phi} \frac{\partial \Psi_M}{\partial p} \tag{6.9}$$

$$[\omega] = \frac{-g}{2\pi a^2 \cos\phi} \frac{\partial \Psi_M}{\partial \phi} \tag{6.10}$$

Thus, the mean meridional velocity depends on the rate at which the stream function changes with pressure, and the zonal average pressure velocity depends on the rate at which the stream function changes with latitude.

The mean meridional circulation is dominated in the solstitial seasons by a single circulation cell in which air rises near the equator, flows toward the winter hemisphere at upper levels, and sinks in the subtropical latitudes of the winter hemisphere (Fig. 6.5). This mean meridional circulation cell is often called the *Hadley cell* after George Hadley, who in 1735 proposed it as an explanation for the trade winds. The mean meridional winds near the surface bring air back toward the equator. During the solstitial seasons the upward branch of the Hadley cell occurs in the summer

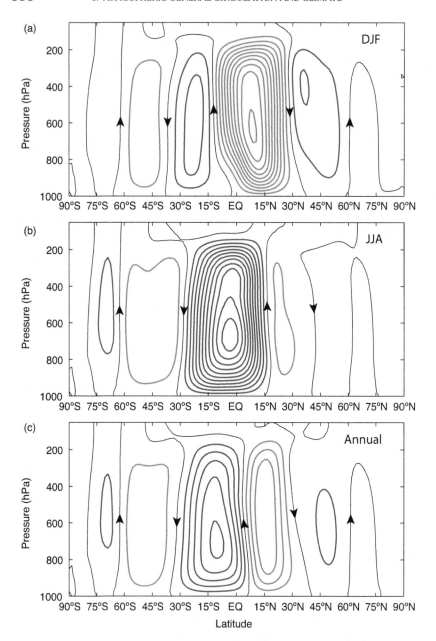

FIGURE 6.5 **Latitude–pressure cross-sections of the mean meridional mass stream function for the (a) DJF, (b) JJA, and (c) annual mean.** Contour interval is 2×10^{10} kg s^{-1} and the arrows on the zero contour indicate the direction of vertical motion. Red is positive and blue is negative. *Based on ERA-Interim data.*

hemisphere. In the annual mean, the rising branch is displaced slightly into the Northern Hemisphere, and the Hadley cell in the Southern Hemisphere is stronger. This asymmetry corresponds to a weak transport of energy from the Northern to the Southern Hemisphere.

In mid-latitudes, weaker cells called *Ferrel cells* circulate in the opposite direction to the Hadley cell. In these mid-latitude mean meridional circulation cells, rising occurs in cold air and sinking in warmer air. Therefore, these cells are thermodynamically indirect, in that they transport energy from a cold area to a warm area. The Ferrel cells are a by-product of the very strong poleward transport of energy and momentum by eddy circulations. Eddies are the deviations from the time or zonal average, and are a key component of the general circulation of the atmosphere. In mid-latitudes, eddies transport energy so efficiently that the mean meridional circulation is thermally indirect, with rising motion in cold air and sinking motion in warm air. The Ferrel cell in the Southern Hemisphere is stronger in the annual mean, because it persists in all seasons. Figure 6.6 shows the mean meridional mass stream function at 700 hPa as a function of season. This shows the greater strength and persistence of the Ferrel cell in the Southern Hemisphere, and also that during the equinoctial seasons, such as April and October, there are two weaker cells, one in each hemisphere.

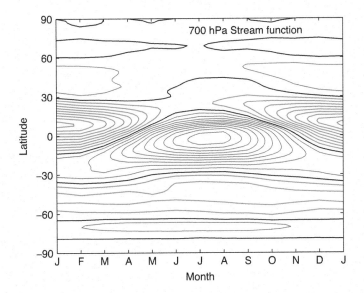

FIGURE 6.6   **Latitude-month cross-section of the mean meridional mass stream function at 700 hPa.** Contour interval is 2×10^{10} kg s^{-1}, northward overturning contours are red, southward turning contours are blue and the zero contour is black. *Based on ERA-Interim data.*

6.3.3 Eddy Circulations and Meridional Transport

The cyclones and anticyclones that are responsible for most of the weather variations in mid-latitudes produce large meridional transports of momentum, heat, and moisture. These disturbances have large wind and temperature variations on scales of several thousand kilometers, which do not appear in a zonal average, but have a profound effect on the zonal mean climate. The fluctuations associated with weather appear as deviations from the time average.

$$x' = x - \bar{x} \tag{6.11}$$

In addition to temporal variations associated with mid-latitude cyclones, the atmosphere exhibits variations around latitude circles associated with continents and oceans that are quasi-stationary and appear clearly in time averages. These are characterized by the deviations of the time mean from its zonal average.

$$\bar{x}^* = \bar{x} - [\bar{x}] \tag{6.12}$$

Northward eddy fluxes of temperature are produced when northward-flowing air is warmer than southward-flowing air, so that, when averaged over longitude, the product of meridional velocity and temperature is positive, even when the mean meridional wind is zero. Using the definitions of the time and zonal averages, the northward transport of temperature averaged around a latitude circle and over time can be written as the sum of contributions from the mean meridional circulation, the stationary eddies, and the transient eddies, which are shown respectively as the three terms on the right of (6.13).

$$[\overline{vT}] = [\bar{v}][\bar{T}] + [\bar{v}^*\bar{T}^*] + [\overline{v'T'}] \tag{6.13}$$

Transient eddy fluxes are associated with the rapidly developing and decaying weather disturbances of mid-latitudes, which generally move eastward with the prevailing flow and contribute much of the variations of wind and temperature, especially during winter. These disturbances are very apparent on weather maps and have typical periods of several days to 1 week.

The positive correlation between poleward velocity and temperature in large-scale atmospheric waves results from the tendency of the temperature wave to be displaced westward relative to the pressure wave, especially in the lower troposphere (Fig. 6.7). This arrangement is associated with a conversion from energy available in the mean meridional temperature gradient to the energy of waves. Cyclone waves whose amplitude is increasing rapidly with time have a large zonal phase shift between their pressure and temperature waves, and thus produce efficient poleward transports of heat and moisture.

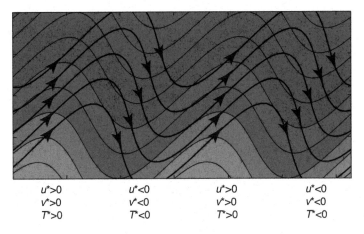

$$u^*>0 \qquad u^*<0 \qquad u^*>0 \qquad u^*<0$$
$$v^*>0 \qquad v^*<0 \qquad v^*>0 \qquad v^*<0$$
$$T^*>0 \qquad T^*<0 \qquad T^*>0 \qquad T^*<0$$

FIGURE 6.7   **Schematic of the streamlines (black with arrows) and isotherms (colors) associated with a large-scale atmospheric disturbance in mid-latitudes of the Northern Hemisphere.** Arrows along the streamline contour indicate the direction of wind velocity. The streamlines correspond approximately to lines of constant pressure, since the winds are nearly geostrophic. The signs of the deviations of the wind components and temperature from their zonal average values are shown with asterisks to illustrate that the NE–SW tilt of the streamlines indicates a northward zonal momentum transport, and the westward phase shift of the temperature wave relative to the pressure wave gives a northward heat transport.

Eddy fluxes by the time-averaged flow are associated with stationary planetary waves. *Stationary planetary waves* are departures of the time average from zonal symmetry and are plainly visible in monthly mean tropospheric pressure patterns (Fig. 6.8). They result from the east–west variations in surface elevation and surface temperature associated with the continents and oceans. Stationary eddy fluxes are largest in the Northern Hemisphere where the Himalaya and Rocky Mountain ranges provide mechanical forcing of east–west variations in the time mean winds and temperatures. The thermal contrast between the warm waters of the Kuroshio and Gulf Stream ocean currents and the cold temperatures in the interiors of the continents also provides strong thermal forcing of stationary planetary waves during winter.

The poleward fluxes of temperature by stationary and transient eddies peak at about 50° of latitude in the winter hemisphere in the lower part of the troposphere (Fig. 6.9). The low-level maximum is associated with the structure of growing extratropical cyclones, in which the phase difference between temperature and pressure is largest in the lower troposphere. The fluxes exhibit a minimum near the tropopause and then increase with height into the winter stratosphere. Temperature fluxes have a large seasonal variation in the Northern Hemisphere, with large values in winter and fairly small values during summer. In the Southern Hemisphere, the seasonal contrast is less, since the Southern Hemisphere is mostly ocean,

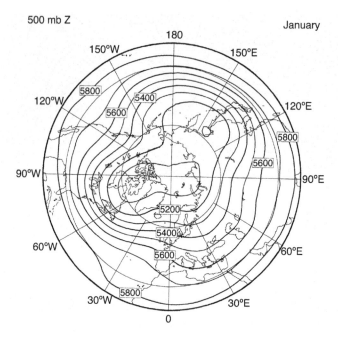

FIGURE 6.8   **Average height of the 500 hPa pressure surface during January in the Northern Hemisphere.** Contour interval 100 m.

which has large heat capacity. Transient eddy fluxes dominate the meridional flux of temperature except in the Northern Hemisphere during winter, when stationary eddies contribute up to half of the flux.

6.3.4 Meridional Water Flux in the Atmosphere

The mean meridional circulation and eddies transport water and play an important role in determining the nature of the hydrologic cycle. Atmospheric eddies move water vapor from the tropics and supply it to middle and high latitudes (Fig. 6.10). The transport of vapor occurs mostly in the lower troposphere where the specific humidity is larger. The left panel of Fig. 6.11 shows the vertically integrated northward transport of water vapor, including the contributions from the mean meridional circulation and from eddies. In the tropics, the mean meridional circulation dominates the total transport, with moisture transported toward the equatorial region by the trade winds. Since the southern branch of the Hadley cell is stronger, the transport of moisture is northward at the equator toward the intertropical convergence zone (ITCZ), whose average position is north of the equator. In the extratropics, the poleward transport is dominated by eddies. The meridional convergence of atmospheric moisture transport in

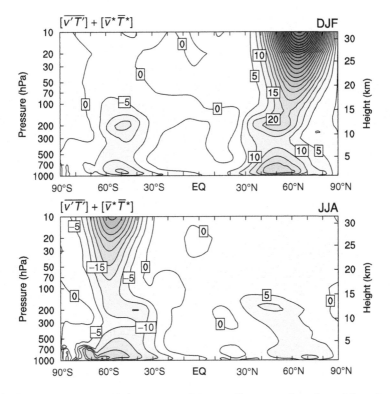

FIGURE 6.9 **Meridional cross-section of the zonally averaged northward flux of temperature by eddies.** Note that in the Southern Hemisphere the poleward fluxes are negative (blue shading) as a result of our arbitrarily defining north as the positive direction. Contour interval is 5 K ms^{-1}. *Data from ERA Interim.*

the tropics is also dominated by the transport provided by the mean meridional circulation, which produces a total convergence of moisture in the equatorial region of about 4 mm day^{-1} (1.5 m year^{-1}) that is supplied from the subtropics (right panel of Fig. 6.11). The eddies draw moisture from the tropics and deliver it to middle and high latitudes. In the Northern Hemisphere, moisture is converged almost uniformly in the region from 40–90°N, but in the Southern Hemisphere, moisture converges more strongly in the belt from 40–70°S because of the presence of the high continent of Antarctica.

6.3.5 Vertically Averaged Meridional Energy Flux

Four types of atmospheric energy are important for determining the meridional transport of energy (Table 6.1). Internal energy is the energy associated with the temperature of the atmosphere, and potential energy

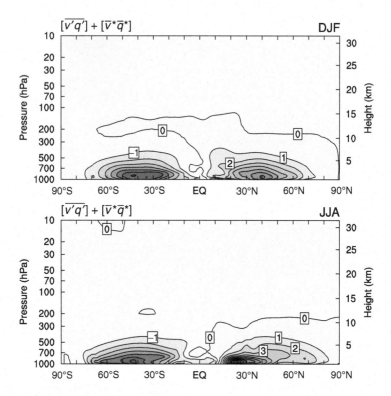

FIGURE 6.10 **Meridional cross-section of the zonally averaged northward flux of water vapor by eddies.** Contour interval is 1 g kg^{-1} m s^{-1}. Northward transport is shaded red. *Data from ERA Interim.*

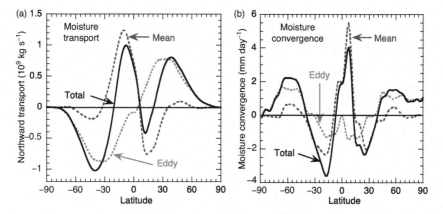

FIGURE 6.11 **Meridional transport of moisture by the atmosphere in 10^9 kg s^{-1} (a) and convergence of meridional transport of moisture in mm day^{-1} (b); totals and contributions by the mean meridional circulation and eddies are shown.** *Data from ERA Interim reanalysis.*

TABLE 6.1   Kinds and Amounts of Energy in the Global Atmosphere

Name	Symbol	Formula	Amount ($J\ m^{-2}$)	Total (%)
Internal energy	IE	$c_v T$	1800×10^6	70
Potential energy	PE	gz	700×10^6	27
Latent energy	LH	Lq	70×10^6	2.7
Kinetic energy	KE	$1/2(u^2 + v^2)$	1.3×10^6	0.05
Total energy	IE + PE + LH + KE		2571×10^6	100

is the energy associated with the gravitational potential of air some distance above the surface. Together internal and potential energy constitute about 97% of the energy of the atmosphere. Although kinetic energy comprises a small fraction of the total energy, it is still very important to understand its generation and maintenance, because the motions are the means by which energy is transported from equator to pole. Motions are also important in converting one form of energy to another. Furthermore, most of the internal and potential energy is unavailable for conversion into other forms. For example, in a dry, hydrostatic atmosphere without mountains, one can show that the ratio of the potential energy to the internal energy is $R/c_v = 0.4$. This simple relation between internal and potential energy reflects the fact that much of the internal energy of the atmosphere is required simply so that the atmosphere may "hold itself up" against gravity, and is not available for generating motion.

Insolation drives the circulation by heating the tropics more than the polar regions. Winds are driven by the density and pressure gradients generated by this uneven heating. The circulation responds not to the total amount of energy in the atmosphere, but to the temperature gradients on constant pressure surfaces. For this reason, the maximum kinetic energy occurs during winter, when the meridional temperature gradients are strongest, and not in the summer, when the total amount of energy in the atmosphere is greatest.

The meridional transport of energy by the atmosphere may be divided into contributions from sensible, geopotential and latent forms that comprise the moist static energy.

$$\text{Moist static energy} = c_p T + gz + Lq = \text{sensible} + \text{potential} + \text{latent} \quad (6.14)$$

Here g is the acceleration of gravity, z is altitude, q is the mass-mixing ratio of water vapor, and L is the latent heat of vaporization. Moist static energy is moved around by the motions of the atmosphere and these transports can be integrated through the mass of the atmosphere to reveal the total meridional flux of energy in various forms (Fig. 6.12). Although

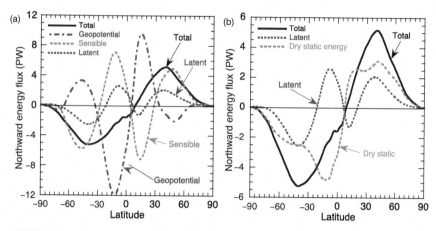

FIGURE 6.12 **Northward transport of energy by the atmosphere in potential, sensible and latent forms.** In (b), geopotential and sensible heat fluxes have been combined to form the dry static energy transport. Units are 10^{15} W. *Data from ERA Interim.*

the total moist static energy transport is smoothly poleward in both hemispheres, the individual components change sign and often strongly offset each other. If the sensible and geopotential transports are combined to form the dry static energy transport, then it can be seen that the transport of latent heat is equally important to the dry static energy transport in mid-latitudes, especially in the Southern Hemisphere. The total transport is negative at the equator, so that the atmosphere transports energy from the Northern Hemisphere to the Southern Hemisphere in the annual mean. This southward energy transport at the equator is related to the fact that the Hadley cell rises in the Northern Hemisphere, as mentioned previously. In Fig. 2.14 we noticed that radiation budget measurements indicated a net northward energy transport at the equator, which requires a northward transport by the ocean to offset the southward transport by the atmosphere. Therefore, it seems likely that northward transport by the ocean at the equator is requiring southward transport by the atmosphere there. The seasonal variation of the northward energy transport is shown in Fig. 6.13. The strong role of the Hadley circulation near the equator is very apparent in the seasonal and latitudinal distributions of northward flux of geopotential, sensible, and latent energy. The Hadley cell influence on the total energy is less obvious because of the cancellation between energy fluxes. Poleward energy transports are largest in mid-latitudes and peak strongly in winter when the radiative forcing of meridional temperature gradients is greatest. The seasonal variation is larger in the Northern Hemisphere, where more land is present.

The cancellation between the transports of different types of energy is particularly strong in the Hadley circulation. The net transport in the

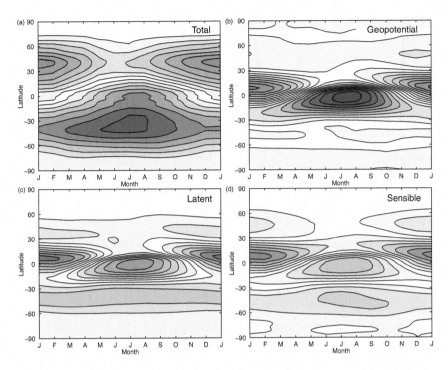

FIGURE 6.13 **Contour plots of the vertically integrated (a) total, (b) potential, (c) sensible and (d) latent northward energy transport as functions of latitude and season.** Red shading indicates northward transport and blue shading southward transport. Contour interval is 1 PW for total and latent and 3 PW for potential and sensible energy.

Hadley cell is only about 10% of the potential energy transport. Mean meridional circulation cells are not a particularly efficient means of poleward energy transport, especially in view of the strong constraints the angular-momentum balance places on these circulation cells. The Hadley cell transports both sensible and latent heat equatorward in the tropics. The equatorward flow near the surface brings warm moist air with it. Heavy precipitation occurs where warm moist air converges in the vicinity of the equator. The release of latent heat and the convergence of sensible heat flux drive strong rising motion in the upward branch of the Hadley cell. In the upward branch of the Hadley cell, latent and internal energy are converted into potential energy. The poleward flow of potential energy in the upper branch of the Hadley cell exceeds the sum of the equatorward flow of latent and internal energy in the lower branch, giving a small net poleward flow of energy. The poleward flow of energy in the Hadley cell can be better visualized by considering the vertical distribution of moist static energy, which is the sum of sensible, potential, and latent energy (Fig. 6.14). Sensible and latent heat per unit mass both peak near the

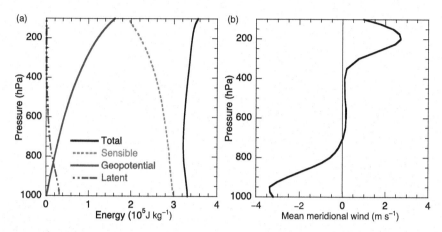

FIGURE 6.14   **Zonal mean energy content (a) and mean meridional velocity (b) at 13.5°N plotted versus pressure.** This shows why the mean meridional circulation transports sensible and latent energy toward the equator, but potential and total moist static energy toward the pole.

surface and so are transported in the direction of the lower branch of the Hadley cell. In the upper troposphere, the Hadley cell transports potential energy poleward, giving a small poleward transport of moist static energy.

6.4 THE ANGULAR-MOMENTUM BALANCE

The general circulation of the atmosphere is heavily constrained by the conservation of angular momentum. *Angular momentum* is the product of mass times the perpendicular distance from the axis of rotation times the rotation velocity. The angular momentum about the axis of rotation of Earth can be written as the sum of the angular momentum associated with Earth's rotation, plus the angular momentum of zonal air motion measured relative to the surface of Earth. Because the depth of the atmosphere is so thin compared to the radius of Earth, the altitude in the atmosphere is unimportant for the angular momentum of the air, and we may use a constant radius, a, to describe the distance from the center of Earth. The latitude has a strong influence on the angular momentum, however, because increasing latitude decreases the distance to the axis of rotation (Fig. 6.15). The distance to the axis of rotation is the radius of Earth times the cosine of latitude. The zonal velocity consists of the velocity of Earth's surface associated with rotation plus the relative zonal velocity of the wind.

$$M = (\Omega a \cos\phi + u)a \cos\phi = (u_{earth} + u)a \cos\phi \qquad (6.15)$$

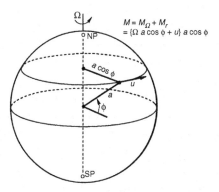

FIGURE 6.15   **The component of angular momentum about the axis of rotation of Earth.** *From Peixóto and Oort (1984) with permission from the American Physical Society.*

The zonal velocity of Earth's surface at the equator, about 465 ms^{-1}, is very large compared to typical zonal wind speeds in the atmosphere.

$$u_{earth} = \Omega a \cos\phi = 7.292 \times 10^{-5}\, \text{rad s}^{-1} \times 6.37 \times 10^{6}\, \text{m} \cdot \cos\phi$$
$$= 465\, \text{m s}^{-1} \times \cos\phi$$

The atmospheric angular momentum associated with Earth's rotation is thus much larger than the angular momentum associated with the zonal winds normally observed in the troposphere. When air parcels move poleward in the atmosphere, they retain the same angular momentum unless they exchange angular momentum with other air parcels or with the surface. Since the distance to the axis of rotation of Earth decreases as a parcel moves poleward on a level surface, the relative eastward zonal velocity of the parcel must increase to maintain a constant total angular momentum. Thus, poleward-moving parcels experience an eastward acceleration relative to Earth's surface.

If a parcel of air moves from one latitude to another while conserving angular momentum, then (6.15) implies a relationship between the zonal velocities the parcel will have at any two latitudes.

$$M = (\Omega a \cos\phi_1 + u_1)a\cos\phi_1 = (\Omega a \cos\phi_2 + u_2)a\cos\phi_2 \qquad (6.16)$$

If a parcel starts out at the equator with zero relative velocity and moves poleward to another latitude while conserving its angular momentum, we have from (6.16) that

$$M = \Omega a^2 = (\Omega a \cos\phi + u_\phi)a\cos\phi \qquad (6.17)$$

which can be rearranged to yield an expression for the zonal velocity at any other latitude u_ϕ,

$$u_\phi = \Omega a \frac{\sin^2 \phi}{\cos \phi} \qquad (6.18)$$

By substituting numbers into (6.18) we find that a parcel of air with the angular momentum of Earth's surface at the equator will have a westerly zonal wind speed of 134 ms^{-1} at 30°N or 30°S. This is much greater than the maximum zonally averaged wind speeds in the subtropical jet stream (Fig. 6.4), and we infer that the poleward angular momentum transport in the upper, poleward-flowing branch of the Hadley cell is more than adequate to explain the existence of a 40 ms^{-1} jet at 30° latitude. The interesting part is explaining why the subtropical jet stream is not stronger than it is. A parcel traveling at a mean meridional velocity of 1 ms^{-1} would require about 30 days to travel from the equator to 30°N, allowing plenty of time for small-scale turbulence or some other slow process to reduce the angular momentum of the parcel. In fact, drag by small-scale eddies is probably not the most important mechanism for reducing the angular momentum of air parcels in the upper branch of the Hadley cell. Rather it is the large-scale eddies in the atmosphere that transport momentum out of the Hadley cell, into mid-latitudes, and downward to the surface.

The zonally averaged meridional flux of angular momentum in the atmosphere can be written

$$[\overline{vM}] = [\overline{v}](\Omega a \cos\phi + [\overline{u}])a\cos\phi + \{[\overline{u'v'}] + [\overline{u^*\overline{v}^*}]\}a\cos\phi \qquad (6.19)$$

and consists of the transport of the zonal mean angular momentum by the mean meridional velocity and an eddy flux term that depends on the covariance between zonal and meridional velocity fluctuations around a latitude circle. This eddy flux term becomes the dominant one near 30° latitude, where the mean meridional velocity is small. The eddy flux of zonal momentum peaks at about the tropopause level at 35° latitude, where it is poleward in both hemispheres and strongest in the winter season (Fig. 6.16).

Northward zonal momentum flux by large-scale atmospheric disturbances is produced when the streamlines of the flow are oriented such that high and low anomalies tilt from southwest to northeast in the Northern Hemisphere as illustrated in Fig. 6.7. It can be seen that when the streamlines are tilted in this manner, the eastward component of wind is greater when the meridional wind component is poleward, and the eastward component of wind is weaker when the meridional flow is equatorward. Therefore, an average over longitude of the product of the deviations of the zonal and meridional components of wind will be positive, indicating a northward flux of zonal angular momentum.

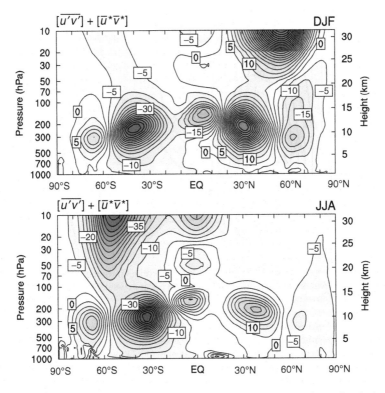

FIGURE 6.16 **Zonal cross-section of the northward transport of zonal velocity by eddies.** Contour interval is $5 \ m^2 \ s^{-2}$, and red shading is northward transport.

The flow of angular momentum in the atmosphere is shown schematically in Fig. 6.17. In the tropical surface easterlies, where the atmosphere rotates more slowly than Earth's surface, eastward angular momentum is transferred from Earth to the atmosphere via frictional forces and pressure forces acting on mountains. This westerly angular momentum is transported upward and then poleward in the Hadley cell. Atmospheric eddies transport angular momentum poleward and downward into the mid-latitude westerlies. Where the surface winds are westerly, the atmosphere is rotating faster than Earth's surface and the eastward momentum is returned to Earth. It is clear that the surface zonal winds cannot be of the same sign everywhere, since eastward angular momentum must flow into the atmosphere where the surface winds are easterly, and must return to Earth where the surface winds are westerly. Thus the tropical easterly winds and mid-latitude westerlies that so regulate the routes of sailing ships are required to satisfy the angular momentum constraint on atmospheric flow.

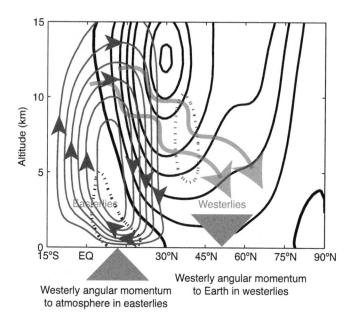

FIGURE 6.17 **Schematic illustration of the flow of angular momentum from the Earth through the atmosphere and back to Earth.** Blue contours with arrows are the mean meridional stream function. Solid black lines are zonal mean wind. Dotted contours indicate negative values of zonal wind and stream function. Wavy arrows indicate poleward and downward angular momentum transport by eddies. Wind and stream function are for January.

6.5 LARGE-SCALE CIRCULATION PATTERNS AND CLIMATE

The circulation of the atmosphere is not zonally symmetric, and east–west variations of winds and temperature are important for regional climates. For example, the subtropical jet of mid-latitudes is not equally strong at all longitudes, but has local maxima associated with the distribution of land and ocean (Fig. 6.18). During January in the Northern Hemisphere the subtropical jet stream has two local wind speed maxima downstream of the Tibetan Plateau and Rocky Mountains over the Pacific and Atlantic oceans, respectively. These maxima in the time-average wind speed have maxima in the transient eddy activity and eddy fluxes of heat and moisture associated with them. They define the so-called *storm tracks*, where vigorous mid-latitude cyclones are most frequently observed. The seasonal migration of these storm tracks plays a key role in the annual variation of precipitation along the west coast of North America discussed in Chapter 5. During July, the wind maximum of mid-latitudes is weaker and farther poleward in the Northern Hemisphere. In the Southern Hemisphere, the mid-latitude jet at about 50°S is present in all seasons, but the

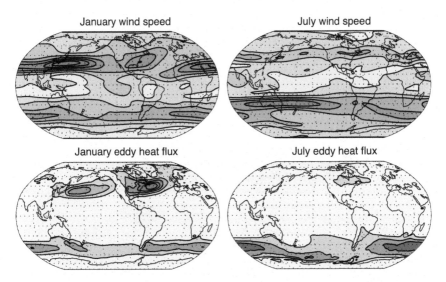

FIGURE 6.18 **Average wind speed (top) and northward heat flux by eddies with periods shorter than about 1 week.** Contour interval on the top is 10 ms^{-1} and contour on the bottom is 5 Km s^{-1}, zero contours are not shown and blue shading indicates southward transport. Robinson projection is used. *Data from ERA Interim.*

subtropical jet is strongest in July (winter) and located in the Western Pacific and over Australia, where strong convection over Indonesia drives a localized Hadley circulation. The extratropical jet and associated transient eddy heat fluxes are concentrated in the Atlantic and Indian Ocean sectors and are weakest in the Pacific sector. The Southern Hemisphere storm track is thus stronger in the Atlantic and Indian oceans in both winter and summer. The seasonal variation of transient eddy heat fluxes is much stronger in the Northern Hemisphere.

The distribution of surface air pressure is an important indicator of the general circulation. In Fig. 6.19, the height of the 1000 hPa surface is shown as a smooth proxy for surface pressure. The global mean value of about 16 m has been removed. Near the surface, air tends to spiral inward toward low-pressure centers, as is clearly indicated for the mid-latitude low-pressure centers over the Pacific and Atlantic oceans during January. By the conservation of mass, this converging air must rise over the low-pressure center. Conversely, the flow near the surface spirals outward from high-pressure centers, so that high pressure is generally indicative of subsiding motion above the boundary layer and suppressed convection. The surface pressure near 30° latitude is generally higher than the surface pressure at the equator. This is consistent with the surface trade winds of the tropics, which generally blow toward the equator from both hemispheres and meet in the ITCZ near the equator, where the surface pressure is low and

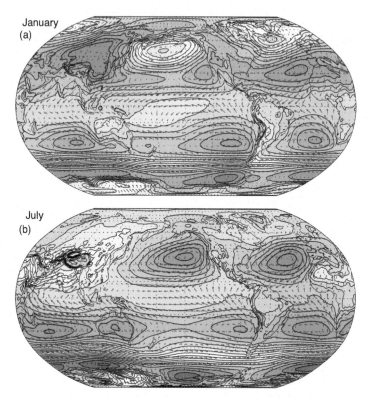

FIGURE 6.19 **1000 hPa height and wind vectors for (a) January and (b) July.** Contour interval is 20 m and largest vector represents a wind of about 15 ms^{-1}. Positive heights are in red. *Data from ERA Interim.*

where deep convection occurs with its attendant latent heat release and large-scale upward motion.

The seasonal variations in sea-level pressure are most apparent in the Northern Hemisphere. During January, the high-latitude oceans are characterized by low-pressure centers with the Aleutian and Icelandic lows centered in the northern margins of the Pacific and Atlantic oceans, respectively. A high-pressure center lies over Asia. During July, the land–sea pressure contrast is reversed in mid-latitudes, with the highest pressures over the oceans and the lowest pressures over the land areas. The oceanic highs are centered over 40°N, compared to their position at about 30°N during January. The dominant low-pressure feature during Northern Hemisphere summer is centered over Asia at about 30°N, and is associated with the Asian summer monsoon.

The dramatic shifts in land–sea pressure distribution are driven by seasonal changes of insolation and the different responses of the land and ocean to heating. Over the oceans the response of surface temperature to seasonal variations of insolation is smaller because the energy is put into a deep layer of ocean with a large heat capacity and because evaporation consumes much of the heat input. Land surfaces have a much smaller capacity for storing heat, and often are not sufficiently wet that evaporation can balance large summertime increases in insolation. As a result, the land surfaces warm up dramatically in summer and cool in winter. The pressure variations along latitude circles in mid-latitudes are associated with the dynamical response to the land–sea temperature and heating contrasts. The low pressures generally occupy the warm regions where the atmosphere is heated, and the high pressures occur where the temperature is low and the atmosphere is being cooled. Land surfaces are warmer than adjacent oceans in summer and are colder than the oceans during winter (see Fig. 1.6).

6.5.1 Monsoonal Climates

The seasonal movement of maximum insolation from one hemisphere to the other causes seasonal shifts in the tropical winds and precipitation, which have dramatic effects on the populations there. In some regions of the tropics, the winds blow consistently in one direction during part of the year and then may weaken or blow from a very different direction for the rest of the year. Such regular seasonal changes in wind speed and direction are called monsoons. The word *monsoon* is derived from an Arabic word meaning season. In many parts of Africa, Asia, and Australia, seasonal changes in wind direction are accompanied by dramatic shifts in precipitation regime between very dry and very rainy. Nowhere is this phenomenon more dramatically displayed than in the Asian monsoon.

During summer the Tibetan plateau is heated by insolation, and much of this energy is transferred to the atmosphere, which warms significantly, leading to a reduction in surface pressure. The low surface pressure encourages a low-level flow of warm, moist air from the ocean to the land (Fig. 6.20), which supports intense precipitation over India and the slopes of the Himalayas during summer, particularly where the summer monsoon winds intersect with the mountains of western India and the Himalayas (Fig. 6.21). A similar encroachment of summertime precipitation occurs over eastern Asia. The heating by insolation and latent heat release drives upward motion in the atmosphere, which is necessary to balance the convergence of air at low levels. The seasonal movement of the main precipitation regions can be seen clearly in the OLR (Fig. 2.11). During winter the situation is reversed. The Himalayas cool dramatically, air flows toward the Indian Ocean from the land at low levels, and India and surrounding lands experience a wintertime drought (Fig. 6.21). The wind shift is particularly strong in the western Indian Ocean, and this causes a strong response in the ocean and the sea surface temperature (Fig. 7.9).

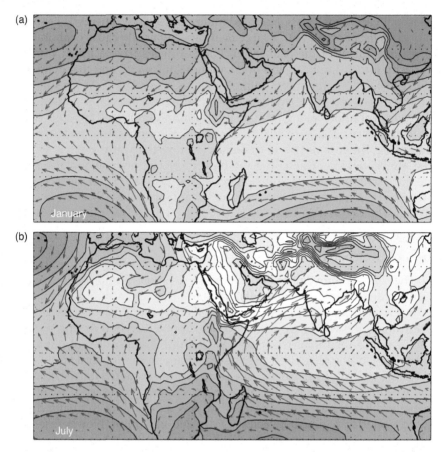

FIGURE 6.20   Same as Fig. 6.19 except for the Afro-Asian monsoon region from 30°S to 40°N and 30°W to 120°E in mercator projection.

The switch from dry to moist conditions occurs abruptly at the time of summer monsoon onset, typically around the middle of June. The date of monsoon onset and the duration and intensity of rainfall during the rainy season vary from year to year. These fluctuations in summer precipitation can produce either flood or drought, and are of great importance for the populace in the regions affected.

6.5.2 Desert Climates

Desert climates occur in land areas where the precipitation is significantly less than the potential evapotranspiration, so that the surface becomes dry. Precipitation will generally occur where humid air is uplifted. Aridity can be caused by a variety of mechanisms, which are all associated

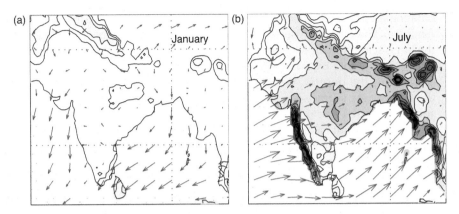

FIGURE 6.21 **Precipitation contours and 1000 hPa wind vectors over Southern Asia during (a) January and (b) July.** Contours of precipitation are 20, 40, 80, 160, 320, 480... mm month^{-1}. *Data from GPCP v6 and ERA Interim.*

with the circulation of the atmosphere. Rainfall can be suppressed by widespread, persistent subsidence associated with the general circulation of the atmosphere. As we have seen, the Hadley circulation implies downward air motions along much of the belt from about 10° to 40° latitude in both hemispheres, where many of the world's great deserts are found (Fig. 6.22). Regions of localized subsidence associated with mountain ranges can also lead to deserts. Persistent subsidence generally occurs on the downwind side of mountain ranges that stand athwart the prevailing wind direction. Examples are the North American Desert, which is downwind of the Cascade, Sierra, or Rocky Mountains, and the Monte and

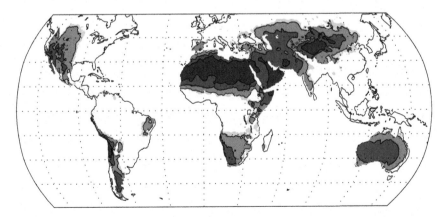

FIGURE 6.22 **Global distribution of dry lands defined as the ratio of precipitation to potential evapotranspiration (P/PET).** Darkest red indicates most extreme aridity. *Data from Feng and Fu (2013).*

Patagonian deserts, which are to the east of the Andes Mountains in South America. The coastal deserts of Peru and Chile result from three causes: the large-scale subtropical subsidence, the localized subsidence associated with easterly flow over the Andes, and the cold sea surface temperatures adjacent to the deserts. When warm air flows over cold ocean surfaces, the air near the surface is cooled by sensible heat exchange and the atmosphere becomes stable to moist convection, because cold, dense air near the surface cannot easily rise through the warmer air above.

Deserts may also be formed in regions that are blocked from a supply of humid air. If air is forced to cross a mountain range, its humidity is greatly reduced as the air stream is uplifted on the windward side. As the air rises, it cools, and the water vapor condenses to form heavy precipitation. On the leeward side, the air sinks and warms, giving the air an extremely low relative humidity. The deserts of central Asia, such as the Gobi, occur in regions that are either very distant from the ocean or separated from the ocean by a major mountain range. The Sahara and Arabian (Rub'al Khali) deserts constitute one of the largest and most arid regions on Earth. This results from their large land mass and their position in the subtropics where mean air motions are downward and insolation is large. Because of their high surface temperature and high albedo, these deserts are regions of net radiation loss in the annual average (Fig. 2.12). The atmospheric circulation provides energy to these regions by import of high-energy air at high levels of the atmosphere. The air then subsides over the deserts as it radiatively cools. The outflow of this air at low levels exports moisture and helps to sustain the great dryness of these regions.

All deserts share a very dry surface and precipitation much lower than the potential evaporation, but can vary greatly in their mean temperatures. For example, Ouallen, Algeria, a town in the Sahara at 22.8°N, 5.53°E, has a maximum monthly mean temperature of 38°C in July, a minimum of 16°C in January, and an annual precipitation of about 22 mm. In contrast, Antofagasta, Chile in the coastal Atacama Desert at 23.5°S, 70.4°W, has an even lower annual precipitation of 2 mm, but has much colder temperatures. Antofagasta has a monthly mean maximum temperature of 20°C in January and a minimum of 13°C in July. The coastal Peruvian and Atacama deserts are remarkable because the low temperatures and frequent stratus cloud and fog associated with the cold temperatures of the adjacent ocean are juxtaposed with an extreme lack of precipitation.

Most deserts have a large diurnal variation in surface temperature, because the surface warms substantially during the day in response to insolation, and cools rapidly at night because longwave radiation emitted by the surface exceeds the downward longwave coming from the dry, cloudless atmosphere. Deserts at high altitude experience an even greater diurnal cycle, because the atmospheric greenhouse effect is weak. Seasonal variations of temperature can also be rather large in mid-latitude arid lands, with freezing temperatures in winter and very hot daytime temperatures in summer.

6.5.3 Wet Climates

Wet climates occur where the precipitation is heavy and exceeds the evapotranspiration for much of the year. Tropical wet climates are supported by the natural tendency of the atmospheric circulation to bring warm moist air to the equator. Islands that fall under the ITCZ are favored by heavy rainfall, and all of the major tropical landmasses have a region with a wet climate near the equator. The major regions of convection and precipitation in the tropics can be seen in the seasonal means of OLR in Fig. 2.11. Along the equator, three minima in the OLR are associated with rainy regions over South America, Africa, and the Indonesian–western Pacific region. These zones of heavy precipitation move north and south with the seasons and tend to occur in the summer hemisphere. The regions of lowest seasonal mean OLR occur where the most extensive upper-level clouds exist (Fig. 3.20), and these clouds are in turn representative of intense convection and rainfall.

The combination of equatorial location, shallow seas, and relatively small landmasses in the region of Malaysia, Indonesia, and New Guinea gives rise to the biggest region of intense precipitation on Earth. Here high sea surface temperatures and strong solar forcing of diurnal land–sea breezes around the many islands foster frequent intense convection and precipitation. This large region of intense convection provides a strong thermal forcing for the large-scale circulation of the tropical atmosphere. The seasonal movement in the mass of convection centered over Indonesia has both north–south and east–west components. During Southern Hemisphere summer, the convection extends southeastward over the South Pacific Ocean and is identifiable as far away as 30°S, 140°W (Fig. 5.4). This feature is called the *South Pacific convergence zone* (SPCZ). During Northern Hemisphere summer, the SPCZ is retracted back across the dateline, and the Indonesian convection extends northwestward into the Bay of Bengal, where it becomes connected with the convection associated with the Asian summer monsoon.

The largest area of tropical rain forest exists in the Amazon Basin of South America. The Amazon Basin receives more than 2 m of precipitation per year. The heavy precipitation over the vast area of the Amazon Basin is favored by the basin's equatorial location and its orientation in relation to the prevailing easterlies of the tropical atmosphere. Northeasterly winds carry large amounts of water vapor into the Amazon Basin from the Atlantic Ocean. The westward flow of moist air is unimpeded by the very gradual slope of the basin until the barrier formed by the Andes Mountains at the western extremity of the basin prevents the water vapor from simply flowing across South America without precipitation. Once convection is initiated over the Amazon Basin, the resulting latent heat release drives upward motion in the atmosphere, rain that falls on the surface can be recycled as further rain and further inflow of humid air from the east can occur at low levels. A circulation of this nature is capable of delivering large amounts of precipitation to the surface of the basin and supports a huge volume of surface runoff to the Amazon River, which has the largest flow of any river in the world. Away from the Equator rainfall is seasonal, however, and is greatest in the summer hemisphere, so that the rainy seasons on opposite sides of the equator tend to occur 6 months apart (Fig. 6.23).

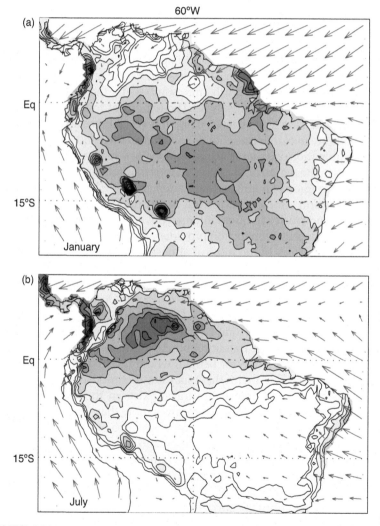

FIGURE 6.23 **Seasonality and annual mean of precipitation and 1000 hPa winds over tropical South America.** Contours of precipitation are 10, 20, 40, 80, 160, 240... mm month^{-1}. *Data from GPCP v6 and ERA Interim.*

6.5.4 Tropical Wet and Dry Climates

Many tropical regions have a wet season and a dry season of varying lengths. An interesting example that covers a very large area is the northern half of Africa, where the climate varies from wet near the equator to extremely arid at 25°N, with the southern boundary of the Sahara Desert roughly along 16°N. Between these two extremes the precipitation is seasonal, with a wet period and a dry period (Fig. 6.24). The precipitation and low-level convergence are tied together and follow the insolation into the summer hemisphere. Figure 6.20 shows that during July the pressure drops over the central Sahara and winds along 15°N come from the southwest, bringing moist air toward the Sahara, just the reverse of during January. During summer, strong solar heating of the surface results in a transfer of heat to the air, which can support mean upward motion. This upward

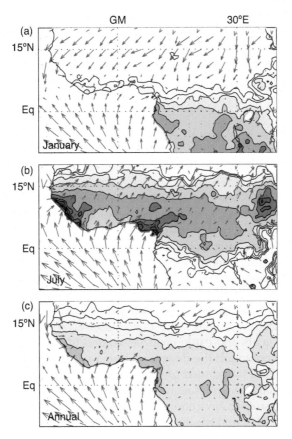

FIGURE 6.24 **Seasonality and annual mean of precipitation and 1000 hPa winds over tropical Africa.** Contours of precipitation are 10, 20, 40, 80, 160, 240... mm month^{-1}. *Data from GPCP v6 and ERA Interim.*

motion can draw moist, low-level air from the south that has its origin in the Gulf of Guinea or in the moist land areas of central Africa. As a result, precipitation in the semiarid lands at the southern margin of the Sahara Desert occurs only during the summer season, when solar heating of the surface is sufficient to drive upward motion and low-level convergence there. In the tropical Western Sahara during July, the minimum sea-level pressure and apparent low-level convergence occur near 22°N, 5°W (Fig. 6.20). However, the precipitation associated with this low-level convergence is quite small, because the air must follow a long trajectory over warm and dry land areas so that it is quite dry when it reaches the area where the low-level winds converge and support shallow rising motion. The low-level wind convergence during January occurs at about 7°N in western Africa, which is much closer to the moisture source in the Atlantic Ocean (Fig. 6.24), and which therefore produces very heavy precipitation. During July, the wind convergence in Western Africa extends to 10°N where the winds become weak in the region of heavy precipitation, but they then accelerate toward the low pressure center farther north. Along the southern coast of western Africa the precipitation peaks twice a year during early summer and late fall. This double maximum is associated with the Sun being exactly over the equator twice a year. Precipitation is stronger in all seasons in the highlands of Ethiopia at around 10°N and 40°E.

The wet season in Africa decreases in duration and reliability as one moves northward from the equator. Figure 6.25 shows the progression from desert, to savanna to tropical wet climate as one travels south from the Sahara toward the equator. As a general rule, the fractional variation of precipitation from year to year increases as the mean annual precipitation decreases, so lands that have low precipitation also have a high probability of unusually dry or wet years. At the southern margin of the Sahara Desert, in a region often called the Sahel, life is particularly sensitive to failures of the summer rainfall, especially when the rains fail for several years in succession as they did in 1910–1913, 1938–1942, and 1969–1973. Drought periods in the Sahel have been related to decadal changes in sea surface temperature, and may be made worse by the local environmental changes induced by humans and their domestic animals. Domestic animals such as goats and cattle need vegetation to eat, and humans need firewood for cooking. If the density of human and animal populations is such that the land is denuded of vegetation, then the desert may advance into regions that were previously semiarid.

Changes in surface environment of arid lands can be very long lasting. When the rains fail and the surface vegetation is removed, the dominant surface covering may become windblown sand, which has a higher albedo than a vegetated surface. Such a surface will reflect more of the insolation,

(a)

(b)

FIGURE 6.25 **Three photographs showing the variation of surface conditions from the subtropical Sahara Desert to the ITCZ.** (a) The Saharan oasis of Ghardaia, Algeria at 32°N. The water table approaches the surface in this depression so that date palms can be grown by irrigation. The annual rainfall is about 75 mm and the surrounding land is barren of vegetation. (b) Sahel region grassland between Agadez and Tanout, Niger at 16°N. The annual rainfall is about 400 mm. (c) Equatorial rain forest on the rim of the Congo Basin near Bondo, Zaire at 4°N. Annual rainfall is about 1800 mm. *Photos courtesy of S. G. Warren.*

(c)

FIGURE 6.25   *(Cont.)*

so that the heating rate necessary to drive low-level convergence may not be attained, and the rains will be more likely to fail. The reverse feedback process is also possible. Random weather events or the remote influence of sea surface temperature changes can lead to a succession of wet years. A few years of enhanced rainfall can foster the development of surface vegetation that will hold down the sand and decrease the surface albedo, thereby increasing the probability of additional wet years. The desert margin is thus a sensitive region, and can be altered by modest forces driving it toward greater or lesser aridity. Paleoclimatic evidence presented in Chapter 8 shows that the Sahara was not always a desert, and that natural changes can transform its climate to a much wetter one.

EXERCISES

1. Discuss why the net energy transport of the Hadley circulation is much less than the poleward transport of potential energy. In what ways are the Hadley circulation and its energy transports related to the climate of the tropics?
2. Draw Fig. 6.7 for the Southern Hemisphere using the same coordinate axes. Make sure that what you draw transports heat and momentum toward the South Pole.
3. Derive (6.19) from (6.15) and (6.13).

4. Calculate the zonal velocity of an air parcel at the equator, if it has conserved angular momentum while moving to the equator from 20°S, where it was initially at rest relative to the surface.

5. The tropical easterlies and mid-latitude westerlies occupy about the same surface area of Earth. Would you expect the surface westerly winds to be stronger, weaker, or about the same as the surface easterlies? Explain your answer with equations.

6. Estimate the rate at which air must subside over the Sahara to balance the heat loss by radiation shown in Fig. 2.11a. (*Hint.* Approximate the vertical gradient of moist static energy with that of potential energy and then use downward advection against this gradient to balance a radiative heat loss, for example, $w\,\partial(gz)/\partial z = R_{TOA} \times (p_s/g)^{-1}$.)

7. It has been speculated that you might be able to bring rainfall to the Sahara by decreasing the surface albedo. Estimate how much the albedo of the Sahara would need to decrease in order to convert its current net annual loss of radiation to a net annual gain of radiation of the same magnitude. This net radiative heating could support mean upward motion that would produce precipitation. How can you account for the likely surface temperature increases and associated longwave energy loss increases that will occur if you increase the solar absorption without providing moisture for evaporation?

8. Using the results of problems 6 and 7 and the tropical humidity distribution in Fig. 1.9, estimate how much albedo decrease you would need to generate net radiative heating to drive mean vertical motion sufficient to produce runoff from the Sahara equivalent to 1 m of precipitation per year. (*Hint*: Equate the vertical transport of water vapor at the top of the boundary layer with the precipitation rate P, for example, $(w\rho_a q)_{1\,km} = P\rho_w$, where ρ_a and ρ_w are the density of air and liquid water, respectively, and q is the mass mixing ratio of water vapor. Compare your result with the net radiation over the Amazon Basin during the rainy season (Fig. 2.11)).

7

The Ocean General Circulation and Climate

7.1 CAULDRON OF CLIMATE

The global ocean plays several critical roles in the physical climate system of Earth. These roles are related to key physical properties of the ocean, such as: it is wet, it has a low albedo, it has a large heat capacity, and it is fluid. Oceans provide a perfectly wet surface, which when unfrozen has a low albedo, and is therefore an excellent absorber of solar radiation. The oceans receive more than half of the energy entering the climate system, and their evaporative cooling balances much of the solar energy absorbed by the oceans, making them the primary source of water vapor and heat for the atmosphere. The world ocean is thus the boiler that drives the global hydrologic cycle. The world ocean also provides the bulk of thermal inertia of the climate system on time scales from weeks to centuries. The great capacity of oceans to store heat reduces the magnitude of the seasonal cycle in surface temperature by storing heat in summer and releasing it in winter. Because seawater is a fluid, currents in the ocean can move water over great distances as well as carry heat and other ocean properties from one geographic area to another. The equator-to-pole energy transport by the ocean is important in reducing the pole-to-equator temperature gradient. Horizontal and vertical transport of energy by the ocean can also alter the nature of regional climates by controlling the local sea surface temperature. The large heat content of the ocean and details of how heat is moved into the deep ocean are very important for the transient warming of the climate system, in response to changing climate forcing, such as the increase in greenhouse gases associated with human activities.

In addition to its direct physical effects on the climate system, the global ocean can affect the climate indirectly through chemical and biological processes. The ocean is a large reservoir of the chemical elements that form the atmosphere. Exchange of gases across the air–sea interface controls the concentration of trace chemical species containing oxygen, carbon, sulfur,

Global Physical Climatology. http://dx.doi.org/10.1016/B978-0-12-328531-7.00007-4

and nitrogen, which are important in determining the radiative and chemical characteristics of the atmosphere. For example, the ocean controls the concentration of carbon dioxide in the atmosphere by exchanging gaseous carbon dioxide across the air–sea interface. Carbon dioxide is converted to organic solids in the ocean, and carbon is then stored by deposition of these solids onto the sea floor. Sulfur-bearing gases released from biological and chemical processes in the ocean enter the atmosphere where they are converted to aerosols that form the nuclei on which cloud droplets form. Evaporation of sea spray also forms salt particles that can form cloud condensation nuclei. The cloud condensation nuclei produced in the ocean can have a substantial influence on the energy balance of Earth, through their effect on the optical properties and extent of clouds.

7.2 PROPERTIES OF SEAWATER

Oceanic currents and the resulting heat transports are determined primarily by the physical properties of the ocean. To specify the physical state of seawater requires three variables: pressure, temperature, and salinity. Because water is slightly compressible, we define the potential temperature and potential density, which are the temperatures and densities at a reference pressure. As described in Chapter 1, *salinity* is the mass of dissolved salts in a kilogram of seawater, and is generally measured in parts per thousand, which we denote with the symbol ‰. The average salinity and temperature of the world ocean are approximately 34.7‰ and 3.6°C, respectively. The effect of ocean on atmospheric composition through biological and chemical processes depends on a more complex mix of physical, chemical, and biological properties. For example, oxygen and nutrient content of seawater are of critical importance for life in the sea. Trace amounts of key minerals may be very important for local biological productivity.

Variations of density on pressure surfaces are important for driving the circulation of the ocean, and depend on the temperature and salinity. Salt content increases the density of water, and seawater expands and becomes less dense as its temperature increases. The salinity of seawater ranges from about 25‰ to 40‰ and the temperature ranges from about –2°C to 30°C. Salinity and temperature variations have roughly equal importance for density variations in the ocean (Fig. 7.1). The density of seawater is almost linearly dependent on salinity. However, the dependence of density on temperature does not have a simple linear behavior. When the temperature of water approaches its freezing point, its density generally becomes less sensitive to temperature. For pure water, for example, the maximum density occurs at 4°C, and the water then expands slightly as it is cooled further. Therefore, fresh water lakes that are cooled from the top continue to overturn convectively until the entire water column reaches

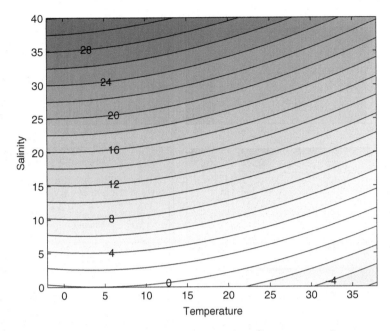

FIGURE 7.1 Seawater density anomaly ($\rho_t - 1000\ kg\ m^{-3}$) at one atmosphere pressure as a function of temperature (°C) and salinity (‰).

4°C, because water that is at 4°C will always be denser than warmer water. When the entire water column is cooled to 4°C, surface water that is cooled further will become less dense than the column and will "float" on the surface. When it reaches 0°C the surface water will freeze and form a layer of surface ice, which provides a floating layer of insulation between the cold air above and the warmer water below. If the lake is deep enough, the water near the bottom will remain at about 4°C, although the air temperature above the surface ice may fall to many degrees below zero. This fact allows fish in high-latitude or high-altitude lakes to survive the winter in the liquid water beneath the surface ice.

For seawater with salinity greater than 24.7‰, the density continues to increase with decreasing temperature until freezing occurs, although more slowly as the freezing point is approached. Therefore, if the salinity is initially uniform, the entire water column must be cooled to the freezing point before ice can form. Sea ice is able to form in high-latitude oceans because the salinity decreases significantly near the surface (Fig. 7.2). Lower salinities near the surface cause a decrease in density that offsets the increase in density associated with colder temperatures near the surface, allowing water near the surface to freeze while warmer water is present below. The low surface salinities result primarily from the excess of precipitation over evaporation in these latitudes. In the Arctic Ocean

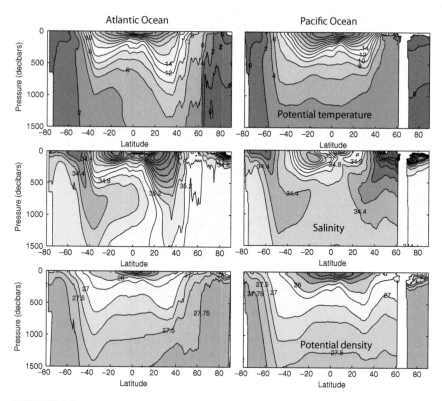

FIGURE 7.2   **Potential temperature (°C), salinity (‰) and potential density (kg m^{-3} – 1000) as functions of pressure (decibar ~1 m of depth) and latitude for the Atlantic (left) and Pacific (right) oceans.** *Data are from MIMOC.*

the supply of freshwater from rivers flowing from the surrounding continents contributes importantly to low surface salinities and therefore to the stable density gradient. The Arctic Ocean has a strong *halocline*, with relatively fresh water near the surface rapidly transitioning to more salty water beneath. Because salinity increases with depth and also increases density, the surface is able to form ice without bringing the temperature of the entire water column to the freezing point. It has been hypothesized that, if the flow of certain key rivers were diverted from the Arctic Ocean to supply irrigation water to continental interiors farther south, the heat balance of the Arctic could be severely distorted, because the normal configuration of a thin layer of surface ice on a mostly unfrozen Arctic Ocean may no longer be stable. Increased salinity of surface waters in the Arctic might lead to either complete removal of most Arctic sea ice or complete freezing of the Arctic Ocean from surface to bottom.

   Only about the first kilometer of ocean between 50°N and 50°S is warmer than 5°C (Fig. 7.2), so that much of the mass of the ocean is between

−2°C and 5°C. The thermal structure of the ocean at most locations can be divided into three vertical sections. The top 20–200 m of water in contact with the atmosphere usually has an almost uniform temperature, which is maintained by rapid mixing through mechanical stirring and thermal overturning. This layer is called the surface *mixed layer* of the ocean. Below the mixed layer, the temperature decreases relatively quickly with depth down to about 1000 m (Figs 1.11 and 7.2). This layer of rapid tempera-ture change, called the *permanent thermocline*, persists in all seasons. It is believed that the permanent thermocline is maintained by heating from above, balanced by a slow upward movement of colder water from below, and by very weak mixing at depth. The cold water in the deep abyss of the oceans is produced at the surface in a few regions of the polar ocean. At the base of the permanent thermocline, the typical temperature is about 5°C, and below this the temperature decreases more slowly with depth, reaching a temperature of about 2°C in the deepest layers of the ocean. The physical properties of the deep ocean show little spatial variability, so that temperature, salinity, and density are almost uniform (Fig. 7.2).

Water is almost incompressible, so the density of seawater is always very close to 1000 kg m^{-3}, even near the bottom of the ocean where the pressure may be several thousand times the surface air pressure. Density of seawater is usually reported as a deviation from 1000 kg m^{-3}, $\rho - 1000$. *Potential density*, ρ_t, is the density that seawater with a particular salinity and temperature would have at zero water pressure, or the density at sur-face air pressure. Potential density increases most rapidly with depth in the first several 100 m of the tropical and mid-latitude ocean (bottom row in Fig. 7.2). This rapid increase of density with depth is supported by the absorption of solar radiation near the surface, which sustains the warm temperatures there. The strong density stratification in the upper ocean inhibits vertical motion and turbulent exchanges, so that the deep ocean is somewhat isolated from surface influences in those regions where this density stratification is present. The strong density stratification is reduced in high latitudes, where in some locations the potential density at the sur-face comes much closer to the densities prevailing in the deep ocean.

The distribution of potential density suggests that water occupying the bulk of the deep ocean came from the Polar Regions, where at certain loca-tions and seasons, surface water becomes dense enough to sink to a great depth. The distributions of other tracers also suggest that slow downward motion of water in high latitudes extends downward and equatorward into the deep ocean, as will be discussed in Section 7.6. Figure 7.2 shows that the potential density in the North Atlantic is greater than that in the North Pacific, especially near the surface where deep water is formed. This is because the Atlantic Ocean is more saline than the Pacific. So deep water can be formed in the Southern Ocean poleward of 50°S, and in the North Atlantic, but not in the North Pacific at the present time. At around 50°N in the Pacific and 50°S, fresh water is transported downward and spreads equatorward below the saline waters of the subtropics. In the Atlantic Ocean, relatively salty water persists below the surface layer of less saline water.

7.3 THE MIXED LAYER

The primary heat source for the ocean is solar radiation entering through the top surface. Almost all of the solar energy flux into the ocean is absorbed in the top 100 m. Infrared and near-infrared radiations are absorbed in the top centimeter, but blue and green visible radiation can penetrate to more than 100 m if the water is especially clear. The depth to which visible radiation penetrates the ocean depends on the amount and optical properties of suspended organic matter in the water, which vary greatly with location, depending on the currents and local biological productivity. The principal component of suspended matter in surface water is *plankton*, which are plants and animals that drift in the near-surface waters of the ocean. The relative abundance of photosynthetic plankton can be estimated from the color of the ocean whose greenness indicates the abundance of chlorophyll (Fig. 7.3). Chlorophyll is most abundant where ocean currents or mixing bring nutrients to the surface where they can be consumed by plankton.

The solar flux and heating rate in the ocean are greatest at the surface and decrease exponentially with depth, in accord with the Lambert–Bouguer–Beer Law, as described in Chapter 3. Under average conditions, the solar flux and heating rate are reduced to half of their surface value by

FIGURE 7.3   The global biosphere – ocean chlorophyll concentration and land vegetation index derived from seaWiFS satellite data for September 1997–August 2000. *NASA image: Mollweide projection.*

a depth of about 1 m, but significant heating can still be present at more than 100 m below the surface. Since the solar heating is deposited over a depth of several tens of meters in the upper layers of the ocean, and cooling by evaporation and sensible heat transfer to the atmosphere occurs at the surface, an upward flux of energy in the upper ocean is required to maintain an energy balance between surface loss terms and subsurface heating. Molecular diffusion is an important heat transport mechanism only in the top centimeter of the ocean. Elsewhere the heat flux is carried by turbulent mixing, convective overturning, and mean vertical motion, which is called *upwelling* or *downwelling* in the ocean. Turbulent mixing in the surface layer of the ocean is greatly aided by the supply of mechanical energy by the winds and their interaction with waves on the surface of the water. In the mixed layer of the ocean, heat transport by convection and turbulent mixing is so efficient that the temperature, salinity, and other properties of the seawater are almost independent of depth.

A schematic diagram showing the processes important in the oceanic mixed layer is presented in Fig. 7.4. The depth of the mixed layer depends

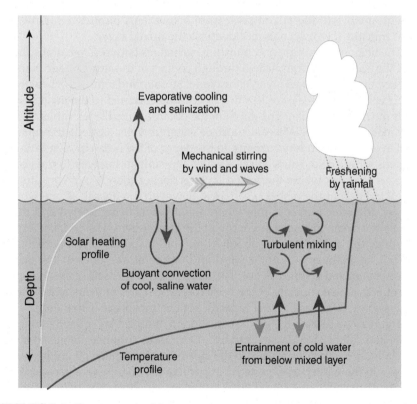

FIGURE 7.4 **Important mixed-layer processes.**

on the rate of buoyancy generation and the rate at which kinetic energy is supplied to the ocean surface by winds. If the surface is cooled very strongly, such as at high latitudes during fall and winter, then cold, dense water is formed near the surface at a rapid rate and buoyancy forces will drive convection, with sinking of cold water and rising of warmer water in the mixed layer. When the surface is cooled only weakly or actually heated, such as during summer when surface solar heating rates are greatest, then the generation of mixing by buoyancy is less and the mixed layer will become thinner and warmer. Buoyancy can be generated by the effect of evaporation on the surface salinity, even when surface temperatures are increasing with time. The density increase associated with increasingly saline surface waters can balance or overcome thermal stratification and encourage mixing. Rainfall represents an input of fresh water at the surface, which acts to decrease the density of the surface waters. Winds blowing over the ocean waves transfer kinetic energy to the water, which results in turbulent water motion as well as mean ocean currents. The supply of turbulent kinetic energy to the upper ocean by winds can induce mixing, even in the presence of stable density stratification. If the intensity of turbulence in the mixed layer is great enough, cool and dense water can be entrained into the mixed layer from below. This implies a downward heat transport, which cools and deepens the mixed layer.

The heat, momentum, and moisture exchanges between the atmosphere and the ocean are accomplished through contact of the atmospheric boundary layer with the mixed layer of the ocean. Storage and removal of heat from the ocean on time scales of less than a year are confined to the mixed layer over much of the ocean. The depth of the oceanic mixed layer varies from a few meters in regions where subsurface water upwells, as along the equator and in eastern boundary currents, to the depth of the ocean in high-latitude regions where cold, saline surface water can sink all the way to the ocean bottom. Regions where the mixed layer is deeper than 500 m constitute a small fraction of the global ocean area, however. In general, as one would expect, the mixed layer is thin where the ocean is being heated and thick where the ocean gives up its energy to the atmosphere. The global-average depth of the mixed layer is about 70 m. The mixed layer responds fairly quickly to changes in surface wind and temperature, whereas the ocean below the mixed layer does not. The thermal capacity of the mixed layer is the effective heat capacity of the ocean on time scales of years to a decade, and is about 30 times the heat capacity of the atmosphere (Chapter 4).

Figures 7.5a,b show the global distribution of mixed layer depth, temperature and salinity for January and July. During the summer the mixed layer is relatively shallow, especially in the Northern Hemisphere, but it becomes deeper in the winter, as expected, but particularly so in several locations. In the North Atlantic, the mixed layer becomes very deep in winter and these are the regions where deep water can be formed. During July (winter), in

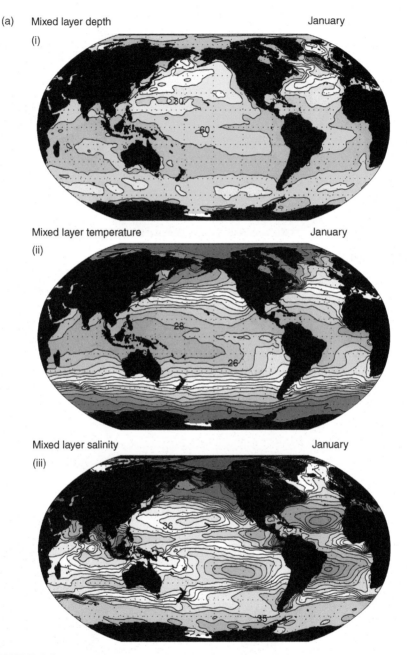

FIGURE 7.5 (a) (i) Map of the depth of the mixed layer for January. Contour interval is 30 dbars, with the reddest color more shallow than 30 dbar and the darkest blue colors the deeper mixed layer depths. (ii) Map of the potential temperature averaged over the mixed layer. Contour interval is 2°C, and warmest contour is 28°C. (iii) Map of salinity averaged over the mixed layer. Contour interval is 0.25‰ and saltiest contour in the Atlantic is 37.25‰. Data from MIMOC.

(b)   Mixed layer depth   July

(i)

Mixed layer temperature   July

(ii)

Mixed layer salinity   July

(iii)

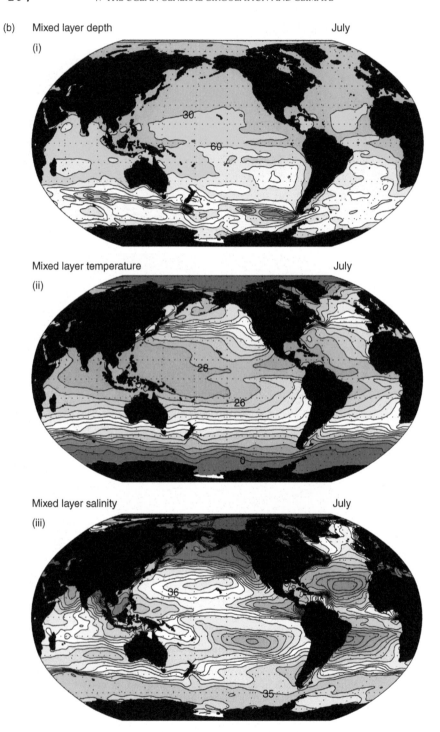

FIGURE 7.5   (*Continued*) (b) Same as Fig. 7.5a, except for July.

the Southern Hemisphere, a band of deepened mixed layer extends from the Indian Ocean to the eastern Pacific in the vicinity of the entrance to the Drake Passage off southernmost South America. This deepened mixed layer marks the subduction of cold, relatively fresh water into intermediate depths that can be seen in Fig. 7.2. The mixed layer does not become as shallow in summer in the Southern Hemisphere because the strong westerlies persist year-round there (Figs 6.4 and 6.19), and continue to stir the mixed layer in summer. As one would expect, the mixed-layer temperature and salinity are very similar to the surface values. The Atlantic is the saltiest ocean. Because of the geometry of the Atlantic Ocean and surrounding continents, some of the evaporation that occurs in the subtropical Atlantic is exported by the atmosphere to other ocean basins, mostly because of westward vapor transport across the Isthmus of Panama and southward transport toward the Southern Ocean along 30°S. As the climate warms this natural export of freshwater will strengthen with the increased moisture of the air, and the subtropical Atlantic will get even saltier.

Figure 7.6 shows examples of the seasonal cycle of the mixed layer from the North Atlantic and Southeast Pacific. The mixed layer is warmest and

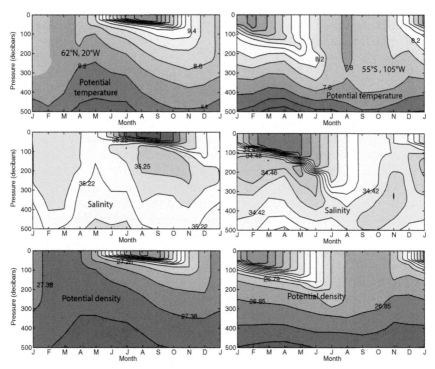

FIGURE 7.6 Plots of potential temperature, salinity and potential density as functions of depth and month of the year in the far North Atlantic (62°N, 20°W) and in the South–West Pacific (55°S, 205°W).

thinnest in late summer near the end of the period of greatest insolation and least intense stirring of the ocean by winds. In late summer, the surface begins to cool, the storminess increases, and the mixed layer deepens and cools. The mixed layer continues to deepen and cool throughout the winter, and by the end of winter, may extend to a depth of several hundred meters and merge smoothly into the permanent thermocline. During most of the rest of the year a seasonal thermocline with steep temperature gradients links the permanent thermocline with the base of the mixed layer. In spring and summer this seasonal thermocline develops and the mixed layer becomes thinner and warmer. Seasonal variations in temperature are confined primarily to the mixed layer and the seasonal thermocline, so temperatures at depths below the deepest extent of the mixed layer experience little seasonal variation. At both these locations, a fresh water layer caps the more saline water below during summer, but during winter salt is mixed upward, making a layer of cold, salty, and very dense water near the surface, especially in the far North Atlantic. These conditions allow dense water to be formed that can sink below the surface and fill the deeper layers of the ocean. Anomalies that are generated by weather or climate events in the winter can be sealed under the strong density stratification of the thin summer mixed layer, and then emerge again to influence the atmosphere in winter.

7.4 THE WIND-DRIVEN CIRCULATION

The transfer of momentum from winds to ocean currents plays a critical role in driving the circulation of the ocean. This is particularly true for the currents near the ocean surface. The wind stress acting on the ocean looks very much like the near-surface winds (Fig. 7.7). In the subtropics, especially in the winter hemisphere, the trade winds apply an equatorward and westward stress to the ocean. In the mid-latitudes, these stresses are reversed, and apply an eastward and poleward force on the surface. In the equatorial East Pacific and the Atlantic during July, the wind stresses are westward and these act to drive upwelling and colder SSTs there (Section 7.5).

The global distribution of large-scale surface ocean currents is shown in Fig. 7.8. These currents are not directly observed, but are inferred from observations of ocean surface height, wind stress, and SST. The basic data have a spatial resolution of 1° in latitude and longitude and many fine scale features are not represented. In the trade wind regions, the near-surface currents flow approximately 90° to the right of the applied wind stress vector in the Northern Hemisphere, and the reverse in the Southern Hemisphere. This is in general accord with the Ekman Theory, and will be introduced in Section 7.5. Near the equator in the Pacific, the current is

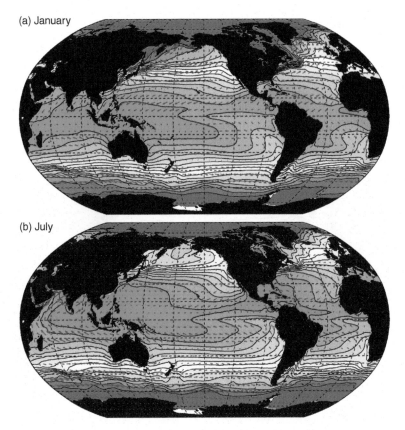

(a) January

(b) July

FIGURE 7.7 **Wind stress vectors and sea surface temperature (SST) for (a) January and (b) July.** *Data from ERAI.*

westward, but north of the equator the North Equatorial Countercurrent flows toward the east, most strongly in northern summer. In the Southern Ocean, the current flows generally eastward in a circumpolar current that squeezes through the Drake Passage between South America and the Antarctic Peninsula. The surface currents are arranged in coherent patterns, with large circulations called *gyres*, occupying the major ocean basins with currents circulating in the same direction as the applied wind stress.

The surface currents in the Indian Ocean show a strong response to the monsoonal wind reversal (Fig. 7.9). During January, the currents are westward north of the equator, which then reverse to eastward in July. Along the coast of Africa and Arabia, the southerly summer winds drive a northward current, which also drifts offshore to cause upwelling of cooler water along the coast. As a consequence of the upwelling and the stronger winds in summer, the SST is warmer in January than in July.

7.4.1 Western Boundary Currents

Some of the most visible current structures are the large clockwise circulations in the northern Pacific and Atlantic oceans. Along the western
boundaries of the Pacific and Atlantic Ocean basins strong poleward-
flowing currents exist in a narrow zone very near the continents. These
currents are called the Kuroshio and the Gulf Stream, respectively, and
may be referred to generically as western boundary currents. Western

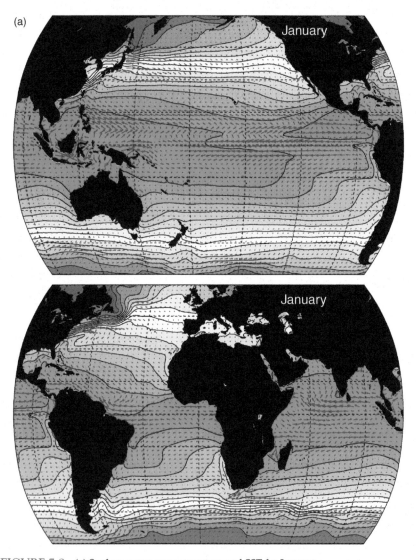

FIGURE 7.8   (a) Surface ocean current vectors and SST for January.

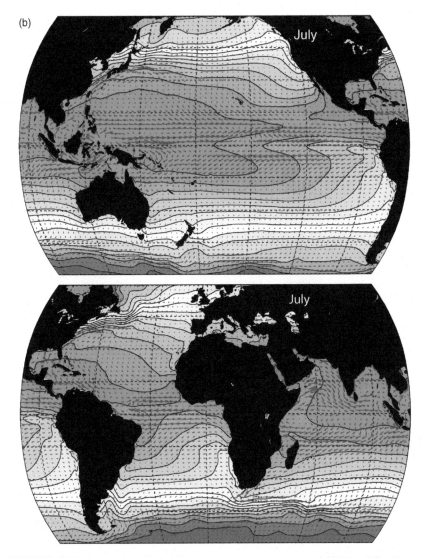

FIGURE 7.8 *(Continued)* (b) Surface ocean current vectors and SST for July. *Current vectors are from the OSCAR data set.*

boundary currents also occur in the Southern Hemisphere along South America (Brazil Current) and along Africa (Agulhas Current). They are generally less sharply defined and extensive in the Southern Hemisphere, perhaps because of the different ocean geometry that allows the Antarctic circumpolar current to flow unimpeded in a continuous eastward current at about 60°S. Western boundary currents carry warm water from the tropics to middle latitudes. The speed of these currents may exceed 1 m s^{-1},

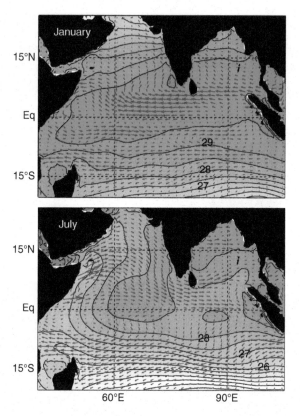

FIGURE 7.9 Surface ocean current vectors and SST (°C) in the Indian Ocean region showing the seasonal reversal of currents associated with the monsoon, and the cooling of water during the summer monsoon associated with the upwelling along the coast of Africa and the greater surface winds during July.

which is quite fast for an ocean current. With the possible exceptions of the Antarctic circumpolar current and some zonal equatorial currents, these currents are the closest oceanographic analogs to the jet streams of the atmosphere, although they flow mostly poleward rather than eastward. The return flow of water from mid-latitudes to the equator is much more gradual and occurs in a broad expanse across the center of each basin.

Figure 7.10 shows the surface current and SST in the western Atlantic region. Currents flow poleward along the western margin of the Atlantic, flowing along the South American coast into the Gulf of Mexico, through the Strait of Florida, and up the East Coast of North America. Around Cape Hatteras, they leave the coast and turn eastward, following the gradient of SST toward the far North Atlantic. Unlike the atmospheric jet streams that generally show a strong seasonal signature, the Gulf Stream continues with almost equal strength in the summer and winter.

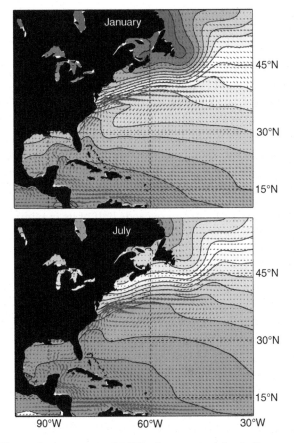

FIGURE 7.10   Surface currents and SST in the western Atlantic Ocean region.

The vertical structure of the Gulf Stream is illustrated in Fig. 7.11, which shows the geostrophic current and potential temperature in the cross-section along 65°W. The geostrophic current is obtained by assuming that the pressure surfaces are level at 2000 dbar, then computing the height above the 2000 dbar surface using the hydrostatic relationship (7.1).

$$z(p) - z(2000\,\text{dbar}) = \int_{p}^{2000} \frac{1}{\rho g}\, dp \qquad (7.1)$$

The density ρ depends on temperature, salinity, and pressure, and g is the acceleration of gravity. Once the heights of the pressure surfaces are known, the currents can be determined from the geostrophic approximation (Holton and Hakim, 2012).

$$u_g = -\frac{g}{f}\frac{\partial z}{\partial y}\bigg)_p \qquad v_g = \frac{g}{f}\frac{\partial z}{\partial x}\bigg)_p \qquad\qquad (7.2)$$

The *Coriolis parameter* ($f = 2\Omega\sin\phi$) measures twice the local vertical component of the rotation rate (Ω) of Earth.

The current has a monthly mean speed of more than 0.5 m s^{-1} and is geostrophically balanced with the sloping thermocline, which implies temperature and density variations along lines of constant pressure. The eastward current at this longitude extends about 1000 m deep to the base of the thermocline and is slightly stronger and about 2° farther north during January than August. Even though averages over a month tend to blur out the sharpness of the jet structure, the current is still narrow; and the width over which the current exceeds half of its maximum strength is about 150 km.

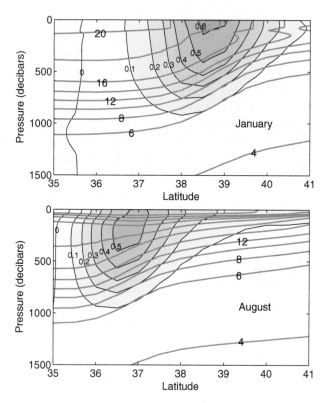

FIGURE 7.11 Eastward geostrophic ocean current (black contours and shading) and potential temperature isolines (red contour lines) for January and August along a meridional cross-section at 65°W where the Gulf Stream travels approximately eastward. Current data are in meters per second and temperatures in degree Celsius. Temperature, pressure and salinity data from MIMOC were used to compute the density.

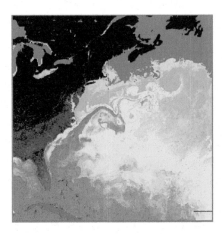

FIGURE 7.12 **SST in the Northwestern Atlantic showing meanders and rings in the Gulf Stream.** Red is warm and blue is cold. *Courtesy of Dr. Otis Brown.*

The Gulf Stream current is not straight or steady, but breaks down into meanders and rings, and eventually loses a clear identity as the flow expands eastward and northward across the basin (Figs 7.12 and 7.13). The current itself and these eddy structures are much smaller than the corresponding jets and eddies in the atmosphere by a factor of about 100.

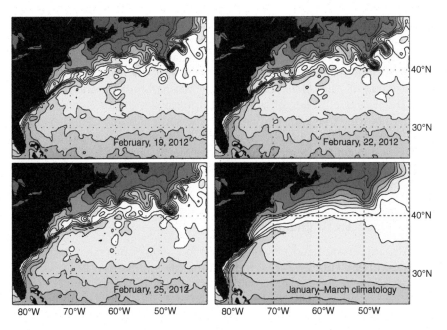

FIGURE 7.13 **Daily sea surface temperature in the Western Atlantic for 3 days in February 2012, plus the 3-month January–March climatology for SST.** *Data from NOAA OI SST high-resolution data set.*

The baroclinic *Rossby Radius of Deformation* is smaller in the ocean as is also the scale of maximum baroclinic instability, whereby eddies in the Gulf Stream derive their energy, as in the atmosphere. Some of the eddy structures in Fig. 7.13 are related to topographic features in the ocean bottom. On each of the three days shown, a vortex with an expression in cold SST exists near 43°N–50°W, and this feature also shows up in the climatology of January through March. This feature is associated with the subsurface topographic feature of the Grand Banks, where the cold Labrador Current from the north meets the Gulf Stream coming from the west and south. Downstream of the Grand Banks at about 43°N–47°W one can see the incorporation of a warm ring of water into colder surroundings over the course of February 19–25, 2012. Further west at about 38°N–68°W, one can see cold water wrapping around a warm intrusion, leading to a warm ring embedded within colder water on February 25, at about 39°N–67°W.

The poleward flux of warm water in the Gulf Stream and Kuroshio currents has a profound effect on the SST and climate of the land areas bordering the oceans, especially the lands immediately downwind of the oceans. The annual average SST shows a strong gradient in the North Atlantic Ocean aligned approximately with the mean position of the Gulf Stream (Fig. 7.13). This strong temperature gradient extends northeastward from the Mid-Atlantic coast of North America to the vicinity of Spitsbergen north of the Scandinavian Peninsula. Some of the heat carried northward by the Gulf Stream is picked up by the Norwegian Current and carried into the polar latitudes. As a result, at middle and high latitudes, the eastern Atlantic is much warmer at the surface than the western Atlantic Ocean and the Pacific Ocean. This asymmetry in the Atlantic sea surface temperature contributes to the milder winter climates of western European land areas compared to eastern North American land areas at the same latitude. Another major contribution to this climate asymmetry is the southward and eastward advection of cold air temperatures above North America in winter. These stationary waves in the atmosphere are caused by mountain ranges and the east–west SST variations.

7.4.2 Eastern Boundary Currents

Also important for climate are eastern boundary currents, which occur in tropical and subtropical latitudes at the eastern margins of the oceans. These currents are generally cold, shallow, and broad in comparison to western boundary currents. The names given to the eastern boundary currents in these geographic areas are the California Current off North America, the Peru or Humboldt Current off South America, the Canary Current off northern Africa, and the Benguela Current off southern Africa. In each

(a)

(b)

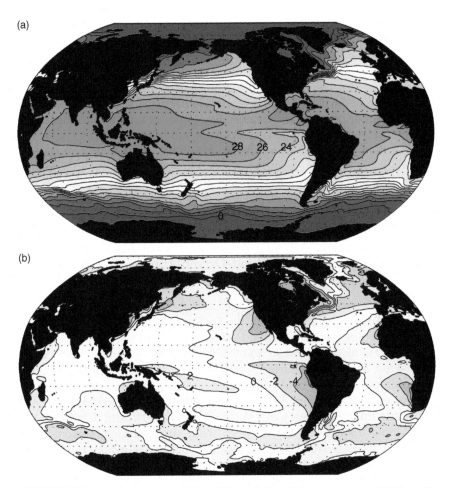

FIGURE 7.14 **(a) Annual mean SST and (b) deviation of the annual mean SST from its zonal average at each latitude Units are °C.** *Data from NOAA OI SST data.*

of these regions, a wind-driven current flows along the coast toward the equator and then turns westward toward the center of the basin. These currents are associated with cold SST, which can be illustrated by plotting the deviation of SST from its zonal average (Fig. 7.14). The SSTs in the subtropics to the west of the continents in the Atlantic and Pacific Oceans are much colder than the zonal average at the same latitude. The coldest water occurs very near the coast and extends westward and equatorward into the oceans.

Upwelling of cold subsurface water, which is driven by alongshore or offshore winds in these regions, produces the low SST near the coast. These low-level winds are associated with the surface high-pressure systems in the atmosphere above the eastern subtropical ocean areas, as shown in Fig. 6.19. The wind systems and the associated currents and cool SSTs are best developed during the summer in the Northern Hemisphere, and are more nearly year-round phenomena in the Southern Hemisphere. It is believed that the geometries of the coastlines in the two oceans, and in particular the northwest-southeast slopes of the coastlines of South America and Africa, cause the eastern boundary currents in the Southern Hemisphere to be better developed and to extend to the equator and then westward along the equator. The cooler than average SST in eastern boundary current regions is often associated with atmospheric subsidence and persistent low stratocumulus cloud and fog (Fig. 3.21).

7.5 THEORIES FOR WIND-DRIVEN CIRCULATIONS

7.5.1 The Ekman Layer, Wind-Driven Transport, and Upwelling

Due to the rotation of Earth, the frictional component of the vertically integrated transport of water in the surface layer of the ocean is not in the direction of the applied wind stress, but 90° to the right of it in the Northern Hemisphere and 90° to the left of it in the Southern Hemisphere. This wind-driven near-surface water transport plays a critical role in determining relatively cold surface temperatures in the eastern boundary current regions and along the equator, and it also plays an important role in driving the subtropical gyres that feed the western boundary currents.

To show the relationship between wind stress driving, currents, and transport, we may consider a homogenous ocean of constant density and pressure, and assume that it is driven by a uniform wind stress with eastward component, τ_x and northward component, τ_y. We seek a steady solution, in which frictional stresses and Coriolis accelerations are in balance (Vallis, 2006).

$$fv = -v\frac{d^2u}{dz^2} \tag{7.3}$$

$$fu = v\frac{d^2v}{dz^2} \tag{7.4}$$

The frictional forces have been described such that the frictional stress is proportional to the shear of the current velocity times a momentum diffusion coefficient, v. Thus, the specified wind stress enters as a boundary condition on the current shear at the surface, and we assume that the current goes to zero at large depths, so that the boundary conditions on (7.3) and (7.4) are

$$v\frac{du}{dz} = \frac{\tau_x}{\rho_0} \\ v\frac{dv}{dz} = \frac{\tau_y}{\rho_0} \left.\right\} \quad \text{at } z = 0; \ u = v = 0 \text{ at } z \to -\infty \quad (7.5)$$

where ρ_0 is the density of the seawater and assumed constant. The solution for the velocities under these conditions is:

$$u_E = \frac{e^{z/z_E}}{\rho_0\sqrt{fv}}\left\{\tau_y\cos\left(\frac{z}{z_E}+\frac{\pi}{4}\right)+\tau_x\cos\left(\frac{z}{z_E}-\frac{\pi}{4}\right)\right\} \quad (7.6)$$

$$v_E = \frac{e^{z/z_E}}{\rho_0\sqrt{fv}}\left\{\tau_y\cos\left(\frac{z}{z_E}-\frac{\pi}{4}\right)-\tau_x\cos\left(\frac{z}{z_E}+\frac{\pi}{4}\right)\right\} \quad (7.7)$$

where, $z_E = \sqrt{\dfrac{2v}{f}}$ is the Ekman Depth and z is measured positive upward.

The steady solution (7.6)–(7.7) describes the Ekman spiral (Fig. 7.15). The current vector has its maximum magnitude at the surface where it is directed at an angle of $\pi/4$ (45°) to the right of the wind stress vector in the Northern Hemisphere ($f > 0$). The current vector turns toward the right with increasing depth, and its magnitude decreases exponentially with depth. The magnitude of the current decreases by a factor of e^{-1} for every increase of depth equal to $z_E = \sqrt{2v/f}$. An appropriate value of $v = 30$ m^2 s^{-1} gives an Ekman depth of ∼ 800 m. If we integrate the currents over the depth range in which the currents are significant, we obtain the integrated transport in the Ekman layer.

$$U_E = \int_{-\infty}^{0} u_E\, dz = \frac{\tau_y}{\rho_0 f}; \quad V_E = \int_{-\infty}^{0} v_E\, dz = \frac{-\tau_x}{\rho_0 f} \quad (7.8)$$

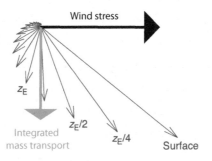

FIGURE 7.15 **Schematic of the Ekman Layer solution (7.6, 7.7 and 7.8) showing the applied wind stress, the current vectors at various depths, and the mass transport integrated through depth.** Current vectors are shown at depth intervals of $z_E/4$.

The net horizontal water transport in the Ekman layer is directed at a 90° angle to the right of the applied wind stress in the Northern Hemisphere, so that if the wind stress is toward the east at the surface ($\tau_x > 0$), the Ekman layer transport is toward the south ($V_E < 0$). If a westward wind stress is applied near the equator, the Ekman layer transport will be northward in the Northern Hemisphere and southward in the Southern Hemisphere, because of the change of sign of f at the equator, so that a net divergence of surface flow will be generated. Conservation of mass requires upwelling along the equator to balance the Ekman layer transport away from the equator. The cold tongue of SST in the eastern equatorial Pacific Ocean during July (Fig. 7.9) is caused largely by this wind-driven upwelling. The cold SST anomalies associated with the eastern boundary currents (Fig. 7.14) are associated with offshore Ekman transport driven by equatorward alongshore surface winds. Offshore Ekman transport near an ocean boundary requires upwelling to replace the exported water. Since water temperatures decrease with depth, upwelling is generally accompanied by cold SST.

Wind stress driving can also cause vertical motions in the open ocean away from boundaries and the equator, if the wind stress has spatial gradients. If we consider the mass continuity equation for an incompressible fluid

$$\frac{\partial u}{\partial x} + \frac{\partial v}{\partial y} + \frac{\partial w}{\partial z} = 0 \tag{7.9}$$

and integrate it over the depth of the Ekman layer, we can derive a relationship between the applied wind stress and the vertical motion at the base of the Ekman layer.

$$w_E(-\infty) - w_E(0) = \int_{-\infty}^{0} \left(\frac{\partial u_E}{\partial x} + \frac{\partial v_E}{\partial y} \right) dz \tag{7.10}$$

Utilizing (7.8) in (7.10) and assuming that the vertical current vanishes at the surface yields an expression for the vertical velocity at the bottom of the Ekman layer in terms of the wind stress applied at the surface. We obtain,

$$w_E(-\infty) = \frac{\partial}{\partial x}\left(\frac{\tau_y}{\rho_0 f} \right) - \frac{\partial}{\partial y}\left(\frac{\tau_x}{\rho_0 f} \right) = \vec{k} \cdot \nabla \times \left(\frac{\vec{\tau}}{\rho_0 f} \right) \tag{7.11}$$

where $\vec{\tau} = \vec{i}\,\tau_x + \vec{j}\,\tau_y$ and $\vec{i}$, $\vec{j}$, and $\vec{k}$ are unit vectors in the eastward, northward, and upward directions respectively. The vertical velocity at the base of the Ekman layer in the open ocean is thus seen to be proportional to the curl of the wind stress vector divided by the Coriolis parameter. Where lateral boundaries are present, the dependence of upwelling on the wind stress is more complex, but wind stress near boundaries can produce large upwelling, even without significant wind stress curl.

7.5.2 Sverdrup Flow and Western Boundary Currents

To understand the large-scale response of the ocean to wind stress forcing it is useful to consider the balance of vorticity in the ocean. *Vorticity* is the curl of the velocity vector and is a measure of the local rotation of the fluid. For the large-scale motions of the atmosphere and the ocean it is the vertical component of absolute vorticity that is of most interest.

$$\zeta_a = 2\Omega \sin\phi + \vec{k} \cdot \vec{\nabla} \times \vec{V} = f + \zeta_r \tag{7.12}$$

The absolute vorticity is the sum of planetary vorticity (f), which is associated with the rotation of Earth, and relative vorticity (ζ_r), which is associated with the fluid motion relative to the surface of Earth. For flow without friction, the absolute vorticity remains constant, unless a parcel of fluid changes its shape. If a parcel of fluid maintains its shape while moving equatorward to a latitude where Earth's rotation is less, then the fluid parcel must exhibit a change in relative vorticity in order to maintain a constant absolute vorticity. Stretching of fluid parcels along the direction of the rotation vector will cause the absolute rotation rate to increase.

The famous oceanographer Harald Ulrik Sverdrup showed that in the interior of the ocean, an approximate balance exists between the meridional advection of planetary vorticity and the stretching of planetary vorticity by divergent motions,

$$\beta v = f \frac{\partial w}{\partial z} \tag{7.13}$$

where $\beta = \partial f / \partial y = 2\Omega \cos\phi / a$. If we integrate (7.13) from the bottom of the ocean to the bottom of the Ekman layer and use (7.11), we obtain

$$\beta V_I = f \vec{k} \cdot \nabla \times \left(\frac{\vec{\tau}}{\rho_0 f} \right) \tag{7.14}$$

where

$$V_I = -\int_{-D_0}^{-z_E} v \, dz \tag{7.15}$$

and it has been assumed that the Ekman layer is thin compared to the depth of the ocean, and that the vertical velocity is zero at the bottom of the ocean, where $z = -D_0$.

If we add the interior meridional transport velocity (7.14) to the Ekman layer meridional transport in (7.8) we obtain,

$$V_I + V_E = \frac{1}{\beta} \vec{k} \cdot \nabla \times \left(\frac{\vec{\tau}}{\rho_0} \right) \tag{7.16}$$

so that the total meridional mass transport is proportional to the curl of the wind stress.

If we consider the ocean circulation between the tropics and mid-latitudes, the wind stress varies from westward in the tropical easterlies to eastward in the mid-latitude westerlies (Fig. 7.7). Thus a wind stress curl opposite to the rotation of the Earth is applied to the ocean, and according to (7.16), we should expect the water transport in the ocean to be equatorward. Physically, the wind stresses are causing the water to rotate about a vertical axis in a direction that is opposite to the rotation of Earth. To maintain a steady state in the face of this application of anticyclonic rotation, water must drift toward lower latitudes. The reduction in absolute vorticity applied by the wind stress is thus expressed as a decrease in the planetary vorticity of fluid parcels as they drift equatorward, and a steady state with constant relative vorticity can be maintained.

According to (7.16), the meridional transport in the ocean will be equatorward everywhere, so long as the wind stress curl is opposite to Earth's rotation. How, then, can the conservation of mass and vorticity be jointly satisfied if the wind stress curl is everywhere negative? How does the water-transported equatorward return to high latitudes and close the circulation of mass and vorticity? The western boundary currents observed in the mid-latitude oceans are the solution to this dilemma.

A simple model can be constructed by adding a lateral diffusion term to the vorticity equation (7.13) that produces a steady gyre circulation with the northward flowing return current intensified along the western margin of the ocean basin, much like the observed northward flow is intensified in the western boundary currents. As the water flows poleward along this western margin, the planetary component of vorticity (f) increases because of its dependence on latitude. If the absolute vorticity of the fluid parcels were to be conserved, then their relative vorticity would have to change to compensate for the increase in planetary vorticity. By flowing along the western margin of the ocean basin in a narrow current, the poleward return flow of the wind-driven circulation is able to collect enough vorticity in the same sense as Earth's rotation to arrive at middle latitudes with a vertical component of absolute vorticity, near that of the planetary vorticity at these latitudes, so that the magnitude of the relative vorticity remains reasonably small and steady. In a simple linear model with lateral diffusion of momentum, the mechanism for collecting the necessary vorticity is through the lateral friction stresses, which generates vorticity of the proper sign, only along the western boundary of the ocean basin. Thus the warm, rapidly flowing western boundary currents of the Atlantic and Pacific Oceans are seen to be a response to the wind stress driving the bounded oceans of the Northern Hemisphere.

7.6 THE DEEP THERMOHALINE CIRCULATION

The term *thermohaline circulation* is used to denote that part of the oceanic circulation that is driven by water density variations, which are in turn related to sources and sinks of heat (thermo) and salt (haline). It is traditional in oceanography to organize the discussion of the oceanic circulation into separate wind-driven and density-driven components, although the circulation of the ocean is not a simple addition of the effects of these two types of forcing. Wind driving influences the sources and sinks of heat and salt for the ocean by transporting surface water from the tropics to latitudes where cooling and evaporation can increase its density to very large values. The heat transport associated with the thermohaline circulation affects the SST gradients that help to drive atmospheric winds. Therefore, wind driving and density driving of the oceanic circulation are very closely coupled and cannot be easily separated. However, it is generally true from a diagnostic perspective that wind driving is the strongest influence on currents near the surface, and variations in density are important for the flow at depth.

Slow circulations below the thermocline are driven primarily by density gradients in the deep ocean. These circulations are difficult to measure directly, since the currents associated with them are very weak, but their nature can be inferred from the distributions of trace constituents of seawater. Away from the surface, temperature and salinity of water masses change very slowly, so that the water masses and their origins can be inferred from the particular combination of temperature and salinity that characterizes them. These layers are often deep and the characteristics within them quite uniform, so that oceanographers refer to them as mode water (McCartney, 1982).

Most gases are soluble in water, so that the concentrations of particular gases can also be used to characterize water sources. The saturation concentration of a gas in seawater is the amount that would exist in solution at equilibrium, if seawater at a particular temperature and salinity were exposed to the gas. Saturation concentrations of gases in seawater increase as the water gets colder. For example, the saturation concentrations of oxygen and carbon dioxide in seawater at 0°C are about 1.6 and 2.2 times their values at 24°C, respectively. The concentration of oxygen in surface water is always slightly greater than its saturation value, probably as a result of efficient mixing of bubbles of air into the surface water and production of oxygen in surface waters by photosynthesis. When surface water sinks into the deeper levels of the ocean its source of oxygen is cut off, and the oxygen is slowly consumed by bacteria as they consume organic matter at depth. Therefore, one may use the depletion of the oxygen concentration below its saturation value as a measure of the time since the water has been at the surface.

Figure 7.16 shows the oxygen saturation versus depth and latitude in the Atlantic and Pacific oceans. In the North Atlantic Ocean we observe that high saturation values extend to great depths, and that these high values extend toward the Southern Hemisphere at depths below ~1500 m. We infer then that significant downwelling of water occurs in the North Atlantic and that this water sinks most of the way to the bottom of the ocean and then spreads southward. In the tropics, oxygen-depleted water exists just a few hundred meters below the surface, indicating that water has been below the surface for a long time. A stability barrier associated with the strong density stratification in the tropics prevents mixing of surface waters with the older water below. The distribution of oxygen saturation in the North Pacific Ocean is very different from that in the North Atlantic. In the North Pacific, we see no evidence of downwelling, and in fact the oxygen at depth is severely depleted, with saturations about 10–15% at latitudes and depths where the oxygen saturation is about 85% in the North Atlantic.

From the oxygen saturation alone we can infer that water from the surface sinks relatively quickly to the deep ocean in the North Atlantic, but that this does not occur in the Pacific. We cannot infer the full circulation of the deep ocean from oxygen alone, nor can we directly infer the subsidence rate, since the rate of oxygen depletion depends on the biological activity at depth, which in turn depends on the rate at which nutrients are supplied to them by deposition from above. The inferences from temperature, salinity, oxygen, and many other tracers suggest a deep-water circulation in the Atlantic in which a large mass of deep water is formed in the northern margin of the ocean, which then flows southward to fill a large fraction of the deep Atlantic (so-called North Atlantic Deep Water, NADW). This water rises toward the surface again in the vicinity of 60°S. Cold, but lower salinity water is formed in mid-latitudes of the Southern Hemisphere and wedges itself between the warm surface water and the North Atlantic deep water below. This sub-Antarctic mode water (SAMW) and closely related Antarctic intermediate water (AAIW) are more evident in the salinity distribution in Fig. 7.2, which shows relatively fresh water formed in southern mid-latitudes is subducted on the northern edge of the circumpolar current during winter as further evidenced by the deepened mixed layer there (Fig. 7.5b). AAIW is freshened from sea ice melting and is more closely identified with the deepened mixed layer, west of South

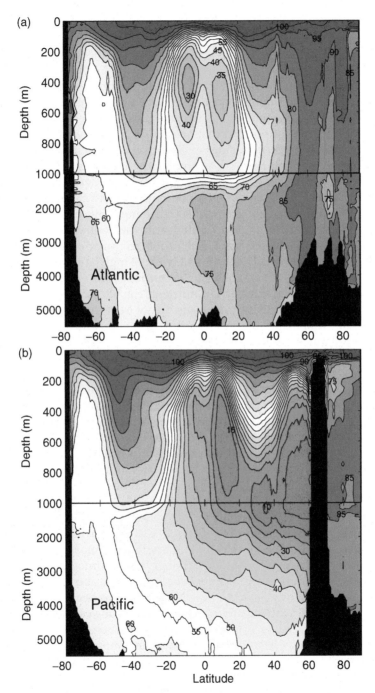

FIGURE 7.16 **Oxygen saturation percentage for the (a) Atlantic and (b) Pacific oceans.**
Note that the vertical scale changes at 1000 m, since variations are much slower below this
depth. *Data from NOAA.*

America. Bottom water is formed around Antarctica, especially in the Weddell Sea where seasonal ice formation adds salt to very cold seawater.

The mechanisms of deep-water formation in the North and South Atlantic are believed to be somewhat different. In the North Atlantic, warm, saline water flows poleward from mid-latitudes, where the Gulf Stream provides an important source of such water. This water is carried farther poleward into the Norwegian and Greenland Seas, where it is exposed to very cold atmospheric temperatures. The cooling of this saline water produces water that is dense enough to sink to great depths. The surface water in high latitudes of the Southern Hemisphere is relatively fresh, because of the excess of precipitation over evaporation in those latitudes, and no western boundary current exists to carry warm, saline water poleward, since the circumpolar current inhibits efficient transport of water from middle to polar latitudes in the southern oceans. Some saline water reaches the surface of the southern polar oceans by flowing southward at intermediate depths and rising at high latitudes in the south Atlantic. This water is also enhanced in nutrients, since it has spent some time at intermediate depths, where nutrients can be dissolved from falling detritus and photosynthetic organisms do not exist to consume the nutrients. In the waters around the Antarctic continent, which are the source regions for much of the Antarctic bottom water, the formation of very dense water is more dependent on sea ice production. Dense water formation is especially efficient in openings in the sea ice, or *polynyas*, where open water and very cold, dry air from come together and foster rapid sea ice production. This process is particularly efficient around Antarctica when cold offshore winds push the ice seaward and new sea ice is formed rapidly in the open water behind the northbound ice. When ice is formed from seawater, salt is rejected from the crystal structure, resulting in the formation of brine, which adds salt to the water immediately under the ice and thereby increases its density. This cold, saline water is dense enough to sink to the bottom of the Southern Ocean.

To infer the rate of downwelling it is necessary to use tracers with known decay times such as carbon-14 (^{14}C) or CFCs. Carbon-14, or radiocarbon is a radioactive isotope produced naturally in the atmosphere by cosmic rays and by the explosion of atomic bombs in the atmosphere. Since the rate of decay of radiocarbon is precisely known, the abundance of radiocarbon can be used to estimate how long water has been below the surface. The rate and spatial distribution of downwelling can also be inferred from transient trace gases such as chlorofluorocarbons (CFCs), which are man-made and have been introduced into the atmosphere starting in the middle of the twentieth century.

By combining the evidence available from tracers of seawater movement it has been convincingly shown that at the present time deep ocean water is formed only at high latitudes in the North Atlantic and Southern Ocean. Only in these locations can water of sufficient density be formed to sink to the deep ocean. From these two locations water spreads out at depth to fill the Pacific and Indian oceans, where the water gradually rises toward the surface. Since the North Pacific is the farthest from either of these two locations, the water at intermediate depths in the North Pacific is the "oldest" ocean water in the sense that it has been the longest time since this water was exposed to the atmosphere. The fact that the oldest water is not at the ocean bottom suggests that the deep water formed in the polar regions slowly rises elsewhere, as would be required by the conservation of water mass. The regions of the ocean where deep water can be formed constitute a small fraction of the total surface area of the ocean. For example, 75% of the ocean has potential density greater than 27.4, but only 4% of the surface water has a density that high. It is estimated that the time required to replace the water in the deep ocean through downwelling in the regions of deep-water formation is on the order of 1000 years. We may call this the turnover time of the ocean. The thermal, chemical, and biological properties of the deep ocean therefore constitute a potential source of long-term memory for the climate system on time scales up to a millennium. Some chemical properties of the ocean take longer than one turnover time to change significantly, so that the potential exists for ocean memory on time scales longer than the ocean turnover time.

Fig. 7.17 shows meridional cross-sections for the zonally averaged global ocean and for sections at 30°W in the Atlantic and 165°W in the Pacific. The turbulent mixing in the deep ocean is very weak and nearly adiabatic, so the flow of deepwater must be approximately along lines of constant potential density (isopycnals). Schematic arrows showing the hypothesized flow of deepwater are indicated for the global ocean in Fig. 7.17. Green arrows indicate the flow of North Atlantic Deep Water that is formed in the far North Atlantic Ocean and flows southward at intermediate depths. This water rises up again in the Southern Ocean, where the isopycnals slope upward and reach the mixed layer (Marshall and Speer, 2012). Antarctic Bottom Water is formed around the edge of Antarctica, especially in the Ross and Weddell seas, where the formation of ice during the winter season adds salt to the water below the ice, making very dense water. In the Southern Ocean the contours of oxygen saturation tend to follow the contours of potential density, supporting the schematic picture of the deep ocean circulation indicated by the colored arrows. The Southern Ocean is thus a key region for both forming bottom water and bringing deep water back to the surface.

The strongest ocean current is the Antarctic circumpolar current. The mass flow through the Drake Passage is about 125 Sverdrups (125 × 106 m^3 s^{-1}) that flows eastward around Antarctica and passes through the Drake Passage between South America and Antarctica. It is driven by the very strong wind stresses associated with the mid-latitude jet in the atmosphere. The wind stress curl in the Southern Ocean drives upwelling poleward of the jet and downwelling equatorward of it. This

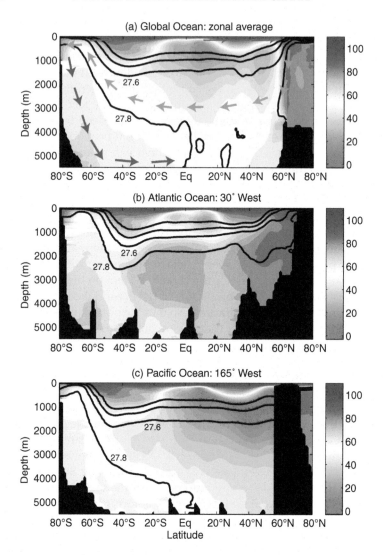

FIGURE 7.17 **Meridional cross-sections for (a) the global ocean – zonally averaged, (b) along 30°W in the Atlantic Ocean, and (c) along 165°W in the Pacific Ocean.** Colors indicate oxygen saturation in percent, solid contours are potential density minus 1000 as in Fig. 7.6. Contours are shown for 27.2, 27.4, 27.6 and 27.8 kg m^{-3}. Arrows indicate the flow of deep water, which tends to parallel contours of potential density. Purple vectors indicate the movement of Antarctic bottom water and green vectors indicate the movement of North Atlantic intermediate water. *Data are from NOAA.*

is consistent with the displacement of the contours of oxygen saturation, with old oxygen-poor water drawn upward poleward of the jet and a deep pool of oxygenated water equatorward of the jet (Fig. 7.17). This wind-driven vertical circulation also increases the temperature gradient and the slopes of the isopycnals, since dense water is drawn up on the poleward side and less dense surface water is pushed downward on the equatorward side. This strong density gradient fosters baroclinic instability of eddies in the ocean, and the Southern Ocean has very vigorous ocean eddies, much as in the mid-latitudes of the atmosphere, but with a horizontal scale of only about 100 km. These eddies produce fluxes of heat and potential vorticity that are important for ocean dynamics and tend to smooth out the meridional gradients that the wind-driven circulation generates.

The Southern Ocean is a key region for climate change, along with the North Atlantic and the tropical Pacific. The Southern Ocean seems to be the key region where deep water returns to the surface, whereas upwelling of less dense water is important in the tropical Pacific and deep water is formed in both the North Atlantic and the ocean around Antarctica. The circulation of the Southern Ocean is important for many aspects of climate variability and change. Since it is the primary region for return of deep water to the surface, the North Atlantic overturning circulation and the Southern Ocean upwelling are closely linked. The return of deep water to the surface means that warming of the Southern ocean will lag behind warming of the globe, and evidence of this effect will be seen in temperature trends calculated in Chapter 9. The circulation of the Southern Ocean interacts with the wind stresses. During an ice age, for example, the strongest wind stresses may move equatorward. If they move equatorward enough, then the importance of the Drake Passage for the circulation of the Ocean may be changed. The Southern Ocean may also be a key player in the carbon cycle variations that occur during ice ages. If sea ice extends equatorward along with the strongest winds, then ventilation of the deep ocean and the return of carbon to the surface may be suppressed, contributing to the reduction of atmospheric CO_2 that supports the ice ages. Reversing these changes may also contribute to the return of CO_2 to the atmosphere during interglacials, the warmer periods between the ice ages that define the large glacial–interglacial cycles of the past 800,000 years.

7.7 TRANSPORT OF ENERGY IN THE OCEAN

The general circulation of the ocean produces horizontal transport of energy from the tropics to the polar regions that is important for climate. However, it is not easy to measure this heat transport directly. It is difficult and expensive to obtain simultaneous current and temperature measurements from the surface to the bottom of the ocean. Such measurements require a ship or a large buoy and a cable with thermistors and current meters that extends from the surface to the ocean bottom, which is a distance of about 4 km on average. The spatial scales of the motions that are important for heat transport in the ocean are often small compared to the great expanse of an ocean basin, so that it is beyond our means to simultaneously measure current and temperature at enough spatial points and frequently enough in time to continuously monitor the product of velocity and temperature that produces most of the heat transport in the ocean. Attempts have been made to measure a series of profiles across a basin at a particular latitude, but these estimates must be assigned a rather large uncertainty.

An alternative to direct measurement of currents and temperatures is to infer the heat transport of the ocean from the energy balance of Earth or of the ocean. In Fig. 2.15, we inferred the annual mean northward heat flux by the ocean as a residual of the top-of-atmosphere and atmospheric heat budgets. This estimate indicates that the maximum meridional energy transport by the oceans in the Northern hemisphere is much smaller than the atmospheric energy transport, and occurs at a lower latitude, peaking in the tropical rather than the middle latitudes. A clear asymmetry can be seen between the Northern and Southern Hemispheres, which is likely related to the shape of the oceans and land areas in the two hemispheres. The poleward heat transport by the ocean in the Northern Hemisphere is about twice that in the Southern Hemisphere.

The oceanic energy flux can also be estimated from the energy balance at the surface of the ocean. Following (4.1) the energy balance at the surface can be written as (7.17).

$$\nabla \bullet \vec{F}_o = R_s - LE - SH - \frac{\partial E_s}{\partial t} \qquad (7.17)$$

The divergence of energy transport in the ocean can be estimated from the ocean surface energy balance if the surface net radiative heating, the evaporative cooling, the sensible cooling, and the energy storage in the ocean can be estimated. All of these terms are discussed in Chapter 4, and maps of the inferred oceanic flux divergence appear in Fig. 4.19. Estimates of meridional energy transport in the ocean obtained from (7.17) are in general agreement with estimates shown in Fig. 2.15, in that they show a maximum transport by the oceans at about 20°N. The surface energy balance method can also provide estimates for individual regions and oceans (Trenberth et al., 2001). The estimates indicate that the Atlantic Ocean transports energy northward across the equator, whereas the Indian Ocean transports energy southward.

7.8 MECHANISMS OF TRANSPORT IN THE OCEAN

The meridional transport of energy in the oceans is clearly important for climate, but because the transports are not measured directly it is uncertain what types of circulations contribute most to the transport. There are three generic types of circulations that are candidates: wind-driven currents, thermohaline circulations, and mid-ocean eddies.

7.8.1 Wind-Driven Currents

The warm western boundary currents such as the Gulf Stream and the Kuroshio and their associated mid-ocean drift currents play an important role in meridional energy transport in the oceans. The swift, warm currents that flow poleward along the western boundaries of the Atlantic and Pacific oceans are capable of carrying large amounts of heat poleward. The equatorward flow of relatively cold water in eastern boundary currents also contributes to the poleward energy flux. In addition to these horizontal gyre circulations, the subtropical oceans have a shallow vertical overturning in the top 700 m that contributes about as much poleward heat transport as the gyre circulations (Bryden et al., 1991). The trade winds drive a poleward drift of warm water at the surface, which is balanced by an equatorward flow of slightly colder water below the surface.

We can estimate the poleward heat flux associated with these circulations by considering the product of the mass flux of water and the temperature difference between the poleward and equatorward flowing at the same latitude. The mass flow is the velocity of the current times its area and the water density. From Fig. 7.11, we estimate that the Gulf Stream is 100 km wide and 600 m deep and has an average current of 0.5 m s^{-1}. Assuming a density of 10^3 kg m^{-3}, we can obtain an estimate of the mass flux in the Gulf Stream of 3.0×10^{10} kg s^{-1}, or 30 sverdrups.

$$\text{Density} \times \text{width} \times \text{depth} \times \text{speed} = 10^3 \text{ kg m}^{-3} \times 100 \text{ km} \times 600 \text{ m} \times 0.5 \text{ m s}^{-1}$$
$$= 3 \times 10^{10} \text{ kg s}^{-1}$$

This estimate of 30 sverdrups agrees with more detailed calculations of the flow through the Florida Straits (Bryden and Hall, 1980). The mass flux in the Gulf Stream increases significantly northward from Florida, as it incorporates more flow from the gyre. If we assume that the Kuroshio has a similar mass flux, then the total poleward mass flux in Northern Hemisphere western boundary currents is 6×10^{10} kg s^{-1}. To calculate the energy flux associated with this mass transport, we need the heat capacity of water (4281 J K^{-1} kg^{-1}) and the temperature difference between the poleward-flowing boundary currents and the equatorward-flowing water at the same latitude. We do not know this temperature difference precisely, and it is very dependent on whether the equatorward flow is above or below the thermocline. It is interesting to consider how big this temperature difference must be in order for the western boundary currents to produce a meridional heat transport of about half the maximum oceanic flux displayed in Fig. 2.15. To obtain an oceanic flux of 1.6 PW requires a temperature difference between the poleward-flowing and equatorward-flowing water of about 6°C.

$$c_w \rho_w v_w \, \text{area}_w \, \Delta T = 4218 \, \text{J} \, \text{K}^{-1} \text{kg}^{-1} \times 6 \times 10^{10} \, \text{kg} \, \text{s}^{-1} \times 6.3 \, \text{K}$$
$$= 1.6 \, \text{PW} \tag{7.18}$$

From Fig. 7.14, we estimate the average surface temperature gradient across the Atlantic at 25°N to be 4–6°C. If the equatorward return flow is primarily in the eastern Pacific or in interior below the thermocline, then it would be easy to produce the required oceanic heat flux with the western boundary currents and an associated colder interior return flow. Similarly, if we look at the vertical structure of the temperature for the subtropical North Atlantic in Fig. 1.11, the temperature drops rapidly by about 10°C from the surface to 500 m depth, so a shallow overturning circulation with poleward flow near the surface and a return flow below will transport a significant amount of heat.

7.8.2 The Deep Thermohaline Circulation

The mass flow of the deep thermohaline circulation is governed by the rate at which deep water can be formed at high latitudes. In the Northern Hemisphere, deep water is formed only in the Atlantic at high latitudes, and the formation rate is quite slow, since it takes several centuries to replace the deep water in the Atlantic. It is estimated that the average rate of deep-water formation in the North Atlantic is 1.5–2×10^{10} kg s^{-1} and in the Antarctic Ocean about 1×10^{10} kg s^{-1}. Since the mass flux of the deep thermohaline circulation is more than a factor of ten smaller than that associated with the western boundary currents, it is safe to assume that it has a much smaller influence on the total northward ocean heat flux than the wind-driven gyres, although it may have a significant effect in the far North Atlantic.

7.8.3 Mid-Ocean Eddies

The Gulf Stream and the Kuroshio spin off long-lived eddies via baroclinic and barotropic instabilities (Fig. 7.12). These are the oceanic analogs to the eddies that produce most of the atmospheric meridional energy transport in mid-latitudes. However, the role of eddies for heat transport in the subtropical ocean is likely much less than in the atmosphere, because of their smaller spatial scales compared with the scale of the oceans. Moreover, the oceanic eddies are best developed well poleward of the latitude of the maximum oceanic transport. The wind-driven and thermohaline circulations are likely to provide much more important contributions to the meridional heat flux in the subtropics. However, in the high latitude Southern Ocean, ocean eddies are very important in determining the overturning circulation and the response of the ocean to changed wind stress.

EXERCISES

1. Use the data in Figs 7.1 and 7.2 to estimate how much the salinity of the surface water of the Arctic Ocean would need to increase before the surface density would equal the potential density at 1000-m depth. How does this compare with the average salinity of the ocean?

2. What depth of seawater would need to freeze in the Arctic Ocean to produce the increase in salinity of problem 1 in the top 100 m of water? Assume that all salt is rejected from sea ice and enters the 100-m layer.

3. With the negative wind stress curl characteristic of today's climate, the wind-driven meridional flow (7.16) is an equatorward drift, which we can hypothesize occurs mostly in the thermocline or above it. How would the net heat transport produced by this drift and its return flow be different if, rather than a warm western boundary jet, the return flow were a slow poleward drift near the bottom of the ocean?

4. Discuss the ways in which the extension of the warm, saline Gulf Stream into the Norwegian and Labrador seas assists in the formation of dense water that can sink to the depths of the Atlantic Ocean.

5. Use Fig. 7.1 to estimate the initial and final density values of a kilogram of water that starts in the tropics with a temperature of 28°C and a salinity of 35‰ and flows on the surface in the Gulf Stream to the Norwegian Sea, where it arrives with a temperature of −1°C. Assume the water conserves its salinity en route and loses heat by sensible heat transfer.

6. As an alternative to problem 5, assume that the kilogram of water starts in the tropics, but is cooled by evaporation along its route rather than by sensible heat loss. Estimate the mass of water that is lost by evaporation en route, if the parcel arrives in the Norwegian Sea with a temperature of −1°C. Calculate the salinity on arrival, assuming that no horizontal mixing or precipitation occurs, and the salinity is well mixed through the top 100 m of the ocean. What is the density on arrival? When you compare the final density with that obtained in problem 5, is the effect of evaporation on the final density significant? Is it important to know the Bowen ratio for the parcel along its route?

7. Suppose that a wind stress is applied to the ocean, taking the following simple form.

$$\tau_x = \begin{cases} A\cos\left(\dfrac{\pi y}{L}\right) & -L < y < L \\ A & |y| > L \end{cases}$$

Derive an equation for the vertical velocity at the bottom of the Ekman layer assuming a constant Coriolis parameter $f = f_o = 2\,\Omega\sin 30°$. Derive an equation for the integrated meridional transport V_I, using (7.14) with the f and β appropriate for 30°N latitude. Determine a numeric value for the maximum w_E and V_I using the following constants: $A = 2$ dyn cm^{-2} = 0.2 N m^{-2}, $L = 1500$ km, $\rho_o = 1025$ kg m^{-3}. Plot τ_x, w_E, and V_I on the interval $-L < y < L$. Assuming that the ocean basin is 5000 km wide, calculate the water mass flux at 30°N associated with the interior flow. Compare this number with the estimate for the Gulf Stream mass flux given in Section 7.8.

8

Natural Intraseasonal and Interannual Variability

8.1 STUFF HAPPENS

Experience shows that weather forecasts are often inaccurate in predicting where and when precipitation will fall and that they get less accurate as the lead time gets longer. In the 1960s, Edward N. Lorenz used simplified versions of the equations that describe the motion of the atmosphere to show that these equations lead to chaos, and prediction of detailed weather information such as the position of weather fronts is likely to be impossible for lead times of greater than about 2 weeks. This is because small errors in the initial field rapidly grow to dominate the forecast fields. Despite this limitation, the quality of forecasts has increased dramatically over the past several decades, so that 10-day forecasts are now often useful. Figure 8.1 shows the correlations between the predictions of a weather forecast model and the observed anomalies in the height fields over the Northern and Southern hemispheres as a function of time starting in 1980. An anomaly correlation of 1 would be a perfect forecast of the 500 hPa height field, so that 3-day forecasts of the height field are very good indeed, but the accuracy drops very quickly for lead times beyond 5 days. It is also apparent that forecasts in the Northern Hemisphere were much better than those in the Southern Hemisphere in the 1980s and 1990s. In the mid-1990s, the Southern Hemisphere forecasts began to improve faster than those in the Northern Hemisphere so that by 2002 the forecasts in the two hemispheres became about equally good. This is because global satellite data began being used much more effectively to determine the initial state of the weather during the 1990s, and satellite observations are much more important in the more ocean-covered and less-populated Southern Hemisphere, where only about 10% of the human population lives.

Although the details of the weather at any instant of time cannot be predicted more than 2 weeks into the future, no matter how much we improve observation systems and models, we can make useful predictions about

Global Physical Climatology. http://dx.doi.org/10.1016/B978-0-12-328531-7.00008-6

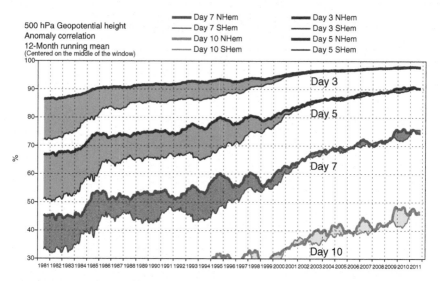

FIGURE 8.1 **Quality of weather forecasts as measured by the anomaly correlation of the 500 hPa geopotential height as a function of time from 1981 to 2012 for forecast lead times of 3, 5, 7, and 10 days.** Quality estimates are given separately for the Northern and Southern hemispheres, and they converge in about 2002. *Courtesy of Martin Janousek, ECMWF.*

the statistics of the weather beyond 2 weeks' time. Examples of useful predictions of weather statistics might be the monthly mean temperature during summer, or the probability of an extreme precipitation event in a particular region during a particular season. The possibility to predict the statistics of weather comes from two principle sources. First, the climate system has modes of variability with time scales longer than 2 weeks that may be predictable. If we can predict these low-frequency variations in the natural climate system, then we can make useful predictions about the statistics of weather. Some longer-term predictability may come from the coupling of the ocean to the atmosphere, since the ocean may evolve on longer time scales that are still predictable. Some of these low-frequency modes of variability will be discussed in this chapter. Second, if we know how the climate is being forced, say by greenhouse gas emissions, we can predict how those greenhouse gas emissions will continue into the future. Then, we can make potentially useful predictions of weather statistics a century or more into the future. This is the subject of Chapter 13. Also, from understanding the basic physics of the climate system, we can predict how it will maintain itself differently in a warmer climate, so that the statistics of rainfall intensity, storm intensity, freezing, and drought might be very predictable.

We often distinguish between unforced variability and variability forced by natural or anthropogenic causes. Unforced variability arises from the internal dynamics without any specific cause. Forced variability can be associated with some change in the boundary conditions of the climate system, such as a volcanic eruption or solar variability on the natural side, or gas or aerosol emissions by human activities on the anthropogenic side. Unforced variability can occur on a variety of time scales from that of a week or two that we normally associate with weather; to intraseasonal variability that might result from internal atmospheric dynamics or interactions between the ocean and the atmosphere; to interannual variability that might result from ocean–atmosphere interactions on time scales of a few years; to natural internal variability that may last up to a thousand years, about the time it takes to turn over the global ocean. Variability that lasts thousands to millions of years may be caused by interactions between variations in Earth's orbital parameters and its cycles of carbon and ice, and will be discussed in Chapter 12. In this chapter, we will discuss natural variability that has been observed with instrumental climate records, which limits consideration to time scales less than about a decade, if we wish to observe enough events to generalize about them. In Chapter 9, variations on longer time scales that can only be inferred from natural recording systems will be considered.

8.2 INTERNAL ATMOSPHERIC VARIABILITY

The climate system is continuously forced by solar heating of the surface in the tropics, and cooled from the atmosphere and from high latitudes by infrared emission. This heating gradient drives the movement of energy upward in convection and poleward in mid-latitude weather systems or ocean currents. Weather systems and convection events often develop from instabilities of the state toward which the heating gradients drive it, which adds randomness to the variability. The atmosphere and the ocean are relatively weakly damped by frictional dissipation, so that these motions increase in amplitude until they are strongly interacting in a nonlinear way, moving energy from one scale to another. In such a system, energy can be collected in large spatial and temporal scales that are of interest from the perspective of climate. The release of energy on relatively small scales leads ultimately to the great wind and current systems of Earth described in Chapters 6 and 7.

8.2.1 Extratropics: PNA, NAO, and SAM

The atmosphere has a concentration of energy in structures with scales of thousands of kilometers or more that is of interest for weather and climate. Energy is put in at large scales associated with the latitudinal gradient of insolation. The large-scale interactions in mid-latitude weather systems move energy from the scale at which baroclinic instability generates waves (~3000 km) to both larger and smaller scales, with a preference for energy to move toward larger scales and collect there. The energy of the mid-latitude westerly winds is delivered from the smaller scale eddies that define the weather in mid-latitudes. The energetic interactions between mid-latitude storm systems and zonal jets are thus intense, and this translates into modes of variability in which jets and their embedded eddy structures move north and south. One can define these structures as those that best describe the observed variance in a mean-square sense, and this can be done efficiently using what Lorenz called *empirical orthogonal functions* (EOFs). This is an objective mathematical technique that can find spatial patterns in data that explain a disproportionate amount of variance. EOFs are spatial structures that take advantage of the correlations between different locations to explain a lot of variance. If they robustly explain a lot of variance, then we tend to associate them with real modes of variability of the climate system, especially if a physical argument can be given for why such a structure should exist.

To introduce the concepts of space and time scales of mid-latitude atmospheric variability, we can consider the 500 hPa height field. The 500 hPa height field is a useful scalar quantity for characterizing extratropical variability, since the wind and temperature can both be related to the geopotential height through the geostrophic and hydrostatic relations. Since 500 hPa is at the mid-point of the mass of the atmosphere, it captures phenomena that extend through the depth of the troposphere. We can divide its temporal variability by filtering into periods shorter than 7 days, periods between 7 and 30 days, and periods longer than 30 days. The total variance and the variance in each of these categories are shown in Fig. 8.2 for the Northern Hemisphere winter half year (October to March), and in Fig. 8.3 for the Southern Hemisphere winter half year (April to September). Also shown is the climatological height field, the average over all days in the half year and all the years in the sample.

Figure 8.2 shows that the variability of 500 hPa height during Northern Hemisphere winter is highest over the oceans and least in mountainous areas. The variability over the oceans is contributed mostly by phenomena with periods longer than 30 days. Variability on intermediate time scales between 7 and 30 days occurs preferentially on the downstream (eastward) sides of the oceans. The high frequency variability, with time scales less than 7 days, occurs on the westward edges of the oceans – one in the Pacific and one in the Atlantic. The high-frequency eddies evolve into the intermediate-scale eddies as they propagate toward the east. The high frequency variance maxima are associated with the jet streams and their associated storm tracks, as previously discussed in Fig. 6.18.

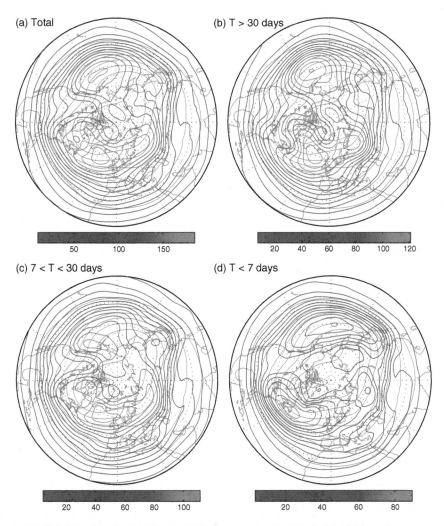

FIGURE 8.2   **Standard deviation (STD, in meters) of 500 hPa height (blue-red contours and color bar) from its climatology for different time scales during the Northern Hemisphere winter half year (October to March).** The climatology is shown as thin magenta contours. (a) Total STD of daily 12Z data, (b) STD of variability with periods longer than 30 days, (c) STD for periods between 7 and 30 days, and (d) STD for periods shorter than 7 days. The contour interval for the climatology is 100 m. The projection is polar stereographic starting at 20°N. *Data from ERA Interim (1979–2013).*

In the Southern Hemisphere during winter (Fig. 8.3), the variability is much more zonally symmetric (large at all longitudes around 60°S), but the variability is greatest upstream of the topographic barrier formed by the Andes Mountains and the Palmer Peninsula and smallest over the mountains. High frequency variability is greatest in a zonally elongated

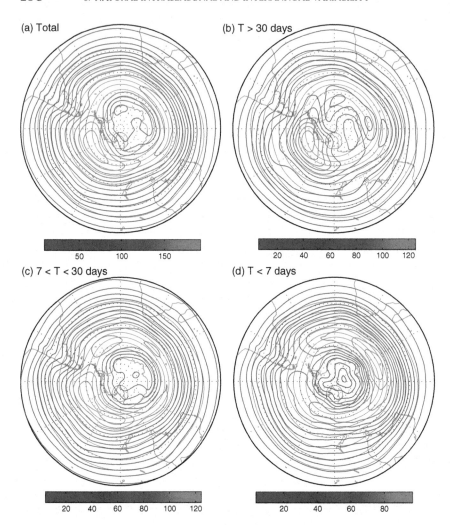

FIGURE 8.3   **Same as Fig. 8.2, but for the Southern Hemisphere during its winter half year (April to September).**

region peaking at about 50°S and 70°E in the Indian Ocean sector. The intermediate scale variability is largest downstream of this in the Pacific Sector between Australia and South America.

We can characterize the structure of the variability with one-point correlation maps of 500 hPa height anomalies. We choose a point where the variability of the time scale of interest is large, then correlate that point with all others. This reveals the characteristic spatial scale and structure of the variability with the time scale of interest.

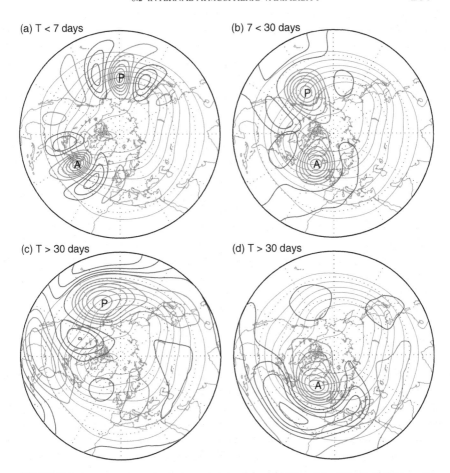

FIGURE 8.4   **One-point correlation maps for the variability of the 500 hPa height on the time scales shown for the Northern Hemisphere during its winter season as in Fig. 8.2.** The point where the letter is located is correlated with the rest of the hemisphere shown. Correlation patterns for two reference points are shown in (a) and (b) and only one in (c) and (d). Positive correlations are red contours, negative correlations are blue, the contour interval is 0.15 and the zero contour is not shown.

We begin by discussing the structures associated with the high-frequency variability in the Northern Hemisphere winter. One point was selected near the maximum of high-frequency variance in each ocean, and both one-point correlation maps are shown in Fig. 8.4a, along with the mean climatological height field. These are trains of waves with typical wavelengths of 50° of longitude (seven waves around a latitude circle or wavenumber 7). They occur near the troughs in the climatological height downstream of Asia and North America where the mean zonal westerlies are strong. They tend to move eastward following the direction of the mean wind, but slower than the mean wind at 500 hPa.

In the Northern Hemisphere winter, the structures associated with 7–30 day variability are larger in spatial scale, however, they do not have such strong wavelike structure as the high-frequencies eddies, but appear mostly as blobs that represent localized up and down movement of the height surfaces without any strong connections to other regions. The variability with periods greater than 30 days does have some characteristic structure. A reference point in the mid-latitude Pacific Ocean captures a large-scale wave train following a great circle route from the western tropical Pacific to the Gulf of Mexico (Fig. 8.4c). It arises from natural variability of the atmosphere, but it is also strongly affected by tropical SST variations associated with El Niño events, as we shall see later. It appears here with the sign opposite to that associated with El Niño events, which produce a low pressure center in the North Pacific. A reference point in the North Atlantic shows a height correlation with a strong north–south dipole pattern. This is known as the North Atlantic Oscillation or NAO, and is the dominant mode of low-frequency variability in the North Atlantic. Note that it is elongated in the east–west direction forming a north–south dipole, whereas the high frequency wave trains are elongated in the north–south direction and form eastward propagating wave trains. These differences in structure can be related to linear wave theory and to the mechanisms by which these modes gain energy (Hoskins et al., 1983).

In the Southern Hemisphere during winter (Fig. 8.5a), we also see wave trains at high frequencies, and the ones with the largest variance are in the Indian Ocean sector where the high frequency variance is greatest. The scale of the wave train in the Indian Ocean sector is larger than that in the Pacific Ocean sector because the winds are stronger in the Indian Ocean. Because of the greater zonal symmetry and stronger zonal winds in the Southern Hemisphere, the intermediate scales also take the shape of east–west-oriented wave trains, but the zonal wavelengths are closer to 90° of longitude, so with a zonal wavenumber of 4 instead of 7 as for the high-frequency wave trains (Fig. 8.5b). If we choose a point near the center of the maximum height variability upstream of South America at 65°S, we also find a low-frequency pattern that looks like a wave train that appears to stretch from New Zealand to the subtropical Atlantic Ocean (Fig. 8.5c). A bit of north–south dipole is also mixed in with this Rossby wave train, as the heights equatorward of the center at 65°S are negatively correlated with it.

If we choose a reference point directly over the South Pole (Fig. 8.5d), then the correlation pattern has a strongly annular pattern with height anomalies of opposite sign at most all longitudes at about 45°S. It has a shape like a donut, with heights in mid-latitudes going up or down while the heights over the pole go down or up. This mode of variability is called the Southern Annular Mode, or SAM. It results from interactions

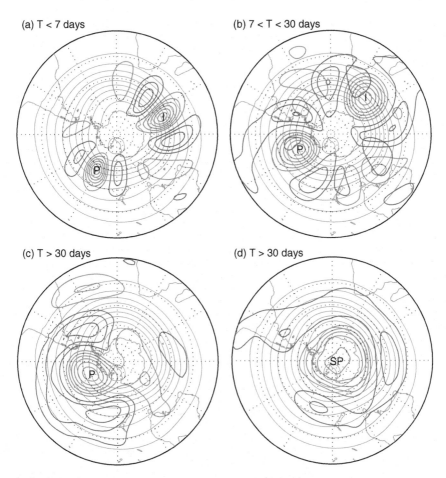

(a) T < 7 days

(b) 7 < T < 30 days

(c) T > 30 days

(d) T > 30 days

FIGURE 8.5   One-point correlation maps as in Fig. 8.4, except for the Southern Hemisphere in its winter half year of April through September.

between the high-frequency eddies and the zonal mean wind. SAM variability is internally generated, but it can be influenced by other things, and responds to forced climate change in global warming simulations.

Although most of the patterns in Figs 8.4 and 8.5 can be understood as trains of Rossby waves, the SAM is unusual in that the zonal-mean variation is dominant. To get another view of this mode of variability we can consider the EOFs of the daily zonal mean wind. The daily zonal mean wind is first computed, then the structures that explain the most variance of the daily zonal wind are computed, weighting by area and mass for the region from 20°S to 90°S and from 1000 hPa to 200 hPa, respectively. The first two such EOFs explain 40% and 21% of the variance thus constrained. Shown in Fig. 8.6 are the regressions of the zonal wind onto the first two

EOFs, so that the plots show the amplitude explained by typical variability of the EOF. The first EOF is equivalent to the Southern Annular Mode, and represents a north–south shifting of the eddy-driven mid-latitude jet (see Figs 6.4 and 6.16), whereas the second EOF represents a narrowing and broadening of the jet. The SAM is interesting in that it explains 40% of the variance, which is about twice that of the second EOF. Another special feature of SAM is that it not only explains a large amount of variance, but an even larger fraction of the variance at very low frequencies. For periods of 30 days or longer, the SAM explains 52% of the variance, whereas the second EOF explains only 22% and the third 8%.

The SAM anomalies are more persistent than those of the second EOF because the high-frequency eddies act to sustain jet shifts. High-frequency eddies propagate out of the jet, and as they do so they transport zonal momentum into the jet, thus sustaining it. The second EOF is not sustained by eddy interactions and its variations have less persistence. This is because, as the jet gets sharper, the eddies are less able to propagate away, so the momentum flux into the jet weakens as it gets sharper. The SAM is very interesting and important within the context of climate change. Its amplitude is large enough that it represents a shift in the latitude of the jet maximum of about 5° of latitude. In addition to its weather impacts, this shift is very important for wind stress driving of the ocean and sea ice. The SAM also appears to respond strongly to human forcing of the climate system, such as changes in the stratosphere due to ozone depletion, and models predict that the SAM will respond to global warming by moving the extratropical jet poleward.

8.2.2 Tropics: The Madden Julian Oscillation

In the tropics, convection releases latent energy that drives circulations with scales ranging from individual convective systems to global-scale circulations. Convection is organized both because convection moistens the atmosphere above the boundary layer and this encourages further convection, and also because convection drives large-scale motions that can supply moisture at low levels, supporting convection where it is occurring and suppressing it elsewhere. This can result in superclusters of intense convection or tropical cyclones, easterly waves, and global-scale tropical variability with time scales of a month or more.

In 1971, Roland Madden and Paul Julian discovered that tropical pressure and wind fields have a preferred mode of variability with a broad time scale centered around 40–50 days, but extending from 25 to 60 days. The Madden–Julian Oscillation (MJO) is strongest in the western Pacific and Indian Ocean regions during Northern Hemisphere winter, but occurs in all seasons and can be seen at all longitudes in the tropics, and also has effects in middle latitudes. It appears to result from a coupling between tropical convection and global-scale wind patterns near the equator. Its amplitude

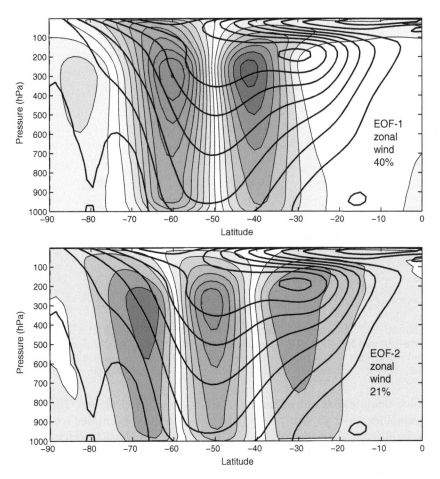

FIGURE 8.6 **The first two EOFs for daily zonal mean wind variations in the Southern Hemisphere for all days of the year.** These structures explain 40 and 21% of the area- and mass-weighted daily variance of the zonal mean wind between 20°S and 90°S, respectively. The heavy black contours are the zonal mean wind climatology for the annual average. Contour intervals for zonal mean wind are 5 ms⁻¹ and contours for the EOFs are 0.5 ms⁻¹. Colors indicate the sign of the EOF anomaly. Top level is 1 hPa.

and persistence are sufficient to cause breaks in the Asian Summer Monsoon and transitions in the state of the tropical coupled ocean–atmosphere system (see ENSO below). It can also modulate the occurrence and intensity of tropical cyclones, typhoons and hurricanes (Fig. 8.6).

To give a sense of the time and space variability of the MJO, we average the OLR from 15°S to 15°N, filter it in time to remove variations with time scales longer than 90 days and shorter than 15 days, and plot it as a function of time (Fig. 8.7). Two representative year-long periods are shown starting on July 1 and ending on July 1 the following year. Negative anomalies

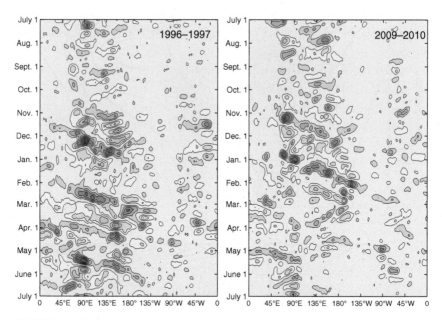

FIGURE 8.7   **Longitude–time plots of band-pass filtered tropical OLR anomalies for 1996–1997 and 2009–2010.** OLR is averaged from 15°S–15°N and periods longer than 90 days and shorter than 15 days have been filtered out. Blue shading is negative, the contour interval is 10 Wm^{-2}, and the zero contour is not shown. *OLR data from Liebmann and Smith (1996).*

(blue shading) indicate enhanced cloudiness and precipitation. The largest anomalies are ± 40 Wm^{-2} – quite large. From these examples we can see several features of the MJO. It is strongest in the Northern Hemisphere winter half year. It is intermittent and does not have a fixed period, but large amplitude variations in OLR occur with characteristic time scales that are longer than a month. The anomalies often move from west to east, starting in the Indian Ocean region and weakening in the central Pacific Ocean. Some other phenomena can be seen that propagate westward with shorter periods.

Wheeler and Hendon (2004) suggested an index of the MJO that is based on the EOFs of OLR and zonal wind at 200 hPa and 850 hPa, all averaged over the latitude band from 15°S to 15°N. Here we remove all variability with time scales longer than 90 days to remove the seasonal and El Niño signatures, then compute the joint EOFs of zonal wind at 200 and 850 hPa. This produces two leading EOFs that explain 20 and 16% of the combined variance of the 200 and 850 hPa zonal wind fields. Then we project these EOF structures onto the original data to compute the daily time series of the amplitudes of these two leading EOFs. These time series are highly coherent with each other at periods between 90 and about 25 days, with the first EOF preceding the second EOF by several weeks. Thus the first two EOFs are expressing a single phenomenon, the eastward moving MJO seen in Fig. 8.7. To show the full structure of the MJO, we take the time

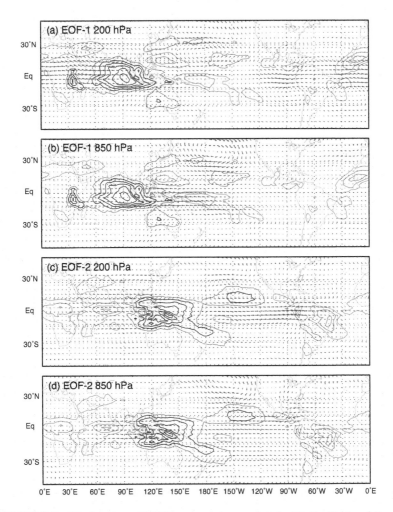

FIGURE 8.8 **Wind vector and OLR regressions associated with the First and Second EOFs of 200 hPa and 850 hPa zonal wind during the Northern Hemisphere Winter Half year.** Contour interval for OLR is 2 Wm^{-2} and the zero contour is not shown. Black contours indicate lower OLR, indicating high clouds and precipitation.

series of the amplitude of these two EOFs and regress them onto the OLR and the wind vectors at 200 and 850 hPa. The result is shown in Fig. 8.8.

The top two panels of Fig. 8.8 show the regressions of the wind vectors and the OLR onto the first EOF. This EOF has a large negative OLR anomaly centered at about 90°E in the Indian Ocean, indicating strong convection and precipitation there. This OLR anomaly is associated with easterly 850 hPa wind anomalies near the equator extending to the east as far as the central Pacific, and westerly anomalies extending west to the African coast. The low-level winds are thus converging where the precipitation is occurring. At 200 hPa, strong easterly wind anomalies extend westward from the region of low OLR and convection. The wind anomalies thus have a strong vertical shear, being westerly near the surface and easterly near the tropopause over the Indian Ocean. At the time of the oscillation represented by EOF-1, a strong anticyclonic wind circulation appears in the North Pacific, centered at about 40°N, just east of the Dateline and most obvious at 850 hPa. The MJO thus has a coherent effect on the mid-latitude circulation during the winter season. The MJO heating anomaly interacts with the subtropical jet stream in the western Pacific to produce this extratropical wave anomaly.

The bottom two panels of Fig. 8.8 showing EOF-2 are similar in many ways to the top panels except that the negative OLR anomaly has shifted to the east and is now centered over Indonesia and the western Pacific, with a convection anomaly extending down the South Pacific Convergence Zone. These two EOFs together represent the eastward propagation of a global-scale wind and precipitation anomaly. At the later stage represented by EOF-2, westerly wind anomalies at 850 hPa extend toward the west over the Indian Ocean to Africa with easterly anomalies above at 200 hPa in the same region. Low-level easterly anomalies and high-level westerly anomalies extend from the convection over Indonesia all the way across the Pacific to South America, so that the wind anomalies on the equator are near global in extent.

8.3 EL NIÑO, LA NIÑA, AND THE SOUTHERN OSCILLATION

The El Niño-Southern Oscillation phenomenon (ENSO) is a coupled mode of variability of the tropical ocean–atmosphere system centered in the Pacific Ocean sector, but coupled to global weather patterns. The El Niño nomenclature comes from the warm sea surface temperatures (SST) phase of ENSO appearing at the west coast of South America around Christmas time. La Niña was introduced as a convenient nomenclature for the cold phase, or feminine side, of ENSO. Southern Oscillation refers to the associated patterns in surface pressure that were discovered by Sir Gilbert Walker in the 1920s (Horel and Wallace, 1981).

To understand the genesis of the ENSO phenomenon, we need to consider the climatology of the coupled ocean–atmosphere system along the equator in the Pacific Ocean sector. Figure 8.9 shows the atmospheric vertical velocity and the oceanic potential temperature along the equator. To understand ENSO, we focus on the Pacific Ocean. The thermocline slopes downward toward the west, with a deep, warm mixed layer in the western Pacific and a shallow mixed layer and shallow thermocline in the east. This sloping thermocline is maintained in large measure by the easterly wind stress on the tropical Pacific Ocean, indicated by the large black arrow in Fig. 8.9. The strength of the easterlies is enhanced in the Pacific by the upward motion

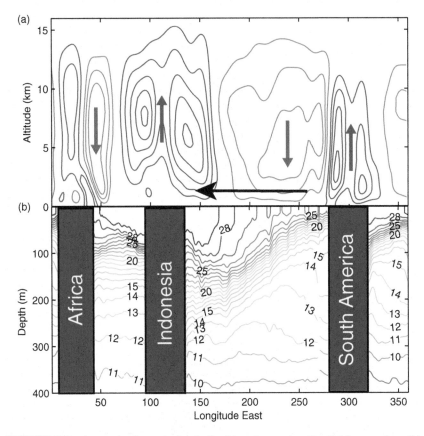

FIGURE 8.9 **Atmospheric vertical velocity (a) and oceanic potential temperature (b) along the equator.** Vertical arrows indicate direction of vertical motion, black horizontal arrow indicates direction of surface winds over the Pacific Ocean.

driven by convection over Indonesia, balanced by the downward motion supported by radiative cooling east of the dateline, where convection is much less probable. One thus sees an immediate connection between the preference for convection to occur in the far west Pacific and the easterlies that drive the westward SST and thermocline depth gradients. The positive feedback between warm SST in the west and cool in the east that drive easterlies that drive cool SST in the east and warm in the west is called "Bjerknes feedback." If these easterlies weaken, then pressure forces would begin to flatten the thermocline and warm surface water would flow toward the eastern Pacific. This would make precipitation more likely to form over the warmed SST in the eastern Pacific and the easterly winds would further weaken. This is the basic idea behind the warm ENSO, or El Niño event, which employs the Bjerknes feedback in reverse.

Figure 8.10 is a schematic diagram showing La Niña, Normal, and El Niño conditions along the equator in the tropical Pacific. The coupled

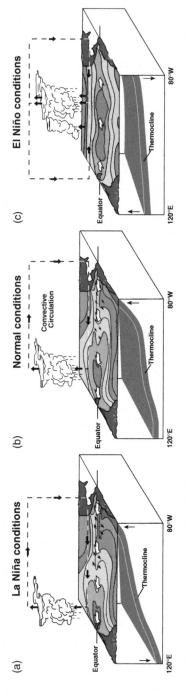

FIGURE 8.10   Schematic diagram showing the atmospheric and oceanic conditions along the equator in (a) La Niña, (b) Normal, and (c) El Niño conditions. *NOAA/PMEL TAO Project.*

system has a lot of available potential energy when the thermocline is steeply sloped, as during La Niña conditions, and an instability can occur then that leads to a rapid and sustained warming of the eastern Pacific with attendant shifts in the precipitation and associated atmospheric circulation patterns. La Niña occurs when the convection is farther west, the thermocline slope is great and the SSTs are cold in the central and eastern Pacific. El Niño is the opposite with rainfall in the central and eastern Pacific and relatively flatter SST and thermocline across the Pacific from west to east. When the convection is farther east, the easterly trade winds weaken and this suppresses the upwelling in the east that normally sustains the sloping thermocline.

The state of ENSO is well characterized by the SST along the equator. Figure 8.11 shows the monthly mean SST averaged between 5°S and 5°N from 1886 to 2013. The mean annual cycle has been removed, but the long-term trend is visible by comparing the early years with the latest years. Particularly large warm events occurred in 1982–1983 and 1997–1998. The largest SST anomalies occur from the coast of South America (~80°W) to the dateline (180°W). Sometimes the SST anomalies emerge first at the coast of South America (e.g., 1966 and 1973) and sometimes they emerge first farther offshore (e.g., 1982 and 1992). Some smaller SST anomalies occur in the Atlantic Ocean and to an even lesser extent in the Indian Ocean.

The characteristic time scale of the ENSO cycle can be estimated from the power spectrum of the SST anomaly averaged over the region from 5°S to 5°N and from 90°W to 150°W. This is often called the Niño-3 index of the state of ENSO. The power spectrum in Fig. 8.12 shows that the strongest variability is at periods from 3 to 5 years, but that significant variability extends across periods from 2 to 20 years.

One of the principal effects of the changed SST on the atmosphere is to move the precipitation in the tropics, which tends to locate over the warmest SST. Figure 8.13 shows the regression of the outgoing longwave radiation (OLR) anomalies onto the Niño-3 index of tropical SST. When the SST in the equatorial East Pacific is anomalously warm, the convection is enhanced in the central and eastern Pacific and decreased over Indonesia and Australia. The anomalies in tropical convection are to the west of the largest SST anomalies because the SST generally increases from east to west and the response of convection to a given SST anomaly is greater over warmer water. The dipole pattern of OLR anomaly represents an eastward movement of the main convection region in the far western Pacific. Although the OLR time series begins in the middle of 1974 and is therefore rather short, it does show some of the consistent impacts of El Niño on tropical precipitation. During November to March El Niño can mean dry conditions over Indonesia and Australia and also northeast South America, but wet over the Gulf of Mexico. During the May to September season El Niño can also mean reduced summer monsoon

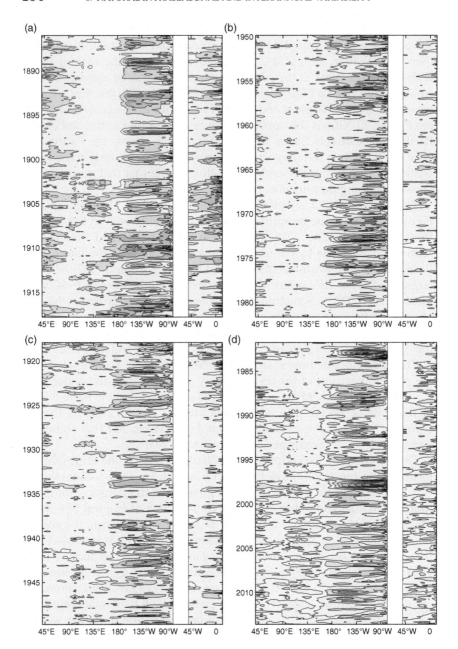

FIGURE 8.11 **Monthly SST anomalies averaged over 5°S–5°N as a function of longitude and time since 1886. Year increases downward in each panel.** The contour interval is 0.5°C and positive anomalies are shaded red. The mean annual cycle has been removed, but the long-term trend is included. The blank area is the land area of Central America. The land area of Africa is not shown. *Data are from the NOAA ERSST data set.*

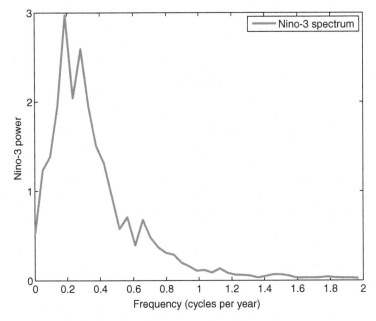

FIGURE 8.12   **Power spectrum of the Niño-3 index, the anomalies of SST averaged over 5°S to 5°N and 90°W to 150°W.** The linear trend was removed before computing the spectrum from the monthly data from 1886 to 2013. The period is just the inverse of the frequency in cycles per year, so 0.2 cycles per year correspond to a period of 5 years. The greatest variance is associated with periods between 3 and 5 years. *Data from the NOAA ERSST data set.*

precipitation in India. Another impact is precipitation reaching the west coast of equatorial South America, where the SST anomalies also have an impact on fisheries.

The extratropical response to ENSO can be assessed by regressing the Niño-3 index of tropical SST into the global field of 500 hPa height. Because of the low-frequency and episodic nature of ENSO variations, we need a long record to get robust results. Here we turn to the Twentieth Century Reanalysis product, which goes back to 1871 (Compo et al., 2011). The 500 hPa field in this reanalysis is mostly a model-calculated response to SST and surface-pressure observations, although it closely resembles the reanalysis products that assimilate all modern observations. We use monthly mean observations to construct the regression patterns shown in Fig. 8.14 for the Northern Hemisphere winter (November to March) and the Southern Hemisphere winter (May to September). The basic patterns are not sensitive to the exact choice of winter half year. Note that the sign is for El Niño conditions with a warm tropical East Pacific Ocean, but would be reversed for La Niña. The entire tropical troposphere warms up

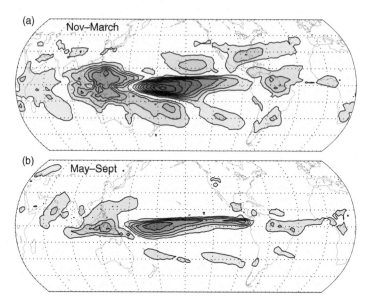

FIGURE 8.13   **Regression of monthly OLR anomalies with the Niño-3 index of SST for the November to March (a) and May to September (b) seasons.** Contour interval is 2 Wm^{-2}, the zero contour is not shown. Negative values are shaded blue and indicate reduced OLR, enhanced high cloud and increased precipitation. Based on NOAA interpolated OLR from June 1974 to 2013.

during an El Niño, so that the 500 hPa height increases everywhere in the tropics from about 20°S to 20°N. This strengthens the subtropical winds a bit and moves them toward the equator. The opposite happens during La Niña.

The response of the extratropical atmosphere to ENSO depends on the anomalies in heating produced as the precipitation shifts to follow the warmest SST values (Fig. 8.13) and the interaction of this anomalous heating with the mean wind fields (Sardeshmukh and Hoskins, 1988). In the November–March season, the extratropical response is somewhat symmetric about the equator, with similar responses in the Northern and Southern Hemispheres. Note the similarity of the November–March pattern with the one-point correlation map in Fig. 8.4c, although it is shown with opposite sign there. Even though the heating anomalies are displaced somewhat into the Southern Hemisphere, the strong subtropical jet in the Northern Hemisphere winter is quite responsive to the heating anomalies so that a nearly symmetric response is generated. The wave forced in the tropics propagates poleward and eastward along a great circle route and interacts with the mean wind and eddies along the way. This downstream wave effect has much significance for North America, as a warm event generates a low over the North Pacific, a high over western North America

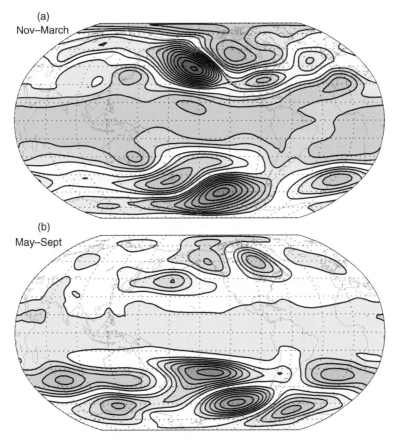

(a)
Nov–March

(b)
May–Sept

FIGURE 8.14 **Regression of monthly 500 hPa height anomalies with the Niño-3 Index of SST for November to March (a) and May to September (b) seasons.** Contour interval is 2 m, the zero contour is not shown and positive values are shaded red. Based on NOAA Twentieth Century Reanalysis and NOAA Extended SST data sets from 1871 to 2013.

and a low over the South East USA. These anomalies are associated with important weather impacts of El Niño, including warmth over western Canada and Alaska, stronger winter storms in Southern California and cooler and wetter conditions in the South East USA.

The response during the May–September season is very different in the Northern Hemisphere because the mean winds are very different and the tropical convection is moved off the equator during summer in association with the Asian Summer Monsoon (Fig. 5.4). The OLR anomalies are slightly smaller during the May to September season and their Northern Hemisphere impact is rather weak, but their influence on the Southern Hemisphere is still strong.

8.4 DECADAL VARIATIONS OF WEATHER AND CLIMATE

Although instrumental climate records are short for defining variations with time scales of decades or longer, it is clear that such variability must exist, perhaps mostly because of the ocean. Decadal time scales are also challenging because on this time scale the natural variability of the climate system becomes difficult to separate from the human influences that have become steadily stronger during the period of instrumental records. Methods are available that may be able to distinguish natural and anthropogenic changes in the observations, but whether they have worked or not is often a judgment call. Methods using climate models with and without human influences can be used to unambiguously separate natural variability from human influences, and these will be discussed in Chapter 13.

Global maps of SST can be used to investigate decadal variability. Global analyses of SST start as early as 1854, but the earlier records are based on extrapolation from relatively sparse data, so that the more recent years are more reliable. To search for natural variability on decadal time scales, we can first start by removing annual cycle and the global mean SST from monthly mean data. Removing the global mean from each month may have the effect of removing global mean trends associated with both the human influence and with globally coherent changes in the observing system. Therefore, we hope that what is left to investigate is the spatial structure of internal climate variability.

One objective way of searching for important patterns in SST variability is EOF analysis. For example, if we start with monthly anomalies of SST over the globe, then the first pattern that comes out is ENSO, and it explains about 17% of the global SST variance, much more than any other mode of variability. Figure 8.15 shows the SST anomaly structure associated with the first EOF of global SST along with its amplitude time series since 1880. The spatial structure is characterized by a large warm anomaly along the equator in the eastern Pacific Ocean and cold anomalies in mid-latitudes of each hemisphere. The extratropical anomalies are caused by the Rossby wave response to the tropical heating anomalies shown in Fig. 8.14. The low pressure centers at around 40° latitude at 150°W in each hemisphere drive cooling of the ocean surface there, mostly during the winter season. This remote forcing of the extratropical SST by waves generated by tropical heating anomalies associated with ENSO has been called the "atmospheric bridge" (Alexander et al., 2002).

The time series of the global ENSO mode (Fig. 8.15b) shows strong variability on time scales of 2–6 years consistent with the power spectrum of the Niño-3 index shown in Fig. 8.12. One can clearly see the strong warm ENSO (El Niño) events in 1998 and 1982, during which the anomalies in SST were about three times the magnitudes shown in Fig. 8.15a. Since

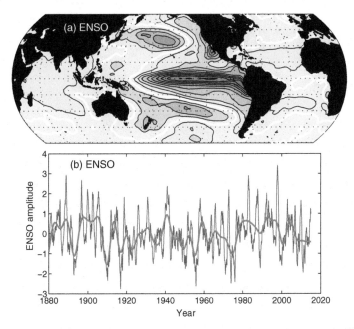

FIGURE 8.15 **The structure of El Niño.** (a) The spatial pattern of ENSO; the regression of monthly SST onto the first EOF of global SST from 1880 through January 2015. Contour interval is 0.1 K, positive anomalies are red and the zero contour is in white. (b) The time structure of ENSO; time series of the amplitude of the spatial pattern in units of standard deviations. The red line is smoothed to remove periods shorter than six years.

1998, the ENSO mode has been mostly negative, indicating colder than normal SST in the equatorial east Pacific. In addition to the 2–6 year time scale of the life cycle of ENSO, one can also see longer period variations in the ENSO mode. These are highlighted by the smooth line in Fig. 8.15b, which has filtered out variations with periods shorter than 6 years. Because ENSO is such a dominant mode of interannual variability, these decadal variations in ENSO can have a significant global influence.

The Pacific Decadal Oscillation (PDO) and the Atlantic Multidecadal Oscillation (AMO) are often discussed. To isolate these we can perform the EOF analysis as above for ENSO, but instead of including the global ocean, consider only the North Pacific and North Atlantic. The PDO was first defined as the dominant EOF of SST north of 20°N in the Pacific (Mantua et al., 1997). This structure and its time series are shown in Fig. 8.16. The PDO looks very much like ENSO, with a bit more emphasis on the extratropical signal and longer time scales. It can be argued that the PDO is nothing more than the low-frequency component of ENSO, and it can be seen that the time series of ENSO and PDO are correlated.

The AMO was first characterized as the anomalies in the areal mean temperature of the North Atlantic (Schlesinger and Ramankutty, 1994).

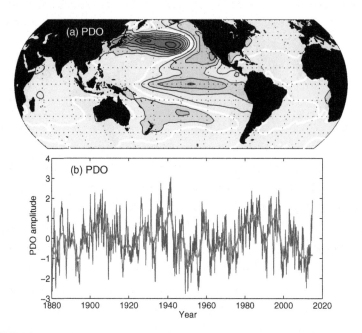

FIGURE 8.16 The structure of the PDO. (a) The spatial pattern of the PDO; the regression of monthly SST onto the first EOF of SST north of 20°N in the Pacific from 1880 through January 2015. Contour interval is 0.1 K, positive anomalies are red and the zero contour is in white. (b) The time structure of the PDO; time series of the amplitude of the spatial pattern in units of standard deviations.

Figure 8.17 shows the first EOF of monthly SST anomalies for the Atlantic Ocean north of 20°S. It consists of two centers of positive SST anomaly in the subtropics west of Africa and in the extratropics east of Newfoundland. The time series of this mode shows very strong decadal variations, with a very interesting transition from negative anomalies in the mid-1970s to the late 1990s toward positive anomalies in the late 1990s to 2015, the latest time shown in Fig. 8.17.

The time series of the amplitudes of the PDO and the AMO can be regressed against anomalies in the Northern Hemisphere 500 hPa height field in winter to show how these patterns relate to meteorology (Fig. 8.18). The 500 hPa height anomaly pattern associated with the PDO is very similar to the response to ENSO (Fig. 8.14) and to the one-point correlation map in Fig. 8.4c. It is the wintertime extratropical response to El Niño. The low pressure over the North Pacific and the associated stronger winds and colder temperatures drive the cool SST anomaly in the North Pacific.

The height anomaly associated with the AMO is a north–south dipole in the North Atlantic that resembles the one-point correlation map of low-frequency variability in Fig. 8.4d, often called the North Atlantic Oscillation. The high-pressure center to the north is associated with weakened

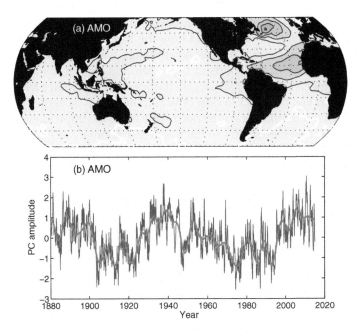

FIGURE 8.17   **The structure of the AMO.** (a) The spatial pattern of the AMO; the regression of monthly SST onto the first EOF of SST north of 20°S in the Atlantic Ocean from 1880 through January 2015. Contour interval is 0.1 K, positive anomalies are red and the zero contour is in white. (b) The time structure of the AMO; time series of the amplitude of the spatial pattern in units of standard deviations.

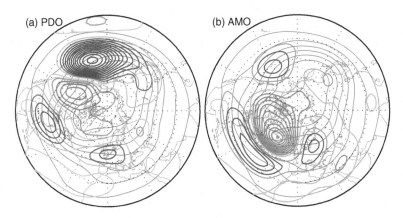

FIGURE 8.18   **Regression of the 500 hPa height anomalies from the Twentieth Century Reanalysis onto the time series of the (a) PDO and (b) AMO time series during the November through March season.** Contour interval is 3 m, positive contours are red. Thin magenta contours indicate the seasonal mean height field. Polar stereographic projection.

winds that generate a positive SST anomaly through reduced sensible heat exchange in winter. To understand how the low anomaly to the south in the AMO pattern (Fig. 8.17b) is associated with warm SST anomalies, one must remember that the winds in the subtropics during winter are easterly in the mean (Fig. 6.19). Therefore the westerly anomalies to the south of the low in Fig. 8.18b act to offset the climatological easterlies there and result in a net reduction of wind speed. This helps to suppress evaporative cooling there and contributes to the warm SST anomaly in Fig. 8.17a.

The strong variability in the PDO and AMO at time scales of a decade or longer make them very interesting, in part because they may help to define the amplitude or structure of temperature change on time scales similar to the onset of human-induced warming. They have decadal time scales presumably because they interact with the ocean, especially in the extratropics where both ocean wave propagation and ocean heat storage possess inherently long time scales. During the winter when the mixed layer is deep, weather anomalies can lay down ocean temperature changes that are buried under the shallow summer thermocline (Figs 7.5 and 7.6) and then reemerge the next winter, giving ocean memory from one winter to the next. It is hoped that with greater understanding, and better data and data assimilation systems, we may be able to use climate models to make useful forecasts of some aspects of the natural variability of weather or climate, perhaps as much as a decade into the future, by using the initial state of the coupled atmosphere–ocean–land system.

EXERCISES

1. Compare the longitude of the maximum in high-frequency eddy variance (T < 7 days) with the longitude of the maxima in intermediate frequency variance (7 < T < 30 days) in Fig. 8.2. Are your results consistent for the Atlantic and Pacific ocean regions?

2. The low-frequency variability in the Southern Hemisphere in Fig. 8.3 is centered upstream of the Palmer Peninsula that juts out from Antarctica toward South America. Can you think of any reasons why this might be the case? How do you think this might influence the variability of sea ice in the Ross, Amundsen, and Bellingshausen seas in that region?

3. The Northern Hemisphere has two wintertime storm tracks, one over the Pacific Ocean and one over the Atlantic Ocean, but the Southern Hemisphere has only one long one centered in the southern Indian Ocean. Why do you think this is the case?

4. Figures 8.4 and 8.5 indicate that the spatial scale of disturbances get larger as the time scale gets longer. Do you know why?

5. Figure 8.6 indicates that a north–south migration of the Southern Hemisphere extratropical jet explains more of the variance of zonal wind than the sharpening or broadening of the jet. Can you explain why?

6. Figure 8.7 shows that the OLR anomalies associated with the Madden–Julian oscillation are strongest in the tropical Indian and Western Pacific regions, but the wind anomalies extend to other longitudes and into the extratropics. Can you explain why?

7. Looking at Fig. 8.9 it is apparent that a strong temperature gradient exists from east to west below the surface. If warm temperature is associated with less dense water, what would you expect the pressure gradient to be 150 m below the surface? What direction would the water at 150 m deep move if there was no wind stress being applied at the surface?

8. Compare and contrast the OLR anomalies associated with ENSO in Fig. 8.13 with those associated with the MJO in Fig. 8.7. It has been argued that a big MJO event can be a precursor to a warm ENSO event. Can you explain why?

9. Theory says that tropical heating anomalies generate Rossby waves near the equator that follow great circle routes as they propagate into the extratropics during winter. Can you connect the centers of the anomalies in Fig. 8.14 and trace out the paths along which eddy energy must be propagating?

10. Assemble a list of arguments for and against the following statement. "The PDO is just the low-frequency, extratropical signature of ENSO, nothing more."

9

History and Evolution
of Earth's Climate

9.1 PAST IS PROLOGUE[1]

The seasons come and go in an orderly cycle, and plants, animals, and humans adapt to this regular rhythm. Though anomalies and extreme weather events occur, it is natural to think of climate as a constant influence to which life has adapted. In the grossest sense, this view of climate is correct. Life has existed on Earth for at least 3.5 billion years, and the climate has been hospitable enough over that great span of time for life to continue. If we look in more detail at the climates of the past by sifting through the evidence recorded by nature, we find that climate is not so invariable and passive as it may seem. Past variations in climate are especially interesting for the climatologist, since they provide clues to the inner workings of the climate system that are difficult to infer in any other way. If past variations in climate can be understood thoroughly, then our chances of anticipating how climate will evolve in the future are greatly increased.

The history of Earth's climate is long and varied, and humans did not experience most of this history, since we appeared only at the last instant of geologic time. Human ingenuity led to the definition and measurement of temperature, pressure, and other climatic variables only during the last several centuries. Climate variables that may be directly measured using modern instruments constitute what we call the instrumental record. In addition to the instrumental record, we have historical data that stretch back farther in time, but are less quantitative. Such records include grape harvests in France, wheat harvests in ancient Egypt, and many bits of climate information contained in written accounts. A large amount of the

---

[1]"We all were sea-swallowed, though some cast again,
And, by that destiny, to perform an act whereof what's past is prologue, what to come,
In yours and my discharge." Antonio in William Shakespeare's *The Tempest*, Act 2, Scene 1.

Global Physical Climatology. http://dx.doi.org/10.1016/B978-0-12-328531-7.00009-8

latter type of information, going back several thousand years, is contained in the ancient Chinese literature. Most of what we know of the deeper history of Earth's climate, before the invention of writing, has come from deducing climate variations from information left behind by various natural recording systems. These recording systems include physical, biological, and chemical information contained in geological information, lake and ocean sediments, terrestrial sediments, ice sheets, and tree rings. Information of this type, which we may call paleoclimatic data, can be used to derive time series of climatic information for many thousands of years into the past.

9.2 THE INSTRUMENTAL RECORD

Much of the early development and use of the thermometer took place in Florence in the mid-seventeenth century; however, the usefulness of early measurements was limited by the lack of a standard scale and calibration. In 1742, Anders Celsius invented the *Celsius temperature scale* (called the *centigrade scale* until 1948), but regular measurements of air temperature came into fashion considerably after its acceptance. Torricelli invented the barometer in 1644. It came more quickly into widespread use, because of the greater ease of its calibration and the perceived relationship between pressure and weather changes, which gives it a predictive capability. For a while, possessing a barometer was a status symbol. The first known reference to rain gauges is Indian literature from the fourth century BCE.

Temperature records as long as 150–200 years are available for only a few locations. Manley (1974) constructed a temperature record for central England going back to 1659. Other temperature time series include those for: Berlin, Germany beginning in 1700; de Bilt, Netherlands in 1706; Germantown, Pennsylvania in 1731; Milan, Italy in 1740; and Stockholm, Sweden in 1756. Sufficient measurements to define the hemispheric or global mean surface air temperature are available for only about the last century. Most of the variance of temperature is associated with seasonal and latitudinal variations, or weather. Any climatic trends are a small signal among much larger magnitude variations, particularly within the rather short period for which the instrumental record is available.

Because of concern over human-induced climate change, considerable effort has been devoted to developing estimates of global surface temperature based on the instrumental record and evaluating temporal trends in those estimates over the past century (Hansen et al., 2010; Morice et al., 2012). To obtain consistent records it is necessary to account for changes in instruments and their locations and surroundings. The land-based record has to be corrected for urbanization. Over the oceans, sea surface temperature measurements changed over time from buckets, to ship intake manifolds to drifting

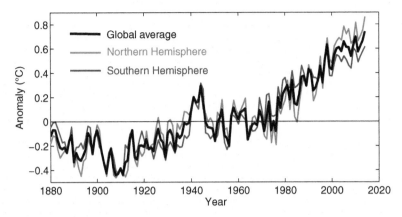

FIGURE 9.1   **Time series of annual mean temperature anomalies from 1880 to 2014 for the Global, Northern Hemisphere and Southern Hemisphere means.** *Data are from the NOAA global surface temperature anomaly data set (reference period 1901–2000).*

buoys, and each of these has distinct biases that need to be accounted for. The instrumental temperature record of global mean land and ocean surface temperature shows generally increasing temperatures since the 1880s, with a warming of about 0.3°C culminating in about 1940 (Fig. 9.1). After 1940, global mean surface temperature appears to have remained roughly constant until about 1970 when it began an increase of nearly 0.8°C until the present. After 1998, the temperature anomalies in the two hemispheres appear to diverge, with warming continuing in the Northern Hemisphere, but slowing in the Southern Hemisphere. It is believed that the change in the warming rate after about 1998 is related to the shifts in the Pacific Decadal Oscillation (PDO) and the Atlantic Multidecadal Oscillation (AMO) shown in Figs 8.16 and 8.17 that occurred at about that time after the large 1998 El Niño event. The change in the rate of warming in about 1998 could be attributed to changes in the natural modes of variability of the coupled ocean–atmosphere system, or could be part of the global-warming trend.

   The spatial structure of warming can also be estimated from instrumental records, but it is more uncertain because some regions of the globe were very sparsely measured until the advent of global satellite measurements in about 1979, and profiling drifting buoys in about 2000. Figure 9.2 shows maps of the standard deviation of annual mean temperature and its trend over the periods 1880–2014 and 1979–2014. Not every feature in Fig. 9.2 is necessarily real, because of changes in spatial sampling and instrumentation over time. Nonetheless, some features of the standard deviation and trend seem robust. The standard deviation of annual mean temperature is greatest over the high latitude land areas, and less over the oceans. The variation over land is contributed by both the summer and winter seasons, but a bit more by the winter season when the variations

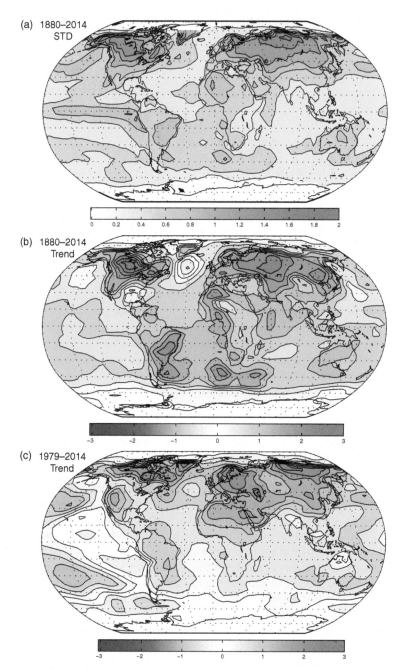

FIGURE 9.2   (a) Standard deviation of annual mean temperature, (b) linear trend over the periods from 1880 to 2014, and (c) linear trend over the period from 1979 to 2014 on the basis of the NOAA combined land and ocean surface temperature data set. Data in polar regions are not sufficient to compute these statistics. Contour interval is 0.2°C for standard deviation and 0.3°C for trend over time.

are larger. The large standard deviation in the equatorial eastern Pacific is associated with ENSO. The trends also show the largest warming over the large continents of the Northern Hemisphere, and generally smaller trends over the oceans. In the North Atlantic and the Far Southern Ocean, the trends are slightly negative, or at least much smaller than the global mean. These are regions where deep mixing occurs, so the effective heat capacity of the ocean is very large.

For contrast, one can consider the trends since 1979, when satellite measurements first started to give global coverage. For this period the standard deviation of annual means has a similar structure (not shown), but the trend structure is significantly different. The greatest warming is also over the northern continents for this shorter record. A stronger cooling over the Southern Ocean is evident, and the contrast between the warming in the Northern Hemisphere and Southern Ocean cooling is greater. In the Pacific Ocean, an east–west asymmetry appears with cooling in the east Pacific and warming in the west Pacific. An asymmetry is also suggested over mid-latitude North America, with greater warming in the Southwest than in the Northeast. Some part of the trend since 1979 is undoubtedly associated with natural variability of the climate system such as ocean–atmosphere coupling in the Pacific, since the changes in the Pacific look like a drift toward more La Niña conditions, but it is also possible that human-induced warming is affecting the natural modes of variability. Some of the dominant modes of natural variability like the PDO and the AMO were shown to have significant trends between 1980 and 2014 in Figs 8.16 and 8.17. A theorem of statistical mechanics says that change first appears in the natural modes of variability of a system.

9.3 THE HISTORICAL RECORD

The historical record consists of written or oral accounts of past events such as agricultural data (records of grape harvest dates in France, wheat yields in Egypt, cherry blossoming dates in Japan, etc.), weather diaries, diaries of a few intellectuals with an interest in weather (e.g., Aristotle, Tycho Brahe, and Johannes Kepler), ship logs, and ancient writings. Other more quantitative information might be about river levels and flooding. The flood levels for the Nile have been recorded for millennia, with especially careful records since 622 CE. Historical climate information is sometimes subjective, often sketchy, and limited to a few geographic areas. Often extreme events find their way into historical accounts, and these may not be representative of the climate of the time.

Although evaluation of the climatic information in ancient art and literature requires great care and skill, much useful and well-supported information can be deduced from historical accounts. Interesting examples include written and archeological evidence of cities that once flourished and then disappeared because of changes in the environment, including the climate. In some cases, it is clear that these changes were in part natural and in part human induced. An example is the city of Ephesus, an ancient Greek city in what is now Turkey. During the fourth century BCE, Ephesus was a major economic power and a center of learning, with an important port and a thriving local agriculture. An amphitheater in Ephesus had a seating capacity of 24,000. Before the city developed, the surrounding hills were covered with oak trees. As the city grew, these hills were cleared and given over to pasture and to cultivation of wheat. At the same time, the climate appears to have become more arid. The combination of changing land use and climate ultimately led to the demise of the city. Erosion led to the filling of the harbor with silt from the surrounding hills. By the ninth century the harbor was unusable, despite dredging the harbor and moving the port. Ephesus was left out of the economic life of the Mediterranean and fell into ruin.

Climatic information can also be contained in art. Rock paintings in the Sahara Desert dating from about 7000 years ago show pictures of the hippopotamus and grazing animals that can no longer exist there. This suggests that the central and southern Sahara were much wetter during the warm epoch following the end of the last glaciation. Paleoclimatic evidence, such as hippopotamus bones dating from the same period and evidence of much higher lake levels, confirms the inference derived from the paintings.

9.4 NATURAL RECORDING SYSTEMS: THE PALEOCLIMATIC RECORD

Another very important class of information on past climates exists for which the recording does not require a human attendant, and from which information for thousands or millions of years into the past can be obtained. This information is cataloged in various types of "natural recording" systems. These give continuous time histories that go back a few million years and are especially good for the last 100,000 years. The sources of paleoclimatic data are outlined in Table 9.1.

The data with the most accurate time chronologies come from analysis of annual tree rings (dendrochronology) or annual layers in lake sediments. The width and structure of the tree rings give some information on the climatic conditions when the tree ring was formed. By correlating tree-ring characteristics with contemporaneous instrumental data on temperature and precipitation, a transfer function can be developed to convert tree-ring characteristics into weather information. Once this transfer function is established and verified, tree-ring data can be used to estimate

TABLE 9.1 Characteristics of Some Paleoclimatic Data Sources

Proxy data source	Variable measured	Continuity of evidence	Potential geographical coverage	Period open to study (year BP)	Minimum sampling interval (year)	Usual dating accuracy (year)	Climatic inference
Layered ice cores	Oxygen isotope concentration, thickness (short cores)	Continuous	Antarctica, Greenland	10,000	1–10	±1–100	Temperature, accumulation
	Oxygen isotope concentration (long cores)	Continuous	Antarctica, Greenland	100,000+	Variable	Variable	Temperature
Tree rings	Ring-width anomaly, density, isotopic composition	Continuous	Midlatitude and high-latitude continents	1,000 (common) 8,000 (rare)	1	±1	Temperature, runoff, precipitation, soil moisture
Fossil pollen	Pollen-type concentration (varved core)	Continuous	Midlatitude continents	12,000	1–10	±10	Temperature, precipitation, soil moisture
	Pollen-type concentration (normal core)	Continuous	50°S to 70°N	12,000 (common) 200,000 (rare)	200	±5%	Temperature, precipitation, soil moisture
Mountain glaciers	Terminal positions	Episodic	45°S to 70°N	40,000	–	±5%	Extent of mountain glaciers
Ice sheets	Terminal positions	Episodic	Midlatitude to high latitudes	25,000 (common) 1,000,000 (rare)	—	Variable	Area of ice sheets
Ancient soils	Soil type	Episodic	Lower and mid-latitudes	1,000,000	200	±5%	Temperature, precipitation, drainage

(Continued)

TABLE 9.1 Characteristics of Some Paleoclimatic Data Sources (*cont.*)

Proxy data source	Variable measured	Continuity of evidence	Potential geographical coverage	Period open to study (year BP)	Minimum sampling interval (year)	Usual dating accuracy (year)	Climatic inference
Closed-basin lakes	Lake level	Episodic	Midlatitudes	50,000	1–100 (variable)	±5% ±1%	Evaporation, runoff, precipitation, temperature
Lake sediments	Varve thickness	Continuous	Midlatitudes	5,000	1	±5%	Temperature, precipitation
Ocean sediments (common deep-sea cores, 2–5 cm/1000 year)	Ash and sand accumulation rates	Continuous	Global ocean (outside red clay areas)	200,000	500+	±5%	Wind direction
	Fossil plankton composition	Continuous	Global ocean (outside red clay areas)	200,000	500+	±5%	Sea-surface temperature, surface salinity, sea-ice extent
	Isotopic composition of planktonic fossils; benthic fossils; mineralogic composition	Continuous	Global ocean (above $CaCO_3$ compensation level)	200,000	500+	±5%	Surface temperature, global ice volume; bottom temperature and bottom water flux; bottom water chemistry
Rare cores, >10 cm/1000 year	As mentioned previously	Continuous	Along continental margins	10,000+	20	±5%	As mentioned previously
Cores, <2 cm/1000 year	As mentioned previously	Continuous	Global ocean	1,000,000+	1000+	±5%	As mentioned previously
Marine shorelines	Coastal features, reef growth	Episodic	Stable coasts, oceanic islands	400,000	–	±5%	Sea level, ice volume

From National Research Council, 1975.

characteristics of the climate on an annual basis for thousands of years into the past, well before instrumental data became available. A reconstruction of precipitation in Iowa based on a 300-year-old tree correctly shows the "dust bowl" period of the 1930s and the lesser drought of the 1950s. It also shows four decades of dryness comparable to the 1930s that occurred prior to the beginning of the instrumental record. Lake sediments from Africa clearly show evidence of changing precipitation and the climate of the surrounding area can be inferred from pollen spores and other biological indicators in the sediment.

The paleoclimatic data source yielding the longest continuous record is the ocean sediment core. Ocean sediments are deposited over time, so that a core drilled into the sea bottom contains a time history of the environment at the time the layers in the core were formed. The time resolution attainable with sediment cores is determined by the sedimentation rate and the degree of stirring of the most recent sediment by bottom-dwelling animal life such as worms, a process called *bioturbation*. Ocean cores measure the history of the ocean ecology as laid down in the sediment. The sediment includes organic matter and the shells of tiny sea creatures (foraminifera, coccoliths, etc.). Each sea creature has a particular niche in the ecology of the ocean. The relative abundance of some species is related to sea surface temperature (SST), so that relative abundance can be used to estimate SST in the past. In addition, the relative abundance of oxygen isotopes in deep-sea sediment cores provides an indication of the global mass of water tied up in terrestrial ice. Because lighter isotopes (^{16}O) are more readily evaporated, they are more likely to be removed from the ocean and incorporated in ice sheets. Therefore, ice ages are marked by a relatively rich mixture of heavy isotopes (^{18}O) in ocean waters. These higher ^{18}O ratios are imprinted in the shells of sea creatures and are collected in sediments on the ocean bottom. The oxygen isotope abundance in ocean sediment can thus be used to estimate the global volume of ocean water that is tied up in continental ice sheets and mountain glaciers. The anomalies in ^{18}O are expressed as the fractional deviation from a reference oxygen isotope ratio, usually that of standard mean ocean water (9.1).

$$\delta^{18}O = \frac{\left(\dfrac{^{18}O}{^{16}O}\right)_{Sample} - \left(\dfrac{^{18}O}{^{16}O}\right)_{Standard}}{\left(\dfrac{^{18}O}{^{16}O}\right)_{Standard}} \times 1000 \tag{9.1}$$

Figure 9.3 shows a time history of $\delta^{18}O$ of the ocean as a function of time into the past, which has been pieced together from ocean sediment cores by Lisiecki and Raymo (2005). It shows that 5 million years ago the ocean

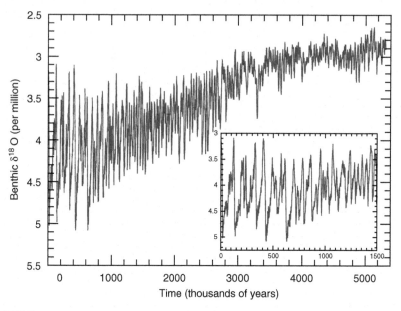

FIGURE 9.3   **Ocean oxygen isotope record from ocean cores for the past 5 million years, present to the left and past to the right.** Large values indicate large volumes of ice on land. Ordinate is plotted reversed so that up means a warm climate with less land ice and down means a glacial age. Inset shows the same record for the past 1.5 million years. *(Data from Lisiecki and Raymo 2005).*

was relatively light isotopically, indicating a relatively small amount of land ice, and the variations were small, indicating a relatively stable land ice volume. About 3 million years ago, the $\delta^{18}O$ began to increase and its variability also increased, indicating more ice and larger variations of ice volume. Until about a million years ago the variations had a characteristic time scale of about 40,000 years, but after that the variations grew even larger and took on a characteristic time scale of about 100,000 years (Fig. 9.4). The periodicities in this time series will be discussed further in relation to the orbital parameter theory of ice ages in Chapter 13.

Similar stratigraphic evidence can be collected in ice cores. Anomalies in stable isotopes such as $\delta^{18}O$ and δD [2] indicate temperature anomalies when the ice fell as snow. Water vapor containing heavier isotopes is more likely to condense. If the water in the ice is isotopically light, this means most of the water vapor containing the heavier isotope has condensed out before it reached the ice sheet and the air was anomalously cold. Since saturation vapor pressure depends on temperature,

[2] D = Deuterium = ^2H

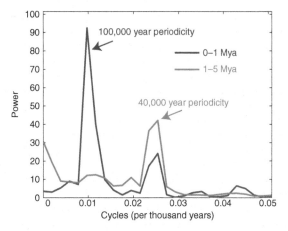

FIGURE 9.4   **Power spectra of the variance of the oxygen isotope record in Fig. 9.4 for the past million years (blue) and for the period from 1 million years ago to 5 million years ago (red).** Note how the strong 100,000-year variations in global ice volume appear only in the past million years, but that the strong variations at 40,000 years appear throughout the record.

the reduction in heavy isotope is related to the amount of cooling since the vapor evaporated. Most of the evaporation occurs in the subtropical oceans, where the temperature is less variable. Aerosols trapped in ice give evidence of past atmospheric dust loading and aerosol chemistry. Air bubbles in ice provide information on the gaseous composition of the atmosphere at the time the bubbles were formed. In the upper parts of ice sheets, annual layers can be identified. Deeper into the ice, and thus farther back in time, the ice becomes compacted and stretched out as the ice flows away from the accumulation regions toward the regions where the ice sheet melts or breaks off into icebergs at a marine boundary. Thus, the ability to resolve rapid changes decreases with depth and age in the core.

9.5 A BRIEF SURVEY OF EARTH'S CLIMATE HISTORY

9.5.1 Early Earth

Earth is believed to have formed nearly 5 billion years ago by the accretion of solid materials and gases that formed the solar nebula. The modern atmospheres of Earth, Mars, and Venus have less abundance of noble gases (e.g., argon, neon) than is typically present in stellar nebula, so it is assumed that any primordial atmosphere collected from the gases in the solar nebula was removed early in solar system history. The primordial atmosphere could have been removed during the collisions of large planetesimals with Earth or been swept away by the more intense solar wind of the early sun. Thus, the modern atmosphere is a secondary one, which has resulted from the release of gases that were mechanically or chemically trapped inside the solid Earth during its formation and were released slowly over time. The process of releasing gases from the interior of a planet may be called *outgassing*, and it continues on Earth today, most obviously in the form of volcanic eruptions. The gases released are primarily water vapor, carbon dioxide, and nitrogen.

We can hypothesize that immediately after the removal of their primordial atmospheres, each of the inner planets consisted of an essentially bare ball of rock, although some gases may have continued to be collected from space during and after the Sun's T-Tauri phase. Assuming that the release of heat from within the planets was negligible, their surface temperatures would have equilibrated at the emission temperature of a sphere with the albedo of bare rock and with a solar constant appropriate to the distance of the respective planet from the Sun. Thus Venus, because of its greater proximity to the Sun, started out at a higher temperature than Earth. Being farther from the Sun, Mars would have started at a lower temperature than Earth. With time, the temperatures of the three planets would gradually increase, because of the greenhouse effect of the water vapor and carbon dioxide that would steadily build up in their atmospheres through outgassing.

On Venus, the surface temperature stayed well above the condensation point for water during its evolution, so that all of the water stayed in the atmosphere. Eventually much of the hydrogen in the atmosphere escaped to space, and a thick atmosphere of primarily carbon dioxide remains. This thick atmosphere gives a very high surface temperature of about 700 K. The process that led to these conditions on Venus is often termed the *runaway greenhouse effect*, but it is also possible that oceans did not form because Venus simply received less water during its formation.

On Mars, it has been speculated that the temperature mostly stayed below the freezing point for water so that the greenhouse effect may never have significantly influenced the surface temperatures there. Because of its relatively low temperature, the partial pressure of water reaches the freezing point before substantial amounts of water vapor can accumulate in the atmosphere. Therefore, the greenhouse effect on Mars may never have really gotten established, or been established only for a time, and most of the outgassed water could have frozen on the surface. A considerable amount of frozen carbon dioxide may also exist on Mars. An alternative to this view of Mars is suggested by some surface terrain with erosional features that appear to have been followed by a running fluid, presumably water, very early in Mars' history.

The distance of Earth from the Sun is such that outgassed water vapor condensed into oceans and the surface temperature remained near the triple point of water, where liquid water, water vapor, and ice can exist simultaneously. The collection of outgassed water vapor in the oceans stabilized the climate and provided the environment that is appropriate for the development of life as we know it. Once the condensation point was reached on Earth, any additional water vapor that was outgassed went into the oceans and the infrared opacity of the atmosphere stopped increasing. The carbon dioxide was dissolved into the ocean and eventually reached equilibrium with carbonate rocks. At this point, the atmosphere was composed mostly of molecular nitrogen. Within about a billion years after the formation of Earth, life developed, leading to green plant photosynthesis, which produces molecular oxygen in the atmosphere. With the development of an oxygen-rich atmosphere came the stratospheric ozone layer, which by protecting the surface from harmful ultraviolet-B radiation, allowed life to emerge from the water and occupy the land surface.

Thus Earth stayed cool enough to avoid the runaway greenhouse effect that occurred on Venus, largely because of its favorable distance from the Sun, which allowed the oceans to form. Life developed quickly in the conditions provided by early Earth, and temperatures favorable for the life forms we are most familiar with ($0 < T < 42°C$) have been maintained at least in some places on Earth, ever since. Theories for the life cycles of stars suggest that, early in Earth history, the solar constant would have been as much as 30% less than its current value. This suggests that some compensating changes in atmospheric radiative properties may have occurred over time to keep the surface temperature in a favorable range for life to progress.

9.5.2 The Last Billion Years

The continents have been drifting as part of lithospheric plates for the last several billion years. Geologists have attempted to reconstruct the positions of the continents for the past 600 million years or so, but little is known about the positions of the continents before that. The continents reached their modern positions about 14 million years ago. Little is known about climate back farther than about a billion years, except for fossil evidence that life existed then and that liquid water was present on the surface (Fig. 9.5).

The climate was cold enough for large ice sheets to form during at least three periods over the last billion years. Evidence indicates ice sheets existed several times during the past billion years, with more extensive evidence for glaciation during the late Paleozoic 300 Mya[3], and fairly detailed

[3]Mya = millions of years ago

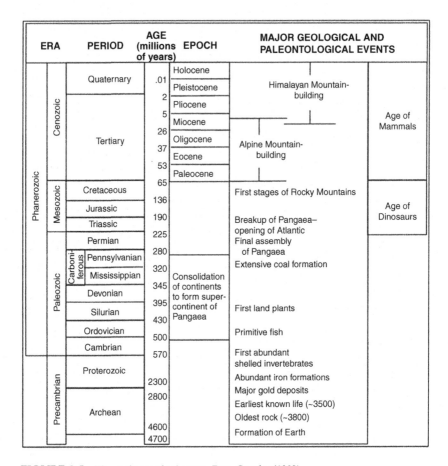

FIGURE 9.5   **The major geologic ages.** *From Crowley (1983).*

documentation for the late Cenozoic glaciation of the last several million years or so, which we are still experiencing.

Evidence suggests that during the Neoproterozoic glaciations about 720 Mya and 635 Mya continental ice reached the sea at low latitudes and the ocean may have been entirely frozen at the surface, which may be termed "Snowball Earth" conditions. At that time the total solar irradiance was about 95% of its current value and greenhouse gases would have needed to be much higher than today to maintain today's climate (Pierrehumbert et al., 2011). The continents were arranged in a single large landmass centered near the equator, which was slowly breaking apart. It is hypothesized that a cool episode was initiated, perhaps by efficient removal of carbon dioxide by weathering from a tropical wet continent, which then led to a full glaciation of Earth through ice–albedo feedback as the polar sea ice expanded to the equator.

On long time scales of thousands to millions of years, weathering and volcanism regulate the CO_2 content of the atmosphere. CO_2 in the atmosphere becomes dissolved in rain to form carbonic acid. When this carbonic acid falls on fresh rock, the rock is dissolved and the carbon flows in streams to the ocean where it is buried as carbonate rock. This carbon is returned to CO_2 when the rock is exposed to heat and pressure within the solid Earth to form metamorphic minerals. The CO_2 is then returned to the atmosphere as outgassing from volcanic activity. Weathering requires liquid water containing dissolved CO_2, so when the surface climate is frozen, weathering is suppressed and outgassed CO_2 can build up in the atmosphere. If the surface is warm and wet, the atmosphere has high concentrations of CO_2, and rock is exposed, weathering proceeds very rapidly. The carbon cycle can thus act as a stabilizing mechanism that eventually drives very cold or very hot climates back toward a more moderate climate.

Other evidence of a frozen ocean is found in iron formations indicating an anoxic ocean, and in carbon isotopic evidence. Carbon that is outgassed from the solid Earth has $\delta^{13}C \approx -6‰$. Photosynthetic life much prefers the lighter ^{12}C isotope of carbon so that organic carbon from marine photosynthetic life is about $-25‰$ lighter than the carbon pool from which it is made. Sequestration of lighter isotopes in organic carbon thus increase the $\delta^{13}C$ of the ocean water. During the Neoproterozoic glaciations the $\delta^{13}C$ in sediments approached the value for outgassed carbon, indicating suppression of photosynthetic life. On top of the glacial layers are thick layers of carbonate rock with $\delta^{13}C \approx -6‰$, indicating the deposition by chemical processes of large amounts of CO_2.

It is estimated that the Earth remained largely ice covered for about 10 million years. The relatively simple life forms present then could have survived in small refugia, such as near geothermal activity, or where wind and sun kept the ocean ice-free. During the snow ball period, outgassing of carbon would continue, while removal of CO_2 by weathering would be suppressed by the cold, dry climate. CO_2 would thus increase in the atmosphere until its greenhouse effect became large enough to overcome the albedo effect of global ice cover. That might require 300 times as much CO_2 as is currently in the atmosphere. Once the ice began to melt, ice–albedo feedback would result in a rapid warming and melting of all the surface ice. At that point, the planet would be very warm, a Hothouse Climate with very high CO_2. Weathering would be very rapid due to the high CO_2 content in a warm climate, and the cap carbonate layer over the glacial sediments would be formed rapidly due to enhanced weathering. This layer has the $\delta^{13}C$ of outgassed CO_2, indicating that the source of the cap carbonate layer was all the CO_2 that had outgassed during the snowball phase.

The late Paleozoic glaciations about 300 Mya appear to have occurred in southern Africa, South America, and Australia, at a time when all these continents were bunched together into a single, large, land mass in high southern latitudes. The glaciated areas were attached to Antarctica and were not far from the South Pole. About 300 million years ago, the continents moved apart and began to drift slowly toward their present positions.

The most studied of the nonglacial climates that occurred between these major glacial ages is the most recent one, the middle Cretaceous, spanning the period from about 120 to 90 million years ago. The continents were in a configuration very different from that seen today. North America and Europe were close together, as were South America and Africa, so that the Atlantic Ocean did not yet exist. Africa had not yet joined Europe, India was a large island in the mid-latitudes of the Southern Hemisphere, and a shallow tropical sea, called the Tethys Ocean, extended from Central America to Indochina (Fig. 9.6). Australia was in middle to high latitudes and still attached to Antarctica. Very little land ice existed during this period, so that the sea level was about 100 m higher than now. This greater mass of ocean water flooded about 20% of the continental areas that are now above water, including large portions of western Europe, northern Africa, and North America.

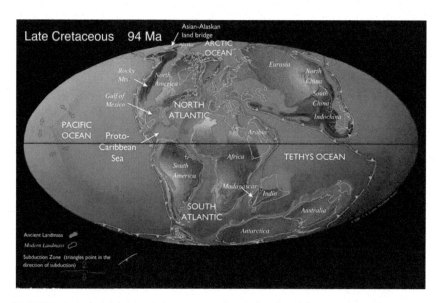

FIGURE 9.6 **Paleomap of the continental positions and sea level during the Creta-ceous period 94 Mya.** *Late Cretaceous paleogeographic map by C.R. Scotese, PALEOMAP Project (Scotese, 2001).*

Abundant fossil evidence exists to suggest that the middle Cretaceous was substantially warmer than the present. Plant habitats moved up to 15° of latitude poleward of their current positions, as did the ranges of many animals. Sedimentary evidence also suggests that coal and other deposits indicating warm, moist climate formed during this period in lands that were then above the Arctic Circle, and dinosaurs ranged there also. Isotopic evidence taken from bottom-dwelling organisms suggests that the deep ocean temperature was between 15°C and 20°C. Such warm temperatures in the deep ocean are inconsistent with the presence of much marine ice in high latitudes. Annual mean Arctic temperature is estimated to have been 10–15°C, compared to current temperatures of around −11°C, a difference of around 25°C.

Unusually large amounts of carbon in the form of coal and oil deposits were laid down during the Cretaceous. Many of the major oil deposits were formed during this period. This high rate of coal and oil formation indicates that the cycling of carbon was very different than it is now. A change in the cycling of carbon might logically be associated with the apparently different oceanic circulation, deep ocean temperatures, and the great expanse of relatively warm, shallow seas. The warmth of this period might have been associated with enhanced levels of carbon dioxide in the atmosphere, which might have been up to five times higher during the Cretaceous. Even with these much higher levels of CO_2, climate models have difficulty keeping temperatures warm enough in the centers of northern continents, where paleoclimatic indicators suggest that freezing did not occur in winter.

The Cretaceous ended with a bang about 65 million years ago. About 75% of the total number of living species became extinct, some of them very abruptly. The dinosaurs were among several species that disappeared entirely, and the ascent of mammals as the dominant group of large vertebrates followed and persists today. The so-called *K–T boundary*, marking the end of the Cretaceous (K) and the beginning of the Tertiary (T) periods, is identified in sedimentary records by a well-defined layer of clay deposits. This layer contains anomalously large amounts of iridium. The iridium content of Earth's crust is lower than the amount in the stuff from which the solar system was formed, because much of Earth's iridium was carried to the core with molten iron during Earth's formative stages. Comets and meteors retain their cosmic abundance of iridium, and the clay layer with its high iridium content could have come from a comet or meteor with a diameter of about 10 km. This iridium anomaly at the K–T boundary can be found over a large portion of the globe and is a rare event in Earth history. The evidence is strong that a large meteorite or comet collided with Earth about 65 million years ago, and evidence of a likely impact site has been found near the Yucatan Peninsula.

The impact of such a large bolide with Earth would be a spectacular event with very serious environmental consequences. The shock would heat the atmosphere in the vicinity of the impact, which would produce large amounts of nitrogen oxides. In the stratosphere, nitrogen oxides would lead to a loss of ozone. The nitrogen oxides would likely leave the atmosphere as highly acidic rains. The impact of the projectile and its penetration of Earth's crust would cast fine particles high into the atmosphere and even into low orbits, where they might persist for several weeks or months. This pall of dust could block out the sunlight needed by photosynthetic organisms and would lead to a cooling of the surface after the heat of the impact had dissipated. All of these effects – shock heating, fires, acid rain, ozone loss, dark and cold – would be stressful to many species and may have been the cause of the rapid species extinctions at the end of the Cretaceous.

9.5.3 The Last 50 Million Years

Changes over the last 50 million years seem to be related mostly to the movement of the continents away from Antarctica. During this period, South America and Australia moved northward from the edge of Antarctica to their present positions as a result of sea floor spreading and movement of the continental plates. As the Drake Passage between South America and Antarctica was opened, the Southern Ocean circulation became circumpolar, and the temperature of the surface and deep waters gradually cooled by more than 10°C. Glaciers developed over this period on Antarctica, and the east Antarctic ice sheet formed about 14 million years ago.

Pollen from ocean cores shows that cool temperate forests existed on the Antarctic continent until 20 million years ago. Ice volume over Antarctica increased to reach its present value about 5 million years ago, and the current Northern Hemisphere polar ice sheets first appeared about 3 million years ago. The overall trend in the last 50–100 million years was for the climate to cool from the warm climate of the middle Cretaceous to our present Quaternary ice age.

9.5.4 The Last 2 Million Years

The last 1.8 million years constitute the Quaternary period, the most recent geologic age and the one during which *Homo sapiens* developed. This period is characterized by the presence of a large amount of land ice, which varied from the amount we have today to much larger amounts during periods of glacier advance. The advance and retreat of this land ice can be inferred from glacial deposits and from the ratios of oxygen isotopes in ocean sediment cores (Fig. 9.3). The last 700,000 years were marked by wide swings that indicate a large shift in the amount of land ice present. These swings had a characteristic interval of about 100,000 years between succeeding periods of maximum glaciation. Glaciers advanced and retreated in both hemispheres simultaneously. Prior to about 700,000 years ago the swings were more frequent and less extreme. The dominant period of the glacial–interglacial swings in the early part of the record is about 40,000 years (Fig. 9.4). We will see in Chapter 12 that some parameters of Earth's orbit vary with these periods.

9.5.5 The Last 150,000 Years

About 125,000 years ago the land ice on Earth reached a minimum amount, comparable to today's interglacial conditions. In the intervening period, the land ice gradually increased, while undergoing some minor oscillations, until the ice volume reached a maximum about 20,000 years ago. The most recent and one of the more abrupt swings in global ice volume occurred in the last 20,000 years, between the last glacial maximum and the current interglacial period (Fig. 9.3). The extent of ice cover at the last glacial maximum can be inferred from a variety of paleoclimatic evidence, including the physical evidence left behind by the huge ice sheets and the $\delta^{18}O$ in ocean cores. A synthesis of this information indicates a giant ice sheet covering much of northern North America, and another large ice sheet in northern Eurasia, which were 3–4 km thick (Fig. 9.7). The weight of these ice sheets was so great that the supporting crust was depressed by nearly 1 km. The plastic deformation of the crust under these massive ice sheets may have been one reason for the rapid removal of ice at the end of each major glacial maximum during the past 700,000 years. As the crust yields under the ice, the altitude of the ice is lowered, thus bringing the ice surface to higher ambient air temperature. Also, glacial lakes may form in the depression made in the crust by the ice sheet, and seawater may flow into the depression and assist in the removal and melting of ice. The crust near the locations of the maximum thickness of the major ice sheets is still rebounding from the removal of the ice sheets. Because so much water was tied up in these large continental ice sheets, sea level was about 120 m lower 20,000 years ago than at present. The $\delta^{18}O$ records and modeling studies suggest that this was about the maximum amount of water that could be moved to continental ice sheets, given the continental positions and climate regime of the late Quaternary. Because sea level was so much lower, the coastlines of the continents were changed. The Bering Sea became a land bridge so that early Americans could walk from Siberia to North America, and Southeast Asia extended to Indonesia (Fig. 9.7).

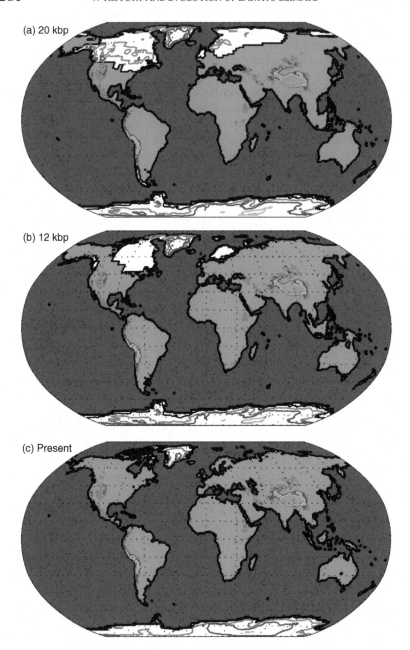

FIGURE 9.7   **Maps of ice sheets and elevation above sea level for (a) 20 kbp, (b) 12 kbp and (c) present.** White indicates perennial ice sheet positions. Red, blue, and green contours represent 2000, 3000, and 4000 m elevations, respectively. *Data from Peltier (1994), archived by the NOAA/NGDC Paleoclimatology Program, Boulder, Co.*

The high-latitude oceans also experienced major changes during the last glacial maximum. The Gulf Stream turned more sharply eastward at about 45°N and warm currents did not extend northward into the Greenland and Norwegian Seas as at present. It is likely that during winter the sea ice extended equatorward of 50°N, and snow cover on land extended ever further equatorward, so that a vast area of ice-covered surface extended from about 45°N to the pole in the Northern Hemisphere during winter. A simultaneous glacier advance was experienced in the Southern Hemisphere, where sea ice around Antarctica was greatly expanded, and mountain glaciers covered parts of Australia, Africa, and South America. During this time, tropical continental regions appear to have been drier than at present, but snowlines on tropical mountains dropped about 1000 m. Mid-latitude continental regions near the ice sheets were also drier for the most part, and windblown dust deposits in these areas indicate dry windy conditions at the equatorward margins of the great ice sheets.

Careful analysis of ice cores from the Greenland and Antarctic ice sheets can reveal much about changes in the climate system over the last major glacial–interglacial cycle. The composition of the atmosphere in the past can be inferred from air bubbles trapped in an ice core. Figure 9.8 shows 800,000-year records from an Antarctic ice core of temperature anomalies estimated from Deuterium abundances in the ice, atmospheric carbon dioxide (CO_2); and methane (CH_4) concentrations measured from bubbles trapped in the ice, and dust in the ice core. The temperature, CO_2, and CH_4 records closely track each other, and the dust is greatest during the coldest periods. About 20,000 years ago, during the most recent glacial maximum, CO_2 concentrations were about 190 ppmv, compared to about 275 ppmv during the warm periods, and CH_4 was about 350 ppbv compared to 700 ppbv during the warm periods. These large changes of atmospheric carbon dioxide and methane between glacial and nonglacial times indicate major changes in the cycling of carbon between ocean, atmosphere, and land, which accompanied the growth and decay of ice sheets. In Chapter 13, we will see how CO_2, CH_4, and other greenhouse gases have increased dramatically in the past century due to human activities, after remaining stable for thousands of years.

Ice cores also contain information on past variations in the amount and chemical composition of aerosols deposited in high latitudes. Non-sea-salt-sulfate and terrestrial calcium are much higher during glacial than interglacial ages. After correcting for the effects of changing accumulation rates, these changes indicate that sulfate concentration in the atmosphere was 20–40% higher during the glacial maxima. The increased atmospheric sulfate is thought to be related to increased production of sulfur-bearing gases by life in the ocean, most probably dimethyl sulfide gas.

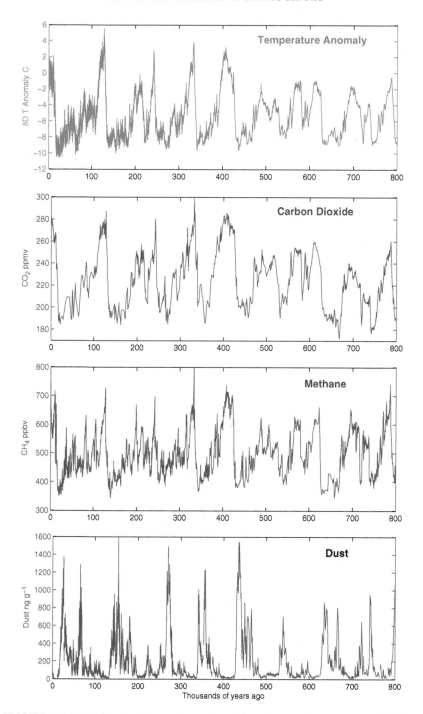

FIGURE 9.8 Estimates of temperature, CO_2, CH_4, and dust from ice cores drilled at Dome C in Antarctica plotted as functions of time before present for the past 800,000 years. *Data from NOAA/NCDC Paleoclimatology Program, Boulder CO, USA; Jouzel et al. (2007), Lambert et al. (2008), Loulergue et al. (2008), and Luthi et al. (2008).*

Variations in atmospheric carbon dioxide of the magnitude observed must be related to changes in the gross biological productivity of the oceans through photosynthesis. An increase in productivity during the ice age would reduce the partial pressure of CO_2 in the water and thereby reduce the CO_2 content of the atmosphere, which is constrained by the ocean concentration on these time scales. If it is assumed that the sulfate is produced biogenetically through dimethyl sulfide (DMS) emissions by phytoplankton, then the sulfate increase is consistent with an increase of gross ocean productivity during the ice age, assuming that the rate at which organisms in the ocean produce DMS rises and falls with gross productivity. Greater production of sulfate may also have increased the number of cloud condensation nuclei, which may have increased the reflection of radiation by clouds and aided in maintaining the cold climate. Comparison of CO_2 from ice cores with temperature estimates from many sources has indicated that the CO_2 concentration lags behind the temperature, with particularly clear evidence during the deglaciation following the Last Glacial Maximum (Clark et al., 2012). This supports the idea that during the Pleistocene the timing of glacial–interglacial transitions were set by orbital parameter variations that changed the temperature and ice volume, and that atmospheric CO_2 responded to the temperature change and thus forms a positive feedback on climate changes arising from other causes.

As discussed in Chapter 7, at the present time much of the intermediate deep water in the global ocean is formed in the North Atlantic Ocean. Because it is formed by cooling water that has spent some time near the surface where nutrients are efficiently consumed, north Atlantic deep water (NADW) is generally deficient in nutrients. In contrast, the source of deep water formed around Antarctica is water that has spent considerable time at intermediate depths collecting nutrients and then rises to the surface in the Southern Ocean. Therefore, Antarctic deep water is comparatively nutrient rich. Because of the different nature of North Atlantic and Antarctic deep water formation, a substantial gradient in deep-water nutrient content exists between the North and South Atlantic and between the Atlantic and the Pacific oceans.

Evidence from ocean sediment cores suggests that the contribution of the North Atlantic to deep-water formation was greatly diminished during the glacial maximum of about 20,000 years ago. Associated with the low nutrient content of NADW are relatively high levels of $\delta^{13}C$, a measure of the ratio of ^{13}C to ^{12}C, and low levels of the cadmium to calcium ratio, Cd/Ca. A time history of past levels of $\delta^{13}C$ and Cd/Ca in deep waters can be obtained from analysis of sediment cores. Records of $\delta^{13}C$ and Cd/Ca from ocean sediments indicate that the rate of deep-water formation in the North Atlantic Ocean was greatly reduced during the last glacial maximum, and nutrients in the deep ocean were more uniformly mixed.

The formation of deep water in the North Atlantic and the associated heat and water transports that were suppressed during the last glacial maximum increased again about 14,000 years ago at a time when the $\delta^{18}O$ in ocean cores indicates that the land ice volume began to decrease rapidly. The proxy data indicate that deep-water formation in the north Atlantic ceased again for a period between about 13,000 and 11,000 years ago. During the same interval, fossil data indicate that the climate in Europe returned nearly to glacial conditions after having warmed considerably during the period from 15,000 to 13,000 years ago. The cold event is known as the *Younger Dryas*, because pollen data indicate that forests that had recently developed in Europe during the aborted warming following the ice age were suddenly replaced again by arctic shrubs, herbs, and grasses, including the herbaceous plant *Dryas octopetella*. It ended abruptly about 11,000 years ago with the return to the interglacial conditions of today. Although the event is strongest in the North Atlantic region, it is seen in many fossil records around the world.

The Younger Dryas event is very evident in time series of $\delta^{18}O$ in ice cores from Greenland (Fig. 9.9). $\delta^{18}O$ in ice cores is a proxy for the air temperature in the vicinity of the ice sheet. When the temperature near the ice sheet is low, it is more likely that the heavier isotope of oxygen would have condensed out before reaching the ice sheet, because of its lower saturation vapor pressure, so lower $\delta^{18}O$ in glacial ice is

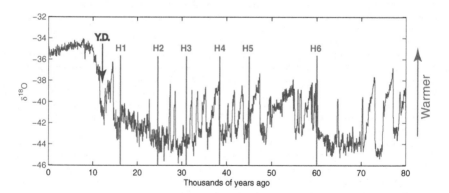

FIGURE 9.9 $\delta^{18}O$ from the NGRIP ice core from Greenland for the past 80,000 years (Andersen et al., 2004). The Younger Dryas event is marked with a blue arrow. The six Heinrich events of exceptional ice rafted debris in Atlantic Ocean sediment cores are shown with red lines (Hemming, 2004).

associated with colder temperatures when the snow fell. Greenland ice cores indicate a local cooling of about 6°C during the Younger Dryas cold event.

Figure 9.9 shows other interesting features of Greenland temperature during the ice ages prior to the current interglacial. Several events similar to the warm event prior to the Younger Dryas cooling occur in the record. A rapid warming of about half way to interglacial conditions occurs, followed by a period of slower cooling back to near glacial maximum temperatures. These events are called Dansgaard–Oeschger (D–O) events after their discoverers, Willi Dansgaard and Hans Oeschger, and they occur about 25 times during the previous glacial period. Particularly strong and long-lasting events are seen at about 38,000 and 48,000 years ago. Temperature changes over Greenland during D–O events were 9–16°C. Signals corresponding to D–O events in Greenland are also recorded in Antarctic ice cores, but with cooling in Antarctica corresponding to the warm events in Greenland, indicating a bipolar seesaw (Barbante et al., 2006). The cooling in Antarctica is much more gradual than the warming in Greenland and lags behind by about 200 years, suggesting that the two hemispheres are connected through the circulation of the Atlantic Ocean.

Also shown in Fig. 9.9 are six Heinrich events. Heinrich events are identified in ocean sediments from the North Atlantic, and are characterized by large amounts of ice-rafted debris. Large amounts of sand from continental rocks indicate that a large number of icebergs from continental glaciers had floated out into the Atlantic and melted, dropping sand and stones into the sediment. Heinrich events tend to occur during times when Greenland is cold, but their relation to D–O events is unclear.

It has been hypothesized that the brief shutdown of the deep thermohaline circulation of the North Atlantic during the Younger Dryas event was associated with the melting of the North American Laurentide ice sheet. The ice sheet melted most rapidly from its southern flank at first and the majority of this melted water flowed down the Mississippi River draining into the Gulf of Mexico. Several great meltwater lakes formed in the depression of Earth's surface left by the retreating ice sheet, and one of these occurred in what is now southern Manitoba. This paleo-lake has been named Lake Agassiz after the geologist, Louis Agassiz, who in 1837 was an early and ardent proponent of the idea that Earth had undergone an ice age. As the ice sheet retreated farther north, a channel was opened that allowed the meltwater lake to drain eastward down the St. Lawrence River to the North Atlantic Ocean. This diversion is dated at about 12,000 years ago from land evidence and by $\delta^{18}O$ records in sediment cores from the Gulf of Mexico, which went from isotopically light to heavy as the low $\delta^{18}O$ meltwater was suddenly diverted away (Fig. 9.10e). The supply of freshwater to the North Atlantic via the St. Lawrence reduced the salinity of the surface ocean waters. Since high salinities are critical to attaining the densities required to form deep water, the supply of freshwater was sufficient to cut off the thermohaline circulation of the North Atlantic. With the thermohaline circulation went its associated heat transport, resulting in large local cooling of the climate. The thermohaline circulation restarted about 11,000 years ago when the meltwater was again directed down the Mississippi, and has continued to operate as the climate warmed to today's interglacial conditions.

The Younger Dryas cool period occurred after the aborted warming of the Bölling-Alleröd from 14.7 to 12.7 kya[4], which followed the Oldest Dryas cold period from 18–15 kya. These temperature variations that are mostly referenced to North Atlantic climate have been shown to have global connections. Figure 9.10 shows a set of climatic indicators from 25 kya to present. Figure 9.10a,b shows temperature and ice accumulation records from Greenland indicating a warming to near present conditions about 14.7 kya (vertical line), which then dropped back to near full glacial conditions during the Younger Dryas event (shading). Figure 9.10c shows the record of temperature over Antarctica, which started warming about 18 kya, then cooled again after Greenland warmed about 14.7 kya, then warmed through the entire Younger Dryas event. These events are characteristic of what is called the *bipolar seesaw* between Greenland and Antarctica, which are connected through the deep circulation of the Atlantic Ocean. During the glacial maximum the Atlantic Meridional Overturning Circulation (AMOC) was suppressed and more of the deep water was formed by the enhanced sea ice extent in the Southern Ocean. As the Earth started to warm around 18 kya, the AMOC remained suppressed until it started again abruptly about 14.7 kya and the north then warmed rapidly. The SST in the Caribbean Sea also increased about 3°C at this time, as is recorded in the sediments of the Cariaco Basin (Fig. 9.10d). Then, however, increased freshwater flooded the Gulf of Mexico in several pulses about 17 and 13 kya, with the second of these marking the beginning of the Younger Dryas and a return to near glacial temperatures. It is hypothesized that the flood of freshwater from the melting ice sheets (Fig. 9.10e) suppressed the AMOC during the Younger Dryas, which then ended very abruptly with a change to near modern temperatures about 12.7 kya. Again the warming in the North Atlantic lagged behind that in the Antarctic, which changed more gradually, since the Antarctic is less dependent on the northward extension of the AMOC than is Greenland and the North Atlantic Ocean. The pattern of changes around the Atlantic

[4]kya = thousands of years ago

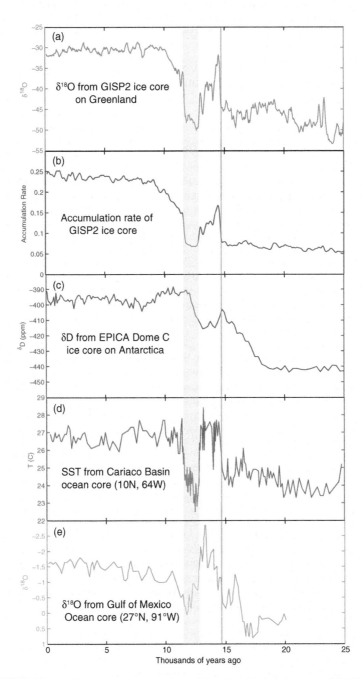

FIGURE 9.10 **Five panels of the deglaciation showing the global nature of the Younger Dryas event (shading) and the Bölling warm event (vertical line) in time series from 25,000 years ago to the present.** (a) $\delta^{18}O$ from GISP2 Greenland ice core indicates temperature near the top of Greenland. (b) Accumulation rate of ice for the GISP2 ice core (Alley, 2000). (c) δD from the EPICA-C ice core indicating temperature near the top of Antarctica (Augustin et al., 2004). (d) SST from an ocean sediment core near Venezuela (Lea et al., 2003). (e) $\delta^{18}O$ from an ocean sediment core in the Gulf of Mexico, indicating freshwater flux from North America and global land ice amount (Leventer et al., 1982).

Ocean indicate that the AMOC started and stopped during the deglaciation following the last glacial maximum.

9.5.6 The Last 10,000 Years

The global climate warmed following the last glacial maximum. Melting of the great ice sheets took about 7000 years from about 14 kya to about 7 kya, with considerable variation in the dates of final melting for different ice sheets. The climate has been comparatively steady for the past 8000 years or so (Figs 9.9 and 9.10). The early Holocene from about 10,000–5,500 years ago was a time when Northern Hemisphere summer climates were warmer than today's. This period was also characterized by remarkable changes in the hydrology of the monsoonal climates of Africa and Asia. Evidence from closed-basin lakes indicates that the period from about 9,000–6,000 years ago was the wettest in northern Africa in the last 25,000 years. It was during this interval that the hippopotamus, the crocodile, and a variety of hoofed animals occupied regions in the Sahara and in Saudi Arabia that are now some of the most arid on Earth. Sediment cores from the Arabian Sea indicate that monsoonal winds were stronger during the early Holocene than at present, and sediments from the Indus and Ganges rivers indicate that precipitation over India was greater than now. All of this evidence is suggestive of stronger summer monsoons during the early Holocene, which has been related to changes in Earth's orbital parameters, as will be discussed in Chapter 12.

Minor advances of mountain glaciers culminated about 5300, 2800, and 150–600 years ago. The last of these falls within the Little Ice Age, circa 1250–1850 CE. Between these cold periods there were relatively warm periods, the best known being around 900–1200 CE and during the present century. During the former period, the Norsemen populated Iceland, Greenland, and North America. The Norse Greenland colonies were abandoned at the onset of the Little Ice Age. Fields in coastal Greenland that had grown crops became perennially frozen. This seemingly drastic change, compared to what has occurred elsewhere, may be because climate changes appear to have larger amplitudes at high latitudes, or because of the sensitivity of Greenland climate to the strength of the deep circulation in the Atlantic Ocean.

9.6 USES OF PALEOCLIMATIC DATA

Paleoclimatic data are very useful in developing a scientific understanding of climate change that will be essential to anticipating the nature of future climate changes associated with human activities and natural processes. The description of past climates derived from paleoclimatic data is useful for three reasons: It gives us perspective on the range of climate changes that are possible and likely, it provides clues about how the climate system works, and it provides a data set for testing theories and models of how climate changes.

Paleoclimatic data give us a perspective on what constitutes a "significant" global climate change. During a full-glacial episode like the one 20,000 years ago, the global mean surface temperature was about 5°C colder than now. Therefore, 1°C in global mean surface temperature represents about 20% of the temperature change between glacial and interglacial conditions. The current climate is very near the warmest that has been observed in the last million years, so we do not have very good analog information for really warm world climates. The warm period about 9000–6000 years ago was nearly the warmest in the last million years and was probably not as much as 1°C warmer than today. To get climates that were substantially warmer than today one must return to the early Pliocene, about 5 million years ago, or to the Cretaceous period, about 65 million years ago, when continental positions and the configuration of the world ocean were very different from today. The dominant time scales for natural climate changes during the current geologic epoch, the Holocene, seem to be in the tens of thousands to hundreds of thousands of years. Although climate changes on a variety of time scales, the largest variations appear to occur on longer time scales. Nonetheless, the record also contains evidence of rapid changes of substantial magnitude on time scales of tens to hundreds of years, such as those associated with the Younger Dryas event and the Little Ice Age.

Paleoclimatic data can provide clues about how the climate system works and how the observed variations were produced. The thick ice sheets over the continents during glacial maxima suggest that continental ice sheets may play an active role in climate change, through their albedo, topography, and effects on ocean circulation and biogeochemical cycles. Temperature changes during an ice age seem to be largest at high latitudes and rather small near the equator. Why is this and what does it tell us about the mechanisms of climate change? The tropical land regions appear to have been drier on average during glacial maxima than during warm periods. What are the reasons for these hydrological and vegetation changes and do they play an active role in climate variability? Air trapped in bubbles in glacier ice shows that the carbon dioxide content of the atmosphere was lower ($\sim$ 190 ppmv) during the last ice age than 200 years ago before the industrial revolution ($\sim$ 280 ppmv). This suggests that global ocean productivity was enhanced during the ice age. Does this information indicate a role for biology in determining climate variability?

Paleoclimatic data provide evidence for testing theories and models of global climate change. Time series of global ice volume can be compared to time series of insolation variations associated with cycles in Earth's orbital parameters. Can we explain the relationship between Earth's orbital parameters and global ice volume? If we put inferred distributions of SST, surface ice, vegetation, and atmospheric composition into our best climate models, do they produce a consistent climate in balance with these conditions? How important are the changes in atmospheric composition for explaining glacial–interglacial cycles of climate change? In Chapter 10 we examine the question of how sensitive the climate is to ice sheets and atmospheric composition, and in Chapter 12 we examine the role of orbital parameter variations and other natural climate forcings in determining climate variability.

EXERCISES

1. Discuss the difficulties with interpreting a curve such as Fig. 9.1 or extrapolating it into the future.

2. One of the earliest examples of primitive humans, *Australopithecus*, lived in Africa between about 4 and 1 million years ago. Hominids began using stone tools about 2 million years ago. Place *Australopithecus* and tool use in their proper place in Fig. 9.5. For what fraction of Earth history have hominids been around?

3. How would the differences of atmospheric CO_2 and CH_4 during the last glacial maximum and the present preindustrial era have contributed to the differences in the climates between then and now?

4. Estimate the rate of freshwater mass production (kg s^{-1}) by the melting of the North American ice sheet, assuming that its melting took 500 years and that it consisted of 10^{19} kg of water. How does this compare with the rate of deep-water formation in the North Atlantic, which is estimated to be 1.5–2×10^{10} kg s^{-1}?

5. Estimate by how much the freshwater flux calculated in problem 5 would decrease the salinity of the surface waters of the far north Atlantic if the flow of the 35% salinity water from the south is 2×10^{10} kg s^{-1}?

6. What fraction of the North American ice sheet at its maximum would need to melt in 100 years in order that the flow of freshwater over this period would decrease the salinity by 2%? How would such a burst of freshwater flux affect the formation of deep water? What would be the response of the climate in the north Atlantic region? How would the rate of melting respond to the climate changes? Can you imagine how the interactions between the melting rate, deep-water formation, and climate variations might cause an oscillation during the decline of the ice sheet?

7. What do you think might have caused the D–O events during glacial times? Why is the warming more rapid than the cooling?
8. Why do you think more dust was deposited in Antarctica during glacial times than during interglacial times?
9. Explain why the North Atlantic and Antarctic glacial age indicators show a bipolar seesaw.

Climate Sensitivity and Feedback Mechanisms

10.1 FOOLS' EXPERIMENTS[1]

Understanding and forecasting of climate change presents a great challenge to geoscientists. The surface climate of Earth is determined by a complex set of interactions among the atmosphere, the ocean, and the land, and these interactions involve physical, chemical, and biological processes. Many of these interactions are poorly understood, and important interactions and processes have probably yet to be discovered. Because it is apparent that humans have the capability to alter climate, it is necessary that we attempt to understand how the climate system works and make quantitative estimates of how our past, present, and future actions will change it. To make progress on a problem with the intimidating complexity of climate change, the proper response of a scientist is to begin by considering simpler questions and then add complexity as understanding is gained. The lessons drawn from these simple models must be taken seriously, but with the full realization that they may not be a faithful representation of nature.

In climate modeling, complexity is determined by the number and kind of processes or interactions among system components that are included. Once a starting point is chosen, one must decide in what order to bring additional processes or interactions into consideration, and with how much detail it is necessary to represent these processes. These decisions can be made on the basis of how important the processes are for the maintenance of climate, or on the basis of their importance for determining the magnitude or structure of a climate change response to a climate forcing of a given measure. The relationship between the measure of forcing and the magnitude of the climate change response defines what we will call the *climate sensitivity*. A process that changes the sensitivity of the climate

---

[1]"I love fools' experiments. I am always making them." —*Charles R. Darwin*

Global Physical Climatology. http://dx.doi.org/10.1016/B978-0-12-328531-7.00010-4

293

response is called a *feedback mechanism*. The strength of a feedback can also be quantified, and will be termed a *positive feedback* if the process increases the magnitude of the response, and *negative* if the feedback reduces the magnitude of the response. To construct a model of climate change, it is logical to order the feedback processes according to importance, and then include the most important feedbacks first. If positive and negative feedbacks of similar strength are possible, then it is important to include both simultaneously, or otherwise a very inaccurate estimate of the overall system sensitivity can be obtained.

For this discussion, it is useful to review the terms climate forcing and climate feedback. Climate forcing is a change to the climate system that can be expected to change the climate. Examples would be doubling the CO_2, increasing the total solar irradiance (TSI) by 2%, introducing volcanic aerosols into the stratosphere, etc. Climate forcings are usually quantified in terms of how many $W\ m^{-2}$ they change the energy balance when imposed. For example, instantaneously doubling the CO_2 changes the energy balance at the top of the atmosphere by about $4\ W\ m^{-2}$. A feedback process is a response of the climate system to surface warming that then alters the energy balance in such a way as to change the temperature response to the forcing. A positive feedback makes the forced response bigger, and a negative feedback makes it smaller. Classic examples of positive feedbacks are ice-albedo feedback and water-vapor feedback. When it warms, ice melts, and this reduces Earth's albedo and causes further warming. When it cools, ice grows, and this increases Earth's albedo, causing further cooling. In water-vapor feedback the saturation vapor pressure is very sensitive to temperature and water vapor is the most important greenhouse gas, producing the strongest positive feedback. The biggest negative feedback is Planck feedback. When the air gets warm, its radiative emission increases following Planck's Law. As the Earth gets warm, it emits more radiation, which cools more strongly.

10.2 OBJECTIVE MEASURES OF CLIMATE SENSITIVITY AND FEEDBACK

The concept of sensitivity can be used to judge which climate processes should be considered first and which can be neglected until later. Sensitivity can also be used as an objective measure of the performance of a model. Quantitatively, sensitivity is the amount by which an objective measure of climate changes when one of the assumed independent variables controlling the climate is varied. For example, we might ask how much the global mean surface temperature, T_s, changes if we vary the TSI, S_0; however, the global mean temperature is a function of more than the TSI. Suppose it depends on a number of variables y_j, which are all parametrically related to S_0, $y_j = y_j(S_0)$. Secondary variables would be temperature, water vapor, other greenhouse gases, cloudiness, ice cover, land vegetation, and many others. Then the chain rule yields,

$$\frac{dT_s}{dS_0} = \frac{\partial T_s}{\partial S_0} + \sum_{j=1}^{N} \frac{\partial T_s}{\partial y_j} \frac{dy_j}{dS_0}$$

(10.1)

where partial and total derivatives are denoted by ∂ and d, respectively, and N is the number of important subsidiary variables. The total change of surface temperature with respect to TSI is the sum of many contributions each made up of the partial derivative of surface temperature with respect to a secondary variable that controls surface temperature, times the total change of that secondary variable with respect to the TSI. As an example, if we increase the TSI, the specific humidity of air will increase because of the dependence of saturation vapor pressure on temperature. The increase of atmospheric water vapor will in turn lead to an increase in surface temperature, because of the greenhouse effect of the added water vapor.

We may simplify the concept of climate sensitivity somewhat by supposing that some *climate forcing, dQ,* with units of W m^{-2} is applied to the climate system, and ask how it is related to a measure of climate change such as the change in global mean surface temperature, dT_s. We define the ratio of the climate response to the climate forcing as one measure of climate sensitivity.

$$\frac{dT_s}{dQ} = \lambda_R$$

(10.2)

As an example, we can consider the global mean energy balance at the top of the atmosphere.

$$R_{TOA} = \frac{S_0}{4}(1 - \alpha_p) - F^{\uparrow}(\infty) = 0$$

(10.3)

If we imagine that some forcing dQ of the energy balance is applied, so that the climate is driven to some new equilibrium state with a new global mean surface temperature, then we can use the chain rule to write

$$\frac{dR_{TOA}}{dQ} = \frac{\partial R_{TOA}}{\partial Q} + \frac{\partial R_{TOA}}{\partial T_s} \frac{dT_s}{dQ} = 0$$

(10.4)

from which, using (10.3) we obtain

$$\frac{dT_s}{dQ} = -\left(\frac{\partial R_{TOA}}{\partial T_s}\right)^{-1} = \left(\frac{S_0}{4}\frac{\partial \alpha_p}{\partial T_s} + \frac{\partial F^{\uparrow}(\infty)}{\partial T_s}\right)^{-1} = \lambda_R$$

(10.5)

In this simple model, the climate sensitivity is controlled by two basic feedback processes: One feedback is measured by the dependence of the

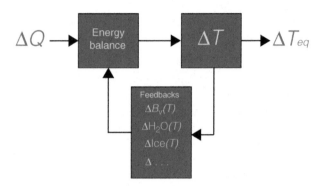

FIGURE 10.1 **Schematic diagram indicating the nature of a feedback process.** An initial climate forcing ΔQ changes the energy balance and leads to an initial temperature change ΔT. This temperature change causes other changes in the climate system; a change in the Planck emission of terrestrial radiation $\Delta B_v(T)$, a change in the water vapor content $\Delta H_2O(T)$, a change in the ice coverage $\Delta Ice(T)$, or some other change $\Delta \cdots$. These changes in turn then influence the energy balance leading to a further change in the temperature. When the temperature change is such that the feedbacks balance the forcing, then the system is in a new equilibrium with a temperature change ΔT_{eq}.

emitted terrestrial radiation flux on surface temperature, and the other is measured by the dependence of the albedo on surface temperature. If the sensitivity parameter λ_R is known, then we can estimate the global mean surface temperature change dT_s expected from a given climate forcing dQ.

$$dT_s = \lambda_R \, dQ \qquad (10.6)$$

Figure 10.1 illustrates the basic concept of a feedback process and distinguishes it from climate forcing. Once the climate forcing, ΔQ, causes some initial temperature change, ΔT, elements of the climate system respond to this temperature change. If the climate forcing is positive, then the temperature change is positive, the emission of terrestrial radiation increases following Planck's law $\Delta B_v(T)$, which acts to increase the radiative cooling of the climate, a negative feedback. Positive feedbacks result because water vapor increases with temperature, $\Delta H_2O(T)$, and ice cover decreases with temperature, $\Delta Ice(T)$. When all the feedbacks are balanced and the system is in energy balance, then a new equilibrium temperature has been established, which is different by the amount ΔT_{eq}.

10.3 BASIC RADIATIVE FEEDBACK PROCESSES

10.3.1 Planck Feedback

We may use the simple global mean model of Chapter 2 to evaluate the strength of the negative feedback associated with the temperature dependence of thermal emission following Planck's Law (3.7). This may be the most important negative feedback controlling the surface temperature of Earth. The emission temperature of a planet is defined by the balance described by (10.3) with the outgoing terrestrial energy emission determined by the Stefan–Boltzmann Law, $F^{\uparrow}(\infty) = \sigma T_e^4$. If we ignore the dependence of albedo on temperature for the moment and assume that the surface temperature and the emission temperature are linearly related, then the sensitivity parameter for this model, which cools like a blackbody, is given by

$$(\lambda_R)_{BB} = \left(\frac{\partial(\sigma T_e^4)}{\partial T_s} \right)^{-1} = (4\sigma T_e^3)^{-1} = 0.26 \text{ K } (\text{W m}^{-2})^{-1} \tag{10.7}$$

The estimate of sensitivity obtained here indicates that about one-quarter of 1°C change in temperature will result from a 1 W m^{-2} forcing of the energy balance. A 1 W m^{-2} forcing of the energy balance can be produced by about a 5.7 W m^{-2} change in the TSI, if the albedo is 0.3. If only Planck feedback is considered, it will take about a 22 W m^{-2} or 1.6% change in TSI to produce a 1°C change in Earth's surface temperature. With such a stable climate, it would be difficult to produce the climate changes observed in the past, so we conclude that some strong positive feedback processes must also operate.

From radiative transfer model calculations, we can estimate that if we double the CO_2 concentration from 300 to 600 ppmv, and leave the temperature and humidity distributions as they are, then the outgoing longwave radiation (OLR) would decrease by about 4 W m^{-2}, which would result in an increase of the net radiation at the top of the atmosphere by the same amount. This is equivalent to a TSI increase of 1.67%, assuming a fixed albedo of 0.3. We can use (10.7) in (10.6) to estimate that about a 1°C global mean surface temperature increase would result from such a climate forcing.

10.3.2 Water Vapor Feedback

One of the most powerful positive feedbacks is associated with the temperature dependence of the saturation vapor pressure of water. As the temperature increases, the amount of water vapor in saturated air increases. Since water vapor is the principal greenhouse gas, increasing water vapor content will increase the greenhouse effect of the atmosphere and raise the surface temperature even further. Thus, the dependence of atmospheric water vapor on temperature constitutes a positive feedback.

Because much of Earth's surface is wet, the humidity of air near the surface tends to remain close to the saturation vapor pressure of air in contact with liquid water. The temperature dependence of the saturation vapor pressure can be obtained from the Clausius–Clapeyron relationship,

$$\frac{de_s}{dT} = \frac{L}{T(\alpha_v - \alpha_l)}$$

(10.8)

where e_s is the saturation vapor pressure above a liquid surface, L is the latent heat of vaporization, T is the temperature, α_v is the specific volume of the vapor phase, and α_l is the specific volume of the liquid phase. From (10.8) we can show that the change in saturation specific humidity divided by the actual value of specific humidity is related to the change in temperature divided by the actual temperature.

$$\frac{dq^*}{q^*} = \frac{de_s}{e_s} = \left(\frac{L}{R_v T}\right)\frac{dT}{T}$$

(10.9)

For terrestrial conditions $(L/R_v T) \approx 20$, so that a 1% change in temperature, which is about 3°C, is associated with about a 20% change in saturation specific humidity, or a 1 K change in temperature corresponds to a 7% change in saturation specific humidity.

It is observed that the relative humidity of the atmosphere, which is the ratio of the actual to the saturation humidity, tends to remain constant, even when the air temperature goes through large seasonal variations in middle to high latitudes. This is not too surprising, since we know that the relative humidity cannot exceed 100% for long and vapor is extracted very efficiently from the ocean when the relative humidity of air is low. As an estimate of the effect of water-vapor feedback on climate sensitivity, we may utilize a one-dimensional radiative-convective equilibrium model (see Section 3.10) in which we assume that the relative humidity remains fixed at observed values while the temperature changes. In this case, we find that the terrestrial radiation emitted from the planet increases much less rapidly with temperature than would be indicated by the Stefan–Boltzmann relationship. Moreover, the terrestrial emission increases linearly with surface temperature, rather than as the fourth power of temperature (Manabe and Wetherald, 1967).

Figure 10.2 shows the surface temperature obtained from a one-dimensional radiative-convective equilibrium model as functions of TSI and atmospheric CO_2 concentration, with two curves showing the result if specific humidity or relative humidity is fixed. In each case, the surface temperature changes faster with the forcing parameter when the relative humidity is fixed rather than specific humidity. In the case of fixed relative humidity (FRH), the water vapor increases about 7% for each degree of warming, so that a strong positive water-vapor feedback is present. The surface temperature increases approximately linearly with TSI, except in the case of fixed relatively humidity for which TSI greater than about 1.1 times today's gives a runaway greenhouse effect in this model. For CO_2

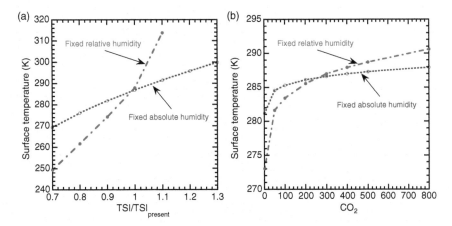

FIGURE 10.2 **Radiative-convective equilibrium calculation of surface temperature as a function of (a) the ratio of total solar irradiance (TSI) to today's value and (b) as a function of CO_2 in ppm.** In each plot one curve shows the result if specific humidity is fixed at today's value and one curve shows the result if relative humidity is fixed at today's value.

the temperature response is not linear because climate forcing from CO_2 varies approximately logarithmically with gas concentration, so that increasing the CO_2 from 150 ppm to 300 ppm produces the same climate forcing as increasing the CO_2 from 300 ppm to 600 ppm.

Figure 10.3 shows OLR as a function of surface temperature obtained from radiative-convective equilibrium calculations such as those shown in Fig. 10.2 in which the model is forced with changing TSI, the relative humidity is assumed fixed and the lapse rate follows the moist adiabat. Except for some unexplained wiggles, the data are fit very well with a line that has a slope of 1.93 W m^{-2} K^{-1}. In the background as dotted lines are shown Planck curves for different emission temperatures, $T_e = T_s - \Delta T$, where ΔT varies from -15 to 75 at intervals of 15 K. The slopes of the black body emission curves are much steeper than the radiative convective equilibria, because of strong water-vapor feedback. The OLR from the model varies approximately linearly over a wide range of surface temperature. From radiative-convective equilibrium calculations with FRH we thus obtain an estimate of the climate sensitivity parameter with only the Planck and relative humidity feedbacks included.

$$(\lambda_R)_{FRH} = \left(\left(f \frac{dF^{\uparrow}(\infty)}{dT_s} \right)_{FRH} \right)^{-1} \approx 0.5 \, K(W \, m^{-2})^{-1} \qquad (10.10)$$

Comparing (10.10) with (10.7) we see that adding relative humidity feedback nearly doubles the estimated sensitivity of the climate, which is also consistent with the differences between fixed specific humidity and FRH in Fig. 10.2. With these two feedbacks included, the response of surface temperature to a doubling of CO_2 concentration is about 2°C.

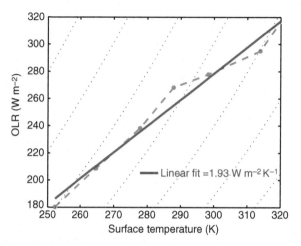

FIGURE 10.3   A plot of OLR versus surface temperature from a set of one-dimensional radiative–convective equilibrium calculations in which relative humidity is fixed and the moist-adiabatic lapse rate is assumed. The blue line is the linear fit to the data indicated by the red dots. The dotted lines in the background indicate black body emission curves. The surface temperature was changed by altering the TSI from 70% to 110% of today's value.

An assumption that the relative humidity remains approximately constant when the climate warms or cools has generally been shown to be an excellent approximation. Figure 10.4 shows zonal cross-sections of the relative humidity for the annual mean, January and July. The relative humidity has strong variations with altitude and latitude. Near the surface, the relative humidity is fairly high, except in latitudes with a lot of dry land areas such as the subtropics, especially in the Northern Hemisphere where the subtropical belt (10–40°N) has a large amount of dry land areas over Africa and Asia. In the mid-troposphere, the relative humidity is also very low in the subtropics, where sinking motion takes cold dry air from upper levels and heats it up adiabatically. The stratosphere is very dry, with very low relative humidity, except over the Antarctic in winter. Although the relative humidity has very strong spatial patterns and seasonal variations, these are related to the circulation, which changes only modestly with climate, in comparison to the saturation vapor pressure, which changes rapidly with temperature.

Figure 10.5 shows the spatial distribution of relative humidity at 500 hPa. Again, large regions of the tropics where the average vertical motion is downward have very low relative humidity. It is in these regions with less water vapor and cloud where the Earth can emit radiation to space effectively (compare the relative humidity in Fig. 10.5 with OLR in Fig. 2.10).

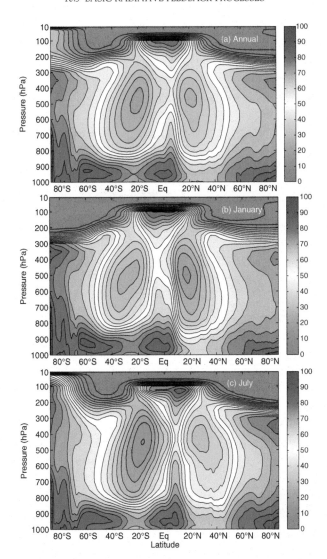

FIGURE 10.4 **Latitude–pressure cross-sections of zonal mean relative humidity for (a) annual mean, (b) January, and (c) July.** Contour interval is 10%. *From ERA Interim reanalysis.*

Emission from these dry zones allows Earth to stay cooler than it would if the water vapor was evenly spread over the surface area of Earth.

10.3.3 Lapse-Rate Feedback

Back in Chapters 2 and 3 we showed that the decrease of air temperature with altitude is critical to the greenhouse effect, since greenhouse

(a) Annual

(b) January

(c) July

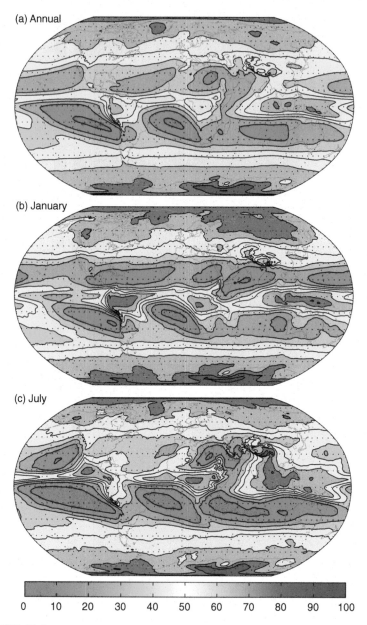

0   10   20   30   40   50   60   70   80   90   100

FIGURE 10.5 **Maps of relative humidity at 500 hPa for (a) annual mean, (b) January, and (c) July.** Contour interval is 10%. *From ERA Interim reanalysis.*

gases and clouds cause the emission of thermal infrared radiation by Earth to originate in the middle troposphere, which is cooler than the surface. Therefore, if we weaken the lapse rate, the surface and the emission temperatures become closer together, the greenhouse effect is weakened, and we should expect the surface temperature to cool. So if the lapse rate decreases with increasing surface temperature, that would be a negative feedback, all else being equal. However, all else is not equal.

Because the saturation vapor pressure is strictly a function of temperature only, if the relative humidity is approximately constant, then the water vapor mixing ratio is a function of only temperature. In that case, when you decrease the lapse rate, the air above the surface warms. If the relative humidity is fixed, these warmed layers will also contain more water vapor. If the emission comes from the water vapor and the water vapor is a function of temperature only, then the water vapor that does the emitting stays at about the same temperature, regardless of the lapse rate. If the lapse rate increases, the water vapor decreases where it cools so that the emission comes from lower in the atmosphere, where the temperature is nonetheless about the same as before the lapse rate change. The effect of decreased pressure broadening on the absorption lines of water vapor has a very modest effect as the emission moves to lower pressures. Figure 10.6 illustrates the effect of lapse rate with a simple schematic. As the lapse rate decreases

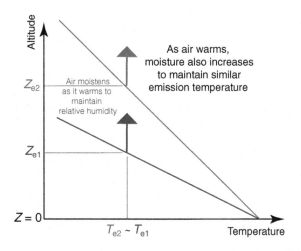

FIGURE 10.6 **Schematic diagram showing how relative humidity feedback offsets the effect of lapse-rate feedback.** Two temperature plots are shown as a function of altitude with larger (blue line) and smaller (red line) lapse rates and the same surface temperature. The emission temperature is marked by the upward-directed arrows. As the lapse rate decreases and the air above the surface warms, the air will also acquire additional water vapor to keep its relative humidity constant. Thus the emission level moves up in altitude and stays at about the same temperature, so that the emission to space is almost independent of the lapse rate, if the relative humidity is fixed.

(red line), the warmed air above the surface gains moisture if the relative humidity is fixed. Then the emission level moves upward and the emission temperature remains nearly constant. Thus the relative humidity feedback cancels the effect of the lapse rate change and the emission to space comes from the same temperature as before the lapse rate changed. For this reason, so long as the relative humidity remains about constant, the effect of lapse-rate feedback will always be greatly muted by the associated water vapor feedback. This was first pointed out by Cess (1975).

10.4 ICE–ALBEDO FEEDBACK

The physical effects of expanded surface ice cover have often been offered as one possible explanation for how the very different conditions of the ice ages could have been maintained. One of the primary physical effects of ice cover is the much higher albedo of ice and snow than all other surface coverings (Table 4.2). The annual variation of surface albedo is controlled largely by snow cover (Fig. 4.4). In the Northern Hemisphere, the seasonal variation of surface albedo is large over a wide range of latitudes from 45°N to the pole, whereas in the Southern Hemisphere it is confined to the latitude where seasonal sea ice is present around Antarctic, from 55°S to 75°S (Fig. 10.7). Vegetation cover can also influence the

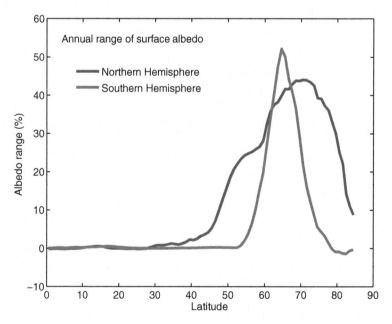

FIGURE 10.7 Surface albedo contrast between winter and summer estimated using January and July. *From NASA-CERES surface radiation budget estimates.*

FIGURE 10.8 **Aerial photo of Arctic Sea Ice.** Dark areas are leads where open water is exposed. Features on the ice floes include pressure ridges and snow drifts. *NASA image; courtesy: the Digital Mapping system team and Operation IceBridge Arctic 2011.*

seasonal variation of albedo, and solar zenith angle is also an important influence on seasonal variations of surface albedo in high latitudes. The albedo of ocean surfaces at high latitudes is typically 10%, whereas a typical albedo for sea ice with a covering of snow at the same latitudes is 60% (Fig. 10.8). The surface albedo contrast between coniferous forest and an ice sheet is also very large. Surface albedo variations are muted by the presence of clouds, since when clouds are present the albedo at the top of the atmosphere is already fairly high and changing the surface albedo has a smaller, but still very significant effect, especially since clouds are not always present.

As the climate cools during the onset of an ice age, the ice would expand into regions previously covered by water or forest in latitudes where the insolation is substantial, particularly in summer. The increased surface ice cover would raise the surface albedo and reduce the amount of solar energy absorbed by the planet. This reduction of solar energy absorption would cause further cooling and thus further ice expansion that might ultimately lead to an ice age. The association of surface glaciation with colder temperatures can constitute a very powerful positive feedback, since it modulates the direct energy input from the Sun.

The albedo feedback effect was incorporated in simple climate models by Budyko (1969) and Sellers (1969). Both of their models produce very sensitive climates, such that it is surprisingly easy to produce an ice age, or even an ice covered Earth, with a relatively modest climate forcing. Their simple models for the ice caps assume that everything about the climate can be characterized by the surface temperature, and that the only independent variable is latitude. The determination of the surface temperature is based on the conservation of energy for the system, so that these models are termed *energy-balance climate models*. The steady-state energy balance has three terms: the absorbed solar energy, the emitted terrestrial energy, and the divergence of the horizontal energy transport by the atmosphere and oceans. Since the growth of polar ice caps and their effect on global climate is the central question that these models are designed to address, and this depends heavily on the sphericity of Earth, these models are functions of latitude on a sphere.

It is convenient to express the three terms in the energy balance as functions of the sine of latitude, $x = \sin \phi$, and the surface temperature T_s.

$$Q_{ABS}(x, T_s) - F_\infty^\uparrow(x, T_s) = \Delta F_{ao}(x, T_s) \tag{10.11}$$

Q_{ABS} is the absorbed solar radiation, $F_\infty^\uparrow$ is the OLR, and ΔF_{ao} is the removal of energy by horizontal energy transport, which would usually be expressed as the divergence of the meridional energy transport by the atmosphere and ocean. The observed climatological zonal average values of these terms are shown in Fig. 2.12. The absorbed solar radiation is written as the product of the TSI, S_0, a function that describes the distribution of insolation with latitude, $s(x)$, and the absorptivity for solar radiation, $a_p(x, T_s) = 1 - \alpha_p(x, T_s)$, which is one minus the planetary albedo, α_p.

$$Q_{ABS}(x, T_s) = \frac{S_0}{4} s(x) a_p(x, T_s) \tag{10.12}$$

The distribution of annual mean insolation is shown in Fig. 2.7. The distribution function $s(x)$ is defined by dividing the annual mean insolation at each latitude by the global average insolation. Its global area average is unity.

$$\frac{1}{2} \int_{-1}^{1} s(x) dx = 1 \tag{10.13}$$

The annual mean insolation distribution function for current conditions can be approximated fairly accurately as

$$s(x) = 1.0 - 0.477 P_2(x) \tag{10.14}$$

where

$$P_2(x) = \frac{1}{2}(3x^2 - 1) \tag{10.15}$$

is the Legendre polynomial of second order in x.

The emitted terrestrial flux is specified as a linear function of surface temperature.

$$F_\infty^\uparrow(x, T_s) = A + BT_s \qquad (10.16)$$

The coefficients A and B can be obtained from a radiative transfer model, or from observational estimates. The transport then can be specified in two different ways that facilitate an easy solution. Budyko assumed a linear form for the transport term, such that at every latitude, the transport relaxes the temperature back toward its global-mean value,

$$\Delta F_{ao}\big|_{\text{Budyko}} = \gamma(T_s - \tilde{T}_s) \qquad (10.17)$$

where

$$\tilde{T}_s = \frac{1}{2}\int_{-1}^{1} T_s\, dx \qquad (10.18)$$

is the global mean temperature.

Sellers (1969) and North (1975) used a diffusive approximation to meridional transport.

$$\Delta F_{ao}\big|_{\text{Sellers}} = \nabla \cdot K_H \nabla T_s = \frac{\partial}{\partial x}(1 - x^2)K_H \frac{\partial T_s}{\partial x} \qquad (10.19)$$

The linear relaxation coefficient, γ, and diffusivity, K_H, are chosen to make the model simulation as close to the observed climate as possible by producing the observed meridional energy transport when the temperature gradient is approximately as observed.

Albedo feedback is introduced by assuming that ice forms when the temperature falls below a critical value and that this causes an albedo increase. It is observed that the annual mean temperature at which surface ice cover persists throughout the year is about $-10°C$ currently. Therefore, we assume that an abrupt transition in the albedo takes place at this temperature.

$$\alpha_p = \begin{cases} \alpha_{\text{Ice free,}} & T_s > -10°C \\ \alpha_{\text{Ice,}} & T_s < -10°C \end{cases} \qquad (10.20)$$

Typical values chosen for planetary albedos with and without surface ice cover are 0.62 and 0.3, respectively.

If we substitute (10.12), (10.16), and (10.17) into (10.11), we obtain the equation considered by Budyko.

$$A + BT_s + \gamma(T_s - \tilde{T}_s) = \frac{S_0}{4}s(x)a_p(x, x_i) \qquad (10.21)$$

Here we have used the fact that because of (10.20), the absorptivity is only a function of sine of latitude x and the position of the ice line, x_i, the point where the temperature equals $-10°C$. We define a new variable I, which is the ratio of the terrestrial emission to the global average insolation.

$$I = \frac{(A + BT_s)}{(1/4)S_0} \qquad (10.22)$$

Substituting this definition into (10.21) yields a simple equation for I.

$$I + \delta(I - \tilde{I}) = s(x)a_p(x, x_i) \qquad (10.23)$$

It is interesting that the linear parameters describing the efficiency of meridional energy transport and longwave radiative cooling occur only in the ratio $\delta = (\gamma/B)$. If δ is large, then meridional transport is relatively efficient compared to longwave cooling, and equator-to-pole gradients in I will be small.

The global area average of (10.23) is

$$\tilde{I} = \frac{1}{2} \int_{-1}^{1} s(x)a_p(x, x_i) dx. \qquad (10.24)$$

which indicates that the global average value of the terrestrial emission divided by the insolation is equal to the global average of the product of the absorptivity and the distribution function for insolation. Thus, if the absorptivity is high where the insolation is also high, the global mean value I will also be high. Since I is linearly related to T_s, the global mean temperature also increases with the average product of insolation and absorptivity. Because the annual mean insolation is least at the poles and greatest at the equator, the albedo increase associated with ice cover will have a greater effect on global mean temperature as the ice cover extends farther toward the equator.

It is interesting to solve for the latitude of the ice boundary as a function of TSI for particular albedo specifications and values of δ. The solution is obtained by specifying the position of the ice line and then solving for I at the ice line latitude using (10.23). Because the albedo specification is discontinuous at the ice line and no diffusion is present in the model, I and T_s are discontinuous and the solution for x_i as a function of S_0 is not unique. A unique solution can be obtained by adding a small diffusive transport and then taking the limit of vanishing diffusion coefficient. This procedure suggests using the average of the albedo on both sides of the ice edge when solving for I at the ice edge (Held and Suarez, 1974). The general structure of such solutions is shown in Fig. 10.9a. Because of the strong nonlinearity introduced by the albedo specification, the model may have between one

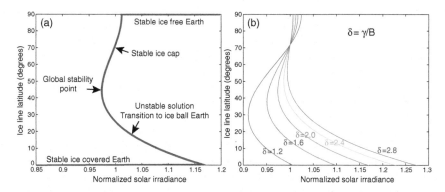

FIGURE 10.9 **Iceline latitude in the Budyko energy balance climate model as a function of normalized TSI.** (a) Schematic diagram showing the stable and unstable solutions for $\delta = 2$, (b) diagram showing the dependence of the ice-line latitude on the ratio of transport to cooling efficiency $\delta = \gamma/B$.

and three solutions for a particular value of the TSI. For sufficiently large TSIs, an ice-free planet is a stable solution to the model. For sufficiently small TSIs, an ice-covered planet is a stable solution. For an intermediate range of TSIs both of these stable solutions are possible.

The diagram has been scaled such that for a normalized TSI of 1.0 the boundary of the polar ice cap is at 72°N, which is approximately the current position of the ice edge. If the TSI is raised or lowered from this value the ice cap will recede or expand. The rate at which the ice cap expands or contracts as the TSI is changed is a measure of the sensitivity of the climate. This rate is equivalent to the slope of the solution curve for the stable ice cap in Fig. 10.9. As the TSI is decreased and the ice cap expands, the model becomes more sensitive until the line is vertical at some critical latitude. The latitude where the slope becomes infinite is labeled the global stability point in Fig. 10.9a. From that point, if the solar irradiance is reduced further the ice cap moves to the equator, yielding the snowball Earth condition. The intervening steady state solution indicated in the diagram shows the ice edge moving equatorward with increasing solar irradiance, which is unphysical, and these solutions are unstable. The point where this solution branch reaches the equator, though, marks the amount by which the solar irradiance must be increased to make an ice-covered planet transition to an ice-free solution. The model thus indicates hysteresis, whereby once the planet is ice-covered by reducing the solar irradiance to 98%, it must then be increased to 117% to get the ice to melt, at which point it melts completely. To get an ice cap to form again, the solar irradiance must be reduced to about 102% (Fig. 10.9a).

The parameters chosen by Budyko (1969) and Sellers (1969) make the ice cap very sensitive to small changes in TSI. For example, Budyko calculated that a reduction in TSI of 1.6% would lead to an ice covered Earth. This result is attractive in that it indicates that a relatively small change in climate forcing could produce the glacial–interglacial transitions that have occurred during Earth history. On the other hand, the global stability point is reached when the ice cap has expanded to about 45° latitude. The inference from this simple model is thus that during glacial maxima, when the average latitude of perennial ice cover was near midlatitudes, the planet is very close to making an irreversible transition to an ice-covered condition. Energy balance climate models are extremely simple, of course, and neglect many important processes, but it is still interesting to see how their sensitivity depends on the parameters needed for their solution.

We may first examine how the sensitivity of the model depends on the parameter δ, which measures the ratio of the meridional transport coefficient, γ, to the longwave radiative cooling coefficient, B. Figure 10.9b shows a set of solution curves for selected values of δ. For larger values of δ the boundary of a small polar ice cap is more sensitive to TSI changes. This is related to the fact that large values of δ produce weak meridional temperature gradients; so that a small change in global mean temperature is translated into a large displacement of the ice line and thereby a large change in absorbed solar radiation. Budyko used a value of $B = 1.45$ W m^{-2} K^{-1}, whereas more recent estimates would place the most appropriate value closer to 2.2 W m^{-2} K^{-1}. The resulting value of δ in Budyko's calculation was 2.6. This high value of δ contributed to the great sensitivity of his model.

The planetary albedo contrast associated with surface ice cover is also a critical parameter controlling the sensitivity of energy balance climate models. The increase of planetary albedo with latitude is an observed fact (Fig. 2.9), but only a portion of this meridional increase of planetary albedo is associated with surface ice. Significant contributions are also made by the increase of cloud cover with latitude (Fig. 3.21) and the increase of the albedo with solar zenith angle, which also increases with latitude (Fig. 2.8). As a result the ice-free albedo increases with latitude, so that in the latitudes where ice is likely to form, the contrast of planetary albedo between ice-free and ice-covered conditions is not as great as one might expect on the basis of the surface albedo change. Much of the surface albedo increase is screened by clouds, which are very good reflectors of solar radiation, particularly at high solar zenith angles. One may take this into account in the energy balance climate model by assuming that the ice-free absorptivity for solar radiation decreases with latitude. A realistic appraisal of ice–albedo feedback is that, although it is an important feedback process, it is not strong enough by itself to explain the ease with which the climate system appears to change from glacial to interglacial conditions.

10.5 DYNAMICAL FEEDBACKS AND MERIDIONAL ENERGY TRANSPORT

In the energy-balance climate models, the efficiency of meridional energy transport and the resulting meridional temperature gradient play an important role in determining the sensitivity of climate. It is interesting to consider how meridional energy transport in the atmosphere and ocean might respond to a developing ice sheet or to some other forcing of the meridional temperature gradient. If we limit ourselves to midlatitudes, the biggest contribution to meridional energy transport comes from eddies in the atmosphere (Chapter 6). The dependence of the eddy fluxes of heat on the mean temperature gradients can be estimated from simple dimensional arguments. We first suppose, as was done by Sellers (1969) and North (1975), that the meridional energy transport depends on the meridional gradient of potential temperature, Θ.

$$[v'\Theta'] = -K_H \frac{\partial \Theta}{\partial y} \tag{10.25}$$

where square brackets indicate a zonal average and y is the northward displacement in a local Cartesian coordinate system. The transport coefficient K_H must have the units of $m^2\,s^{-1}$, so we can think of K_H as the product of a characteristic wind speed V and distance L_e.

$$K_H \propto V \times L_e \tag{10.26}$$

The eddy heat fluxes are produced by disturbances that result from baroclinic instability of the zonal average state of the atmosphere, so we use the theory for baroclinic instability to select the characteristic wind speed and length scales. The length scale is chosen to be the Rossby radius of deformation, which is approximately the scale of disturbances that grow most rapidly in response to the baroclinic instability of the zonal mean flow (Holton and Hakim, 2012).

$$L_e \approx L_R = \frac{N d_e}{f} \tag{10.27}$$

where d_e is the depth scale for the disturbance, f is the Coriolis parameter, and N is the buoyancy frequency, which is related to the vertical gradient of the potential temperature.

$$N^2 = \frac{g}{\Theta} \frac{\partial \Theta}{\partial z} \tag{10.28}$$

The appropriate wind speed scale is the difference in zonal mean wind speed experienced by the eddy across its vertical extent, d_e.

$$V = d_e \left| \frac{\partial u}{\partial z} \right| \tag{10.29}$$

The vertical shear of the zonal mean wind is related to the meridional gradient of potential temperature by the thermal wind approximation, which is accurate away from the equator.

$$\frac{\partial [u]}{\partial z} \approx -\frac{g}{f[\Theta]} \frac{\partial [\Theta]}{\partial y} \tag{10.30}$$

Combining (10.25) through (10.30), we obtain an expression for the meridional flux of dry static energy by transient eddies as a function of the mean state variables.

$$[v'\Theta'] \approx -\left(\frac{g}{[\Theta]} \right)^{3/2} \frac{d_e^2}{f^2} \left(\frac{\partial [\Theta]}{\partial z} \right)^{-1/2} \left| \frac{\partial [\Theta]}{\partial y} \right| \frac{\partial [\Theta]}{\partial y} \tag{10.31}$$

The conservation of potential temperature can be used to scale the characteristic vertical velocity and characteristic vertical-length scale to the horizontal velocity and length scales.

$$W = V \left| \frac{\partial [\Theta]}{\partial y} \right| \left| \frac{\partial [\Theta]}{\partial z} \right|^{-1} \tag{10.32}$$

Using this scaling of the vertical velocity, we can obtain an expression for the vertical heat flux analogous to (10.31).

$$[w'\Theta'] \approx -\left(\frac{g}{[\Theta]} \right)^{3/2} \frac{d_e^2}{f^2} \left| \frac{\partial [\Theta]}{\partial y} \right|^3 \left(\frac{\partial [\Theta]}{\partial z} \right)^{-1/2} \tag{10.33}$$

According to these scaling arguments, the magnitude of the meridional eddy heat flux is proportional to the square of the meridional temperature gradient and to the square root of the static stability (10.31). The depth scale d_e is as yet unspecified. Strongly unstable disturbances occupy the entire troposphere, so that it is reasonable to use the scale height H as the depth scale. For more weakly unstable flows the depth scale itself is proportional to the meridional temperature gradient, so that the total heat flux is even more highly sensitive to the meridional temperature gradient (Held, 1978). Because of the quadratic or higher dependence of the eddy heat flux on the meridional temperature gradient, eddy fluxes act to suppress large deviations of the meridional temperature gradient, so that, to the extent that the meridional temperature gradient is controlled by eddy fluxes, we expect the meridional gradient of atmospheric temperature to be rather insensitive to altered thermal driving of the equator-to-pole temperature difference. Altered meridional thermal drive will result in a large change in baroclinic eddy activity and eddy heat transport, which will suppress the response of the temperature gradient to the altered forcing. The strong sensitivity of the eddy activity and heat flux to meridional thermal driving has been verified by calculations with numerical models of the atmosphere.

We expect that the oceanic heat flux is also sensitive to the meridional temperature gradient, since we expect that both the thermal driving and the wind stress driving of oceanic circulations would increase with the temperature gradient. We do not have any simple quantitative theory for how oceanic fluxes depend on the meridional temperature gradients that corresponds to the simple dimensional arguments given above.

Dynamical models of the free atmosphere suggest that baroclinic eddies will act to maintain a nearly constant meridional temperature gradient by changing their meridional energy flux by a large amount when a small change in meridional temperature gradient is imposed. It is very interesting that paleoclimatic data show that the surface temperature has a very different behavior than expected from these dynamical considerations. During ice ages, the tropical surface temperatures changed only a few degrees from today's values, whereas the polar temperatures cooled by 10°C or more. This "polar amplification" of climate change is an important feature that needs to be fully understood. It points to the importance of surface and boundary layer processes in determining the surface temperature and its latitudinal gradient, or to a strong latitude dependence of feedback processes. Much of the enhanced temperature change in high latitudes appears to be confined near the surface, so that polar amplification of surface temperature change and the hyperdiffusive theory for heat flux described previously are not inconsistent. The meridional gradients of mean tropospheric temperature change much less than the meridional gradients of surface temperature in climate models, so that much of the polar amplification of warming is confined to the lower troposphere.

10.6 LONGWAVE AND EVAPORATION FEEDBACKS IN THE SURFACE ENERGY BALANCE

The energy and water exchanges at the surface are very important for climate (Chapter 4), and it is of great interest to understand how these energy exchanges might be sensitive to temperature. We can gain some intuition about this through relatively simple analysis.

The net longwave heating of the surface is equal to the downward longwave from the atmosphere minus the longwave emission from the surface.

$$F_{\text{Net}}^{\downarrow} = F^{\downarrow}(0) - \sigma T_s^4 \tag{10.34}$$

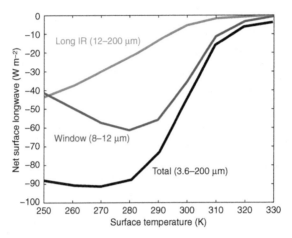

FIGURE 10.10 **Net longwave at the surface as a function of surface temperature calculated from a radiative transfer model.** Separate curves are shown for the total longwave, the window region (8–12 μm) and the longest IR wavelengths that are of importance where the rotational lines of water vapor produce strong absorption and emission (12–200 μm).

Here we have assumed that the surface absorbs and emits like a blackbody. We can calculate the dependence of net radiation on surface temperature under clear-sky conditions using a radiative transfer model with some assumptions. We assume that the temperature decreases linearly with height away from the surface with a lapse rate of 6.5 K km^{-1} until the height where the temperature reaches a minimum of 200 K, and that the atmosphere is isothermal with a temperature of 200 K above that level. We assume that the relative humidity is distributed linearly in pressure according to

$$RH = RH_o \left(\frac{p/p_s - 0.02}{1 - 0.02} \right) \qquad (10.35)$$

where $RH_o = 77\%$ gives a reasonable approximation to the observed global mean relative humidity. The net longwave flux at the ground computed with these specifications of lapse rate and relative humidity and with current values of CO_2 and ozone content is shown in Fig. 10.10. The general character of this plot is not very sensitive to the details of how the relative humidity and temperature profiles are specified. For temperatures less than about 280 K the net longwave loss from the surface is about 90 W m^{-2} and is not sensitive to temperature. This is because there is very little water vapor in the atmosphere at these temperatures and the downward longwave from the atmosphere and the upward longwave from the surface increase at about the same rate, so that the balance remains the same. For temperature greater than about 280 K, as the temperature increases the net longwave loss from the surface decreases. In this temperature range, the infrared opacity of the water vapor begins to increase rapidly, if the relative humidity is fixed. This means that the downward longwave emission from the atmosphere increases more rapidly with temperature that the upward blackbody emission from the surface. At tropical temperatures of about 300 K, the surface longwave loss is decreasing rapidly, changing from about -75 W m^{-2} at 290 K to about -15 W m^{-2} at 310 K, or about $+3$ W m^{-2} K^{-1}. From the atmosphere's perspective, this change in the surface longwave exchange is a loss of energy, since the amount of longwave energy that it emits downward is increasing faster than the amount it receives from the surface. Much of the reduction in net longwave cooling of the surface comes from wavelengths in the range 8–12 μm, called the water vapor window. This is because of the so-called continuum absorption by water vapor that becomes increasingly effective at high vapor pressure. The nature of this absorption is not fully understood, but it is well measured and likely associated with water vapor molecules sticking together at high vapor pressures, and thus forming a complex molecule that can efficiently absorb and emit window radiation. This closes the water-vapor window at high temperatures and makes the cloud-free atmosphere virtually opaque to thermal infrared radiation. According to these calculations, upward and downward longwave radiation at the surface would become almost equal for surface temperatures in excess of ~320 K. The net surface longwave loss decreases most rapidly with temperature in the vicinity of 300 K, which is approximately the mean temperature of the tropical oceans. This constitutes a positive feedback in the surface energy budget that is most effective at tropical temperatures, since as the temperature increases the net longwave cooling of the surface decreases. If no other feedback processes were operating, the tropical sea surface temperature (SST) might be quite sensitive and variable.

The approximate energy balance of the tropical ocean surface is about 200 W m^{-2} solar heating, balanced by 120 W m^{-2} evaporative cooling, 50 W m^{-2} net longwave loss, 10 W m^{-2} sensible cooling, and 20 W m^{-2} energy export by ocean currents. The evaporative cooling of tropical ocean waters is the largest cooling term. It is much greater than the longwave cooling and is sensitive to surface temperature. We can estimate the sensitivity of evaporative cooling to surface temperature if we are willing to make several critical assumptions. We begin with the aerodynamic formula for evaporative cooling approximated using (4.32) and (4.35).

$$\text{LE} = \rho_a L C_{\text{DE}} U q_s^* \left((1-\text{RH}) + \text{RH} \left(\frac{L}{R_v T_s^2} \right) (T_s - T_a) \right) \qquad (10.36)$$

If we assume again that the RH distribution and the temperature profile remain constant as we increase the surface temperature, then the evaporative cooling depends primarily on the temperature dependence of the saturation specific humidity at the surface. We can also assume that the characteristic wind speed and the exchange coefficient are independent of the surface temperature. Making these assumptions and neglecting the weaker linear and quadratic temperature dependencies compared to the exponential temperature dependence of saturation humidity, one can show from (10.36) and (4.35) that the logarithmic derivative of evaporative cooling with respect to surface temperature is only weakly dependent on temperature itself.

$$\frac{1}{LE}\frac{\partial LE}{\partial T_s} \approx \frac{L}{R_v T_S^2} \approx 6\% \; K^{-1} \tag{10.37}$$

This is a result of the exponential dependence of saturation humidity. The derivatives of exponential functions are themselves exponential functions. The numerical estimate in (10.37) is obtained for $T_s = 300$ K, which is a typical tropical oceanic surface temperature. If we multiply (10.37) by a typical evaporative cooling value of LE = 120 W m^{-2}, then we obtain an estimate of the sensitivity that is independent of specific choices of the parameters in the aerodynamic formula (10.36).

$$\frac{\partial LE}{\partial T_s} \approx 7 \, W \, m^{-2} \, K^{-1} \tag{10.38}$$

For FRH, the rate at which evaporative cooling increases with temperature is large compared to the rate at which the longwave cooling of the surface decreases.

Figure 10.10 can be used to calculate the dependence of net longwave cooling on temperature for cloud-free conditions. The maximum rate of change of net longwave heating of the ground occurs at about $T_s = 305$ K and has a numeric value of

$$\frac{\partial F_{Net}^{\downarrow}(z = 0)}{\partial T_s} \approx 3 \, W \, m^{-2} \, K^{-1} \tag{10.39}$$

The magnitude of the rate of increase of evaporative cooling with temperature is more than double the rate of decrease of the longwave cooling of the surface in our simple analysis here. The ocean surface gradually loses its ability to cool by longwave radiative emission as the temperature increases above 300 K, but the evaporative cooling increases more rapidly with temperature than the longwave cooling declines with temperature. The sensitivity parameter for tropical sea surface

(TSS) conditions obtained by including only these two feedbacks is rather small.

$$(\lambda_R)_{TSS} = \left\{ \frac{\partial LE}{\partial T_s} - f \frac{\partial F^{\downarrow}_{Net}(z=0)}{\partial T_s} \right\}^{-1} \approx 0.3 \, K(W \, m^{-2})^{-1} \tag{10.40}$$

If surface longwave heating and evaporation vary with temperature as these simple estimates suggest, tropical SST will be relatively stable. This stability, which is rooted in the rapid change of saturation vapor pressure at tropical surface temperatures, may explain why tropical SST has changed relatively little in the past. It may also contribute to the apparent polar amplification of surface temperature changes by providing a tropical damping of temperature changes.

Evaporative cooling of the tropical ocean surface is very sensitive to the thermodynamic properties of the atmospheric boundary layer, which are determined through a complex interaction among boundary-layer turbulence, mesoscale convection, and large-scale flow. One may illustrate the potential importance of relative humidity variations by differentiating (10.36) with respect to relative humidity for fixed temperature.

$$\left(\frac{\partial LE}{\partial RH} \right)_{T_s} \approx \frac{-LE}{(1-RH)} \% \approx -8 \, W \, m^{-2} \, \%^{-1} \tag{10.41}$$

For tropical conditions, the effect on the evaporation rate of a 1 K increase in temperature at FRH is approximately equal and opposite to the effect of a 1% increase in relative humidity at fixed temperature. The processes that control relative humidity in the tropical boundary layer are extremely important to the sensitivity of climate.

The rate of evaporation increase in (10.37) is not likely to be achieved in the global climate system because of the limitations of the atmospheric energy balance. Globally, evaporation must increase only at the rate at which it can be balanced by radiative cooling of the atmosphere (6.3), and the radiative cooling rate of the atmosphere only increases at 1–2% K^{-1} in climate models. In climate models, the relative humidity at the surface over the oceans increases with warming so that the evaporation increase is limited to what the atmosphere can dispose of energetically.

10.7 CLOUD FEEDBACK

According to the numbers presented in Table 3.2, clouds double the albedo of Earth from 15% to 30% and reduce the longwave emission by about 30 W m^{-2}. Because of the partial cancellation of these two effects, the effect of clouds on the global net radiative energy flux into the planet is a reduction of about 20 W m^{-2}. The effect of an individual cloud on the local energy balance depends sensitively on its height and optical thickness, the insolation, and the characteristics of the underlying surface. For example, a low cloud over the ocean will reduce the net radiation substantially, because it will increase the albedo without much affecting the longwave flux at the top of the atmosphere. On the other hand, a high, thin cloud can greatly change the longwave fluxes without much affecting the solar absorption, and will therefore increase the net radiation and lead to surface warming.

If the amount or type of cloud is sensitive to the state of the climate, then it is possible that clouds provide a very strong feedback to the climate system. Since the fractional area coverage of cloud, A_c, is about 50%, and the net effect of this cloud on the energy balance is about -20 W m^{-2}, we can make the crude estimate that

$$\frac{\partial R_{TOA}}{\partial A_c} \approx \frac{\Delta R_{TOA}}{A_c} \approx \frac{-20\text{W m}^{-2}}{0.5} = -40\,\text{W m}^{-2} \qquad (10.42)$$

Thus, a 10% change in cloudiness has the same magnitude of effect on the energy balance as a doubling of CO_2 concentration. Increasing the cloudiness by 10 would offset the effect of CO_2 doubling. Decreasing the cloudiness by 10% would double the effect of CO_2 doubling. If the cloud type changed in such a way as to decrease the shortwave cloud forcing by 5% and increase the longwave cloud forcing by 5%, this would also have the same effect as doubling the CO_2 concentration. This could be achieved by decreasing the area coverage and raising the mean cloud top height. Therefore, both the cloud amount and the distribution of cloud types are extremely important for climate. The effect of cloud feedback on the sensitivity of climate is highly uncertain, but we will discuss estimates from current global climate models in Chapter 11.

10.8 BIOGEOCHEMICAL FEEDBACKS

The climate of Earth has been shaped by life, and it is certain that biology plays a role in the sensitivity of climate. There are many ways that plants and animals can influence climate sensitivity. Perhaps the strongest and most direct means by which organisms affect climate is through their control of the composition of the atmosphere. The concentration of CO_2 in the atmosphere is determined by a complex set of processes, but a key process is the uptake of CO_2 by plankton in the surface ocean and plants on land. It is estimated that up to half of the global cooling during the last glacial maximum was contributed by the reduction in atmospheric CO_2

concentration This reduction in atmospheric CO_2 must have been produced by an alteration in the biology and chemistry of the oceans, since on these time scales atmospheric CO_2 is controlled by the partial pressure of CO_2 in surface waters.

Sea salt particles and other biological materials contribute cloud condensation nuclei over the oceans. Another important source of condensation nuclei for cloud droplets is the production of sulfurous gases such as dimethyl sulfide by tiny organisms in the surface waters. In the atmosphere, these sulfur-bearing gases are converted into sulfuric acid particles, which form the nuclei around which cloud droplets form by condensation of water vapor. If more of these particles are present, more cloud droplets will form. Because the droplets also tend to be smaller when more condensation nuclei are available, the cloud droplets tend to remain in the atmosphere longer before coagulating and falling out as rain. Also, if the cloud water is spread over more droplets, the albedo of the cloud will be higher (Fig. 3.14). We expect that larger production of sulfur gases that are the precursors of cloud condensation nuclei will lead to higher oceanic cloud albedos and thereby to a cooling of Earth. If the rate at which the organisms release sulfur gases is dependent on temperature, we have the elements of a feedback mechanism. The magnitude of this possible feedback mechanism has not been quantified, and the sign of the feedback is also unknown.

Lovelock (1979) has argued that it may be appropriate to think of Earth as a single complex entity involving the biosphere, atmosphere, ocean, and land, which can be called *Gaia*, the Greek word for "Mother Earth". It is further hypothesized that the totality of these elements constitutes a feedback system, which acts to optimize the conditions for life to exist here. Active control to maintain relatively constant conditions is described by the term *homeostasis*.

Watson and Lovelock (1983) have offered Daisyworld, a simple heuristic model for how life might maintain a very stable climate. They imagine a cloudless planet with an atmosphere that is completely transparent to radiation. The only plants are two species of daisies: one dark (say, black) the other light (say, white) in color. One species reflects less solar radiation than bare ground; the other reflects more. The daisy population is governed by a differential equation for each species.

$$\frac{dA_w}{dt} = A_w(\beta x - \chi) \tag{10.43}$$

$$\frac{dA_b}{dt} = A_b(\beta x - \chi) \tag{10.44}$$

where: A_w, A_b, and A_g are the fractional areas covered by white daisies, black daisies, and bare ground, respectively; β is the growth rate of daisies per unit of time and area; χ is the death rate per unit time and area; and x is the fractional area of fertile ground uncolonized by either species.

The growth rate is assumed to be a parabolic function of the local temperature T_i, which has a maximum value of one at 22.5°C and goes to zero at 5°C and 40°C.

$$\beta_i = 1.0 - 0.003265(295.5 \ K - T_i)^2 \tag{10.45}$$

The planet is assumed to be in global energy balance, such that

$$\sigma T_e^4 = \frac{S_0}{4}(1-\alpha_p) \tag{10.46}$$

The planetary albedo α_p is the area-weighted average of the albedos of bare ground, white daisies, and black daisies.

$$\alpha_p = A_g\alpha_g + A_w\alpha_w + A_b\alpha_b \tag{10.47}$$

It is necessary to decide how the local temperatures T_i are to be determined. One extreme is to have all local temperatures equal to the emission temperature T_e. This corresponds to perfectly efficient horizontal transport of heat. The other extreme is for each surface type to be at its own radiative equilibrium temperature. This would correspond to no transport of heat between regions with different surface coverings, so that the local temperature is that necessary to balance the local absorption of solar radiation.

$$T_i^4 = \frac{S_0}{4\sigma}(1-\alpha_i) \tag{10.48}$$

Combining (10.48) with (10.46) yields an expression that satisfies both the local and global energy balance.

$$T_i^4 = \frac{S_0}{4\sigma}(\alpha_p - \alpha_i) + T_e^4 \tag{10.49}$$

We can generalize (10.49) by writing,

$$T_i^4 = \eta(\alpha_p - \alpha_i) + T_e^4 \tag{10.50}$$

where $0 < \eta < (S_0/4\sigma)$ represents the allowable range between the two extremes in which horizontal transport is perfectly efficient, $\eta = 0$, and horizontal transport is zero $\eta = (S_0/4\sigma)$.

Steady-state solutions for Daisyworld can be obtained for specific values of the parameters. Figure 10.11a shows the solution for daisy area and temperature as a function of solar luminosity (TSI) for the case in which the daisies are neutral, because their albedo is the same as the

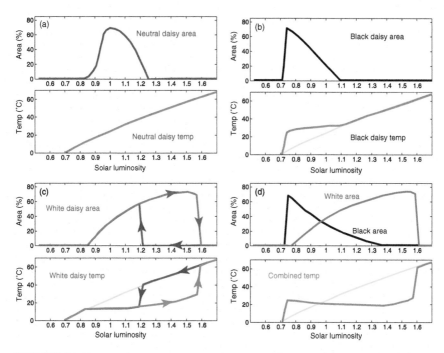

FIGURE 10.11 **Steady-state responses of Daisyworld.** Areas of daisies and global surface temperature are plotted as functions of normalized solar luminosity. The solutions were obtained with a TSI of $S_0 = 1368$ W m^{-2}, a bare ground albedo of $\alpha_g = 0.5$, a daisy death rate of $\gamma = 0.3$, and a heat exchange parameter of $\eta = 2.0 \times 10^9$ K$^4 = 0.12$ $(S_0/4\sigma)$. (a) For a population of 'neutral' daisies with $\alpha_i = 0.5$; (b) for a population of black daisies with $\alpha_b = 0.25$; (c) for a population of white daisies with $\alpha_w = 0.75$ (note that this shows the evolution for both increasing and decreasing solar constant and that the system exhibits hysteresis); (d) for a population of both black and white daisies. Yellow line repeats the neutral daisy temperature from panel (a). *Adapted from Watson and Lovelock (1983). Reprinted with permission from Munksgaard International Publishers Ltd.*

daisy-free land. In this case, the daisy area increases within the temperature range where daisies can survive, but the daisies do not affect the climate. Figure 10.11b shows what happens if the daisies have an albedo that is 0.25 less than the daisy-free land – black daisies. In this case, as soon as the planet is warm enough for daisies to live, their area increases, which lowers the albedo and warms the planet. If the TSI is raised further, less daisy area results and this keeps the planetary temperature fairly constant until no daisies can exist and the temperature follows the same curve as it does for neutral daisies. For white daisies with an albedo of 0.75, a similar stabilization of temperature occurs as soon as daisies can exist and continues until it is too warm for daisies (TSI $\sim$ 1.55), at which point the temperature rises abruptly as the daisies disappear (Fig. 10.11c). If the TSI is then lowered, white daisies do not appear until it becomes cool enough for daisies to exist (TSI ~ 1.2) when daisies appear again and the temperature declines. The system thus exhibits hysteresis. White daisies can maintain cool temperatures to a fairly high TSI, because they reflect lots of sunlight, but once they are gone they cannot reappear until the TSI is much lower than when they disappeared. If both black and white daisies are present, the black daisies are favored by low TSI, the white daisies by high TSI and together they keep the climate fairly constant over a wide range of TSI (Fig. 10.11d).

Because of the strong feedbacks within the system, the global mean emission temperature is remarkably stable over a wide range of TSIs within which daisies can exist. These stable regions are bounded by values of TSI where daisies appear or disappear almost discontinuously and temperature takes a corresponding jump. The stable region exists because changes in temperature cause changes in daisy population that produce a strong negative feedback on the temperature. For example, as the insolation increases, white daisies are favored because they are cooler than the emission temperature. The increase in white daisy population increases the planetary albedo, which reduces the response of the emission temperature to the TSI increase.

Daisyworld is certainly not a good representation of Earth, but it serves to illustrate how biology, simply by doing what is natural for it, can influence the global climate and its stability. Biological mechanisms do operate on Earth that could play the role of the daisies on Daisyworld, although it is likely that their effect would be more subtle and muted. For example, the role of biology in controlling CO_2 might act like the white daisies, since, once photosynthetic life developed, CO_2 was drawn out of the atmosphere. Models suggest that the development of land plants led to a reduction in atmospheric CO_2 in the Paleozoic that helped to offset warming associated with increasing TSI (Berner, 1993). In the Daisyworld model, biological processes are hypothesized to be strongly stabilizing, but CO_2 changes on glacial–interglacial time scales during the quaternary have done just the opposite by contributing significantly to the sensitivity of climate. Clouds might be the earthly counterpart of the white daisies, with their biological connection coming through the production of cloud condensation nuclei by life in the sea. If the release of dimethyl sulfide (DMS) is proportional to the temperature, then the DMS-cloud albedo connection could constitute a strong negative feedback (Charlson et al., 1987). The sign and magnitude of the climate feedback associated with biological production of cloud condensation nuclei are both uncertain; however, since ocean biology is the primary source of cloud condensation nuclei over the ocean, it seems very likely that the cloud contribution to albedo would be less without life.

EXERCISES

1. Derive (10.5) from (10.3).
2. Derive (10.9) from (10.8).
3. If equilibrium is defined by $T_s = \sqrt[4]{N+1}\ T_e$ from Section 3.8, then what is the sensitivity parameter λ_R in (10.6)? First derive a formula, then compute a numerical value for $N = 3$. Compare with (10.7) and explain the difference.
4. Assume that transient eddies are the primary heat transport mechanism determining vertical and horizontal potential temperature gradients in midlatitudes, and that the eddy transports respond to the mean gradients as indicated by (10.31) and (10.33). If the radiative forcing of the meridional gradient produces a larger equator-to-pole heating contrast, how will the vertical gradient of potential temperature (static stability) in midlatitudes respond? Will the change in static stability be relatively large or small compared to the change in meridional temperature gradient? Assume that potential temperature is relaxed linearly toward a radiative equilibrium value and that the relaxation rate of vertical and horizontal gradients is the same.
5. Use the Budyko energy-balance climate model (10.23) to solve for the TSI as a function of iceline latitude. Use (10.14) and the parameters $B = 1.45$ W m^{-2} K^{-1}, $a_0 = 0.68$, $b_0 = 0.38$, and $a_2 = 0.0$. Plot the curves for two values of γ; $\gamma = 3.8$ W m^{-2} K^{-1} and $\gamma = 2.8$ W m^{-2} K^{-1}.
6. According to the energy-balance climate models with ice–albedo feedback, how does the sensitivity of climate relate to the meridional temperature gradient in equilibrium? As the meridional temperature gradient increases, does the climate become more or less sensitive, assuming an ice cap is present?
7. Discuss how the sensitivity of climate might be different if Earth's surface were 70% land and 30% ocean.
8. How do you think the behavior of the Daisyworld model might be different if the planet's sphericity were taken into account? Would it be more or less stable?
9. Program a time-dependent version of Daisyworld for the computer, but add an equation for the population of rabbits whose birth rate is proportional to the area covered by daisies. How does the approach to a steady state depend on the initial conditions and the parameters of the model? Add foxes.

CHAPTER

11

Global Climate Models

11.1 MATHEMATICAL MODELING

We can better understand and predict climate and its variations by incorporating the principles of physics, chemistry, and biology into a mathematical model of climate. A climate model can range in complexity from the simple energy-balance models described in Chapter 2, whose solution can be worked out on the back of a small envelope, to very complex models that require the biggest and fastest computers and the most sophisticated numerical techniques. This chapter will discuss more complex global climate models, since these are the models that can produce the most realistic simulations of climate and are the tool on which we place our hope of predicting future climates in sufficient detail to be useful for planning purposes. These models are also useful for understanding the climate system and how various processes interact to determine the structure that climate change will take.

One can distinguish two general types of climate models at the present time. The first type specifies most of the key climate forcings and then uses a model of the physical climate system to compute the response of the climate. For brevity, we will refer to this type of model as "GCM," which we can take to be an acronym for *global climate model*. This type of model is also sometimes called an atmosphere–ocean general circulation model (AOGCM), since the fluid motions of both the atmosphere and ocean are explicitly calculated. A hierarchy of progressively less complicated climate models exists. These less detailed models are also useful for gaining understanding. Another level of complexity beyond the GCM is the *Earth system model (ESM)*, which may include elaborations way beyond the physical climate system, including the carbon cycle, soil evolution, vegetation that adapts to climate, and may even include human infrastructure and decision making as predicted variables within the model. Current ESMs include the carbon cycle to interactively predict atmospheric CO_2, and can also include chemical and biologic models to predict other

Global Physical Climatology. http://dx.doi.org/10.1016/B978-0-12-328531-7.00011-6

trace gases, aerosols, and cloud condensation nuclei. Models that allow the vegetation cover to react to climate change have also been developed. Most of the discussion here will focus on GCMs, rather than ESMs.

11.2 HISTORICAL DEVELOPMENT OF CLIMATE MODELS

The modem GCM has its roots in the computer models that were first developed to predict weather patterns a few days in advance. L. F. Richardson (1922) was the first to promote the idea that future weather could be predicted by numerically integrating the equations of fluid motion using the present weather as the initial condition. He attempted a hand calculation of a weather forecast while serving as an ambulance driver in France during World War I. The forecast was spectacularly inaccurate, because his initial conditions contained a large spurious wind convergence. The first successful numerical forecasts used a set of equations that are greatly simplified compared to Richardson's and for which the solution is less sensitive to the initial conditions.

Numerical weather prediction was proposed as one potential use for the electronic computer developed by John von Neumann in the late 1940s. The first successful numerical weather forecast with an electronic computer was conducted at the Institute for Advanced Study in Princeton, New Jersey (Charney et al., 1950). This model had only one atmospheric layer and described the region over the continental United States. It included only the fluid dynamical evolution of an initial velocity distribution and none of the physical processes that drive the climate. The first numerical experiment that included radiation and dissipation was conducted by Norman Phillips (1956) using a simple model with only two levels in the vertical. Later, more detailed simulations of the atmospheric general circulation included a more accurate formulation of the equations of motion, more spatial resolution in the horizontal and vertical, and more realistic specifications of the physical processes that drive the atmospheric circulation, such as radiation, latent heat release, and frictional dissipation (Smagorinsky, 1963; Manabe et al., 1965). With these and succeeding models, it has become possible to simulate the general circulation of the atmosphere with reasonable fidelity. The quality of the simulations obtained with atmospheric models has benefited greatly from intensive experimentation associated with providing practical weather forecasts, starting in 1955 with the first routine numerical weather forecasts that were conducted by the U. S. Joint Numerical Weather Prediction Unit. Associated with the practical effort to provide weather forecasts is a substantial effort to collect routine weather observations at the surface and at upper levels of the atmosphere. These observations are capable of describing the atmospheric state in sufficient detail to justify newly initialized forecasts about every 6 h, which are able to provide some useful information about the weather more than a week in the future. The initial states for weather forecasting that are constructed by assimilating all useful observations into a numerical weather prediction model also serve as the primary tool for diagnosing and understanding the atmospheric general circulation, a key part of the climate system, as discussed in Chapter 6. A key contribution to understanding the general circulation and validating climate models has been through *reanalysis* of weather data using state-of-the-art weather models to produce the best record of the atmospheric state during the recent past (Kalnay et al., 1996; Dee et al., 2011).

Numerical models of oceanic general circulation have been constructed by applying the same basic techniques used for atmospheric models. Development of oceanic general circulation models has lagged behind atmospheric models both because the observational base is less in the ocean and because the computational problems are greater for oceanic simulations. No longstanding effort has been undertaken to observe and predict the state of the global ocean on an operational basis, so oceanographers have not had a base in numerical weather forecasting on which to build an oceanic climate modeling enterprise until very recently. Operational forecasts with oceanic initial conditions were employed to make forecasts of El Niño starting in the 1980s. An observed climatology of the ocean that is sufficiently detailed and accurate for climate purposes is still being developed, particularly for the deep ocean. The spatial scale of the significant motion systems is small compared to the width of an ocean basin, and observations of currents, temperature, and salinity at depths below the surface must be acquired from ships and are expensive. Therefore, the number of observations taken in a year has typically been less than is necessary to define the state of the ocean, until perhaps quite recently, when profiling drifting buoys have been deployed across the global ocean and satellites have measured the sea level height variations. The speed of many important ocean currents is small and therefore difficult to observe with great precision, so observing systems focus on temperature, salinity, and ocean surface height. Also, a very long integration time is necessary for an oceanic model to achieve a stable climatology. Although the atmosphere will spin up to climatology from rest in a few weeks or months, the deep circulation in the ocean requires many centuries to spin up from rest or to respond to changed conditions. This long integration time and the high spatial resolution required for realistic simulations of ocean circulation mean that numerical experimentation with ocean models can require substantial and sophisticated computer resources. The number and complexity of the physical processes required for ocean simulation are much less than in the atmosphere, but the circulation is very sensitive to mixing by tides and other processes that are not well understood or modeled.

Early simulations of the atmosphere were obtained by fixing the sea surface temperature. Similarly, early ocean general circulation models assumed fixed wind stress, air temperature, air humidity, precipitation, and radiative forcing. Together these determined the flux of momentum, heat, and freshwater at the surface, which are the key driving forces of the ocean circulation. In a fully general model of the climate, the exchanges of heat, moisture, and momentum between the ocean and the atmosphere must be internally determined so that the state of the coupled climate system can evolve in a consistent manner. Initial experiments with fully coupled atmosphere–ocean models often produced climates that were very different from reality, even though the atmosphere and ocean components each produced a reasonable climatology when run independently with realistic fixed boundary conditions. The greater freedom of fully coupled atmospheric–ocean climate models to determine their own climate has revealed many deficiencies in atmospheric and oceanic general circulation models, and these deficiencies are being addressed through current research efforts. Improved climate simulation and prediction will also require land-surface processes to be treated with greater accuracy and detail.

The schematic parts of an Earth system model are shown in Fig. 11.1. An Earth system model includes not only the physical climate system

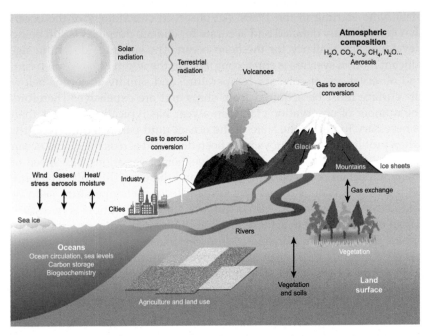

FIGURE 11.1 **Schematic of the Earth system.** The basic elements include the physical elements of the atmosphere, hydrosphere, cryosphere, and land surface, plus the biosphere, all of which interact strongly with humans. *Figure by Beth Tully.*

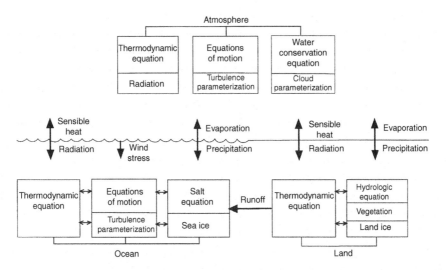

FIGURE 11.2 Schematic diagram of the physical components of a global climate model. This is divided into atmosphere, ocean, and land.

but also the biosphere and human influences, and predicts atmospheric composition, vegetation changes, and many other things not typically included in early climate models. The biogeochemical feedbacks, which determine the atmospheric composition and the nature of the land vegetation, are included. For example, increased CO_2 in the air can cause plants to open their leaf stomata less, which can affect the release of water by plants, which can in turn affect the soil moisture and temperature. These processes were not included in first-generation climate models, but are increasingly being included in Earth system models. All of the components and the processes represented within them are critical to the understanding and simulation of climate.

The component parts of a global climate model that includes only physical processes are shown schematically in Fig. 11.2. This is organized according to the three primary physical components of the climate system: the atmosphere, the ocean, and the land surface. The cryosphere is split between the oceans, where sea ice is important, and the land surface, where snow, glaciers, and ice sheets may form.

11.3 THE ATMOSPHERIC COMPONENT

The atmospheric component of a global climate model is basically the same as a model constructed for the purpose of weather forecasting. Improvements of numerical weather prediction systems have been fostered by efforts to predict weather more than a week in advance. For these longer predictions, the similarity of the model's "climatology" with reality is very important. The errors in the climatology of the model make a large contribution to the errors in these medium range forecasts. Efforts to make longer weather forecasts and to accurately simulate climate are thus synergetic and together are leading to improvements in atmospheric models.

The basic framework of a numerical model of the atmosphere or ocean is a spatial grid work on which the equations of physics are represented. Sometimes so-called spectral methods are used wherein the fields of velocity, temperature, and pressure are represented with continuous functions in order to efficiently solve the equations of motion. These methods allow very accurate calculation of the derivatives needed for following the flow of heat and momentum, but the fields must be interpolated to grid points in order to calculate radiative heating and the effects of sub-grid–scale processes such as convection. The globe is thus always divided into geographic regions or grid cells. The size and number of the grid cells are limited by the amount of computer power available and the time period over which the model must be integrated. Numerical weather prediction models that focus on deterministic 10-day forecasts currently use about 15 km horizontal resolution and about 150 vertical levels. Models that are run many times to make ensembles of forecasts to assess probabilities of weather events are generally run at lower resolution, but forecasts are done many times with slightly different initial conditions. Since climate models may be required to simulate decades or centuries, and include many more processes than a weather model, the spatial resolution is generally much less, currently about 100 km. The spatial resolution and the number of experiments that can be conducted are generally limited by the speed of the fastest computers, which steadily increases. Because the time step must also be reduced when the grid spacing is reduced, to increase the spatial resolution by a factor of two, say from 100 km to 50 km, requires about a factor of 10 increase in computing power.

Vertical resolution is also important to the quality of an atmospheric simulation. Climate models currently have typically about 20–50 levels. The levels are not equally spaced in height or pressure, but tend to be most closely spaced near the lower boundary and near the tropopause, where rapid changes take place and greater resolution is needed. The total number of latitude, longitude, and height grid points represented in an atmospheric model may thus be around 10 million.

11.3.1 The Conservation Equations

The basic dynamical framework of a GCM includes the solution of the equations describing the conservation of momentum, mass, and energy for a fluid. In addition, a water conservation equation for air is needed to predict the amount of water vapor in the atmosphere. Water vapor is critical for the formation of clouds and rain, and is one of the most important variables in the calculation of radiative heating. Numerical models of the atmosphere use the equations of motion in a simplified form called the *primitive equations*. Use is made of the fact that the atmosphere is thin in comparison to its horizontal extent in order to approximate the vertical momentum balance with the hydrostatic balance. Models that do not make the hydrostatic assumption are being developed. Earth is assumed to be spherical and some other small terms in the momentum equations are neglected. If the hydrostatic balance is assumed, the vertical coordinate can be height or pressure. Since it facilitates the inclusion of topography and makes it easier to construct a numerical solution that conserves mass, the *sigma coordinate*, which is pressure normalized by surface pressure, is often used as the vertical coordinate in atmospheric primitive equation models.

The sources and sinks of momentum and heat are associated with phenomena that occur at scales that are much smaller than the grid resolution and so cannot be explicitly simulated. These are often called sub-grid–scale phenomena, and their effects must be specified from knowledge of the state of the atmosphere at the grid scale. This process of including the effect of unresolved phenomena according to the large-scale conditions is called *parameterization*.[1] Key parameterizations in atmospheric models include radiation, the effects of unresolved turbulence and gravity waves, and the effects of convection on the heat, moisture, and momentum budgets. The behavior of the climate model is critically dependent on these parameterized processes.

11.3.2 Radiation Parameterization

Radiative heating from the sun and infrared cooling to space provide the basic drive of the climate system (Chapters 2 and 3). Radiative transfer models that are incorporated in current climate models do not consider each individual absorption line but rather treat groups of lines or band systems in a statistical or empirical manner. They consider the atmosphere and the embedded clouds to be horizontally homogeneous within a grid box. Despite these simplifications, it is generally believed that the transmission of radiation, at least under clear sky conditions, is accurately treated in climate models. Although there are some discrepancies between various parameterization schemes, the basic physical processes and the methods necessary to treat them are reasonably well understood. The largest uncertainties in radiative flux calculations are associated with clouds and the manner in which the amount and nature of the clouds in the model are determined. Early general circulation models typically specified cloud radiative properties as externally

---

[1]The word parameterization is used because one or more adjustable constants or parameters must be introduced to relate sub-grid–scale processes to large-scale conditions.

determined quantities that were often zonally invariant with a fixed seasonal variation. More recent climate models predict cloud amounts and optical properties, which can then interact with the other elements of the climate system. These newer models treat cloud microphysics in some parameterized way and carry cloud water and ice as predicted variables. The predicted cloud water and ice provide a means of connecting the water budget, radiation calculation, and heat budget in a manner that is consistent between the parameterized processes and the large-scale conservation equations.

11.3.3 Convection and Cloud Parameterization

Clouds affect radiative transfer, and the convective motions associated with clouds produce important fluxes of mass, momentum, heat, and moisture. The spatial scales at which cloud properties are determined are generally much smaller than the grid size at which the atmosphere is represented in a climate model, yet the vertical fluxes of heat and moisture associated with sub-grid–scale convection are often greater than those of the large-scale flow. The timing and intensity of precipitation in the tropics and over land areas during summer are controlled as much by unresolved mesoscale phenomena as by the large-scale flow. The atmospheric state averaged over the area of a typical climate model grid box may be stable to moist convection even when intense convection is occurring somewhere in the grid box. Cloud and convection parameterization seek to address the mismatch between the spatial resolution of climate models and the spatial scale of convective motions and clouds.

At least three important effects in a climate model are associated with the formation of clouds: (1) the condensation of water vapor and the associated release of latent heat and rain; (2) vertical transports of heat, moisture, and momentum by the motions associated with the cloud; and (3) the interaction of the cloud particles with radiation. A cloud parameterization in a climate model should treat each of these effects consistently, but this is not yet the case with every climate model. The release of latent heat during intense convection and the radiative effects of the associated clouds drive the Hadley circulation and other important components of the general circulation of the atmosphere, so that the coupling of deep convection to the large-scale flow is critical to a proper simulation of climate.

Most climate models include at least two types of clouds: convective clouds and large-scale supersaturation clouds. Large-scale supersaturation clouds occur when the relative humidity in a grid box at some model level exceeds a critical value. This can be implemented by assuming that condensation occurs in the grid box when the relative humidity reaches some threshold like 80%. Another alternative is to assume that sub-grid–scale temperature variability occurs within the grid box and that the fraction of the grid box where the temperature variability causes the relative humidity to reach 100% is cloud covered.

Convective clouds are associated with the buoyant ascent of saturated air parcels in a conditionally unstable environment. The simplest convective parameterization is moist adiabatic adjustment. If the lapse rate exceeds the moist adiabatic lapse rate (Fig. 1.10), the moisture and heat are readjusted in a vertical layer such that the air in the layer is saturated, the lapse rate equals the moist adiabatic lapse rate, and energy is conserved. Excess moisture is assumed to rain out, but momentum is not transported. In this parameterization, the entire grid box is assumed to behave like a convective element, whereas in reality convection occurs at much smaller spatial scales. Other parameterization schemes attempt to predict the effect of an ensemble of convective cloud elements on the air properties averaged over the grid cell, so that the grid cell is only infrequently saturated or unstable. In the future, when computing power is sufficient that climate models can be run with grid resolution finer than about 1 km, it is hoped that the parameterization of convection will be unnecessary, or at least less of a strong constraint on model behavior. Since cloud properties depend so importantly on the availability of cloud condensation nuclei and ice nuclei, state-of-the-art climate models used for predicting future climates will need to include explicit modeling of the abundance and chemistry of these aerosols.

11.3.4 Planetary Boundary-Layer Parameterization

In the atmospheric boundary layer, rapidly fluctuating phenomena with vertical and horizontal space scales much smaller than the grid spacing of climate models determine the fluxes of heat, momentum, and moisture between the surface and the atmosphere. These phenomena include turbulence, gravity waves, and rolls or other coherent structures that cannot be resolved by climate models and must therefore be parameterized. The simplest and most often used parameterizations are the bulk aerodynamic formulas and similarity theories briefly described in Section 4.6. These formulas allow the boundary layer to be treated as a single layer through which the fluxes can be calculated using the mean variables resolved by the model. They can be made dependent on vertical stability and surface roughness. These models may be elaborated by adding a prediction equation for the boundary-layer depth.

If the model has sufficient resolution to have several levels within the boundary layer, then eddy diffusion formulations may be used in which the vertical eddy fluxes are assumed to be proportional to the vertical derivative of the mean for the model grid box.

$$\overline{w'T'} = -K_T \frac{\partial \overline{T}}{\partial z} \qquad (11.1)$$

The simplest approach is to make the flux coefficient, K_T, a constant, but more realistic parameterizations include height and stability dependencies.

Higher-order closure schemes include prognostic equations for the turbulent kinetic energy in the boundary layer and more complex equations than (11.1) for the vertical eddy fluxes. These models require more than a few computational levels in the planetary boundary layer (PBL) and carry a heavier computational burden. It can be argued that statistical closure schemes do not describe the essential physics of the coherent structures that appear to produce much of the vertical flux in the boundary layer and thereby control the response of this flux to mean conditions. In many cases, the boundary layer contains clouds, and in such cases moist processes are critical to the behavior of the boundary layer. Some planetary boundary-layer parameterizations incorporate a separate parameterization for boundary-layer clouds. Oceanic stratocumulus and trade wind cumulus clouds are important examples where the interaction of the boundary layer with low clouds is essential. In regions of deep convection, the boundary layer properties and the cloud fluxes interact very strongly, so that the PBL parameterization and the convection parameterization must be compatible. The PBL parameterization must also be compatible with the land-surface parameterizations that may incorporate the effect of plant canopies on the fluxes of momentum, heat, and moisture between the surface and the atmosphere.

11.4 THE LAND COMPONENT

The land component of a climate model must contain the surface heat balance equation (4.1) and a surface moisture equation that is at least as sophisticated as the models described in Fig. 5.7 and includes a model for snow cover. Land topography can readily be incorporated in a climate model, but the fidelity with which Earth's surface topography can be included is limited by the resolution of the model. Many important mountain ranges are much narrower than the grid spacing of a GCM, so that smoothed representations of topography must be used.

Experiments with GCMs that include these basic prescriptions for the land surface indicate a substantial climate response to the soil moisture and to the surface albedo. Low soil moisture generally results in a warmer surface and less local evaporation and rainfall. These changes are consistent with the observation that much of the precipitation in such areas as the central United States during summer and the Amazon Basin during the rainy season is re-evaporated rainfall rather than water vapor that has been carried into the region by the atmospheric circulation. The changes in the surface fluxes of latent and sensible heat affect the thermal forcing of atmospheric flow and may thereby cause further weather and climate anomalies. Extreme events, such as the 2003 European drought have been shown to be exacerbated by preexisting dry soil (Fischer et al., 2007), and much of the interannual variability of summertime temperature over land is related to feedbacks between soil moisture and surface temperature.

In land areas, the storage of precipitation in the soil and the subsequent release of this moisture to the air are strongly dependent on the soil type and the vegetative cover. In addition, the absorption of solar radiation and the emission of longwave radiation are sensitive to the geometry and physical state of the vegetative cover. Turbulent transports of heat and moisture between the atmosphere and the soil are also affected by the physical structure of the plant canopy, particularly for forested areas. Parameterizations of surface processes that include all of these effects have been developed in recent years and are being incorporated in climate models. These parameterizations treat the vegetative cover as a variable resistance that moderates the flow of moisture from the soil to the atmosphere. Over land areas, the diurnal cycle of convection is very important, and parameterizations often initiate convection too early or too late and this affects the mean climate over land.

11.5 THE OCEAN COMPONENT

The oceanic component of a global climate model is also built on a framework of the equations of motion describing the general circulation of the ocean, although informative climate simulation experiments can also be conducted with fixed sea surface temperatures, or with an ocean model that is a simple mixed layer that stores heat and releases water vapor, but does not transport heat. Ocean general-circulation models differ from atmospheric models in that the fluid is water, rather than air, and the geometry of the ocean basins is more important and complex. The fundamental driving mechanisms for the ocean circulation include wind stress at the upper surface, radiative and sensible heat fluxes through the surface, and density variations caused by changes in salinity and temperature. Salinity changes are induced at the surface by precipitation, evaporation, runoff, salt rejection during sea ice formation, and the addition of freshwater during sea ice melting. These salinity variations are carried to the interior of the ocean by fluid motions that are forced in part by the density variations associated with salinity and heat.

The spatial resolution with which the oceanic circulation can be resolved is again limited by the computing time required to complete a simulation.

A typical model at the present time might have horizontal resolution of about 10 km with around 50 vertical levels. When you consider that critically important features such as the Gulf Stream and Kuroshio currents are less than 100 km in width, the horizontal resolution currently used in the ocean component of climate models is just at the margin of being adequate. Most models that are run for climatological purposes require parameterized diffusion to simulate the effect of eddies on the general circulation. Mixing by tidal motions and turbulence are small in the ocean, but have a very strong effect on the resulting ocean climate. In atmospheric models, the corresponding eddies are much larger in scale and can be explicitly modeled without parameterization. Ocean model runs with higher spatial resolution reveal many more details and features that are not well represented in simulations with coarser resolution. In some regions such as the Southern Ocean, resolution of transient ocean eddies is necessary to obtain a realistic ocean climate.

11.6 SEA ICE MODELS

Sea ice plays a critical role in climate by increasing the albedo of the ocean surface, inhibiting the exchanges of heat, moisture, and momentum between atmosphere and ocean, and altering the local salinity during sea ice freezing and melting. Thermodynamic processes lead to freezing and melting of seawater, and dynamic processes, such as driving by winds and currents, cause mechanical deformation and transport of sea ice. In most climate models, at least a thermodynamic model of sea ice is employed. Transport of sea ice by winds and currents is treated with varying degrees of complexity, ranging from mixed-layer ocean models with no ice transport at all, to more complete models that calculate the movement of ice in response to both winds and currents.

The simplest model predicts the thickness of a layer of sea ice based on thermodynamic considerations. When the temperature of the upper layer of the ocean reaches the freezing point of seawater ($-2°C$) and the surface energy fluxes continue to remove heat from the water, then surface ice is assumed to form. Heat transport through this layer of ice is assumed to be described by a flux-gradient relationship.

$$F_I = -k_I \frac{\partial T}{\partial z}$$

(11.2)

A layer of snow may be present on top of the sea ice through which heat is also conducted. The thermal conductivity for ice, $k_I \approx 2\ \mathrm{W\ m^{-1}\ K^{-1}}$, is much larger than the thermal conductivity of snow, $k_s \approx 0.3\ \mathrm{W\ m^{-1}\ K^{-1}}$, so that a layer of snow greatly increases the thermal insulation provided by the sea ice. Snow cover also increases the albedo. If all heat fluxes,

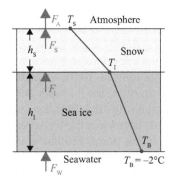

FIGURE 11.3 Schematic diagram of a simple sea ice model with a layer of snow over a layer of ice and the fluxes of heat through the layers. The temperature profiles in the two layers are linear if penetration of solar radiation is ignored and a steady state is assumed. Symbols are as defined in the text.

including solar heating, are assumed to be applied at the surface rather than distributed over some depth, then in steady state the flux through the snow and ice would be equal at every depth. Constant flux implies a linear temperature profile, so that the heat flux becomes proportional to the temperature difference across the ice layer. Under these steady conditions, the temperature profile within the layer of snow and ice would be as shown in Fig. 11.3. The fluxes across the ice and snow layers can then be written in terms of the temperature difference across the layers.

$$F_I = k_I \frac{T_B - T_i}{h_I}, \quad F_s = k_s \frac{T_i - T_s}{h_s} \tag{11.3}$$

where T_s is the surface temperature, T_B is the temperature at the base of the sea ice layer, T_i is the temperature at the interface between ice and snow, and h_I and h_s are the depth of the ice and snow layers, respectively. The interface temperature can be eliminated by applying the requirement that in a steady state the heat flux must be the same through the snow layer and the ice layer. The flux through the sea ice can then be expressed in terms of the temperature difference across the snow–ice layer and the depth of each layer.

$$F_I = \frac{k_I k_s}{k_I h_s + k_s h_I} (T_B - T_s) = \gamma_{SI} (T_B - T_s) \tag{11.4}$$

Here γ_{SI} is the combined thermal conductance of the snow–ice layer. For a thick layer $\gamma_{SI} \approx 1$ W m^{-2} K^{-1}, and for a thin layer $\gamma_{SI} \approx 20$ W m^{-2} K^{-1}. The surface temperature can be determined from the requirement that the net heat flux at the surface be zero, which is consistent with our previous assumption of no heat storage in the ice. The net heat flux includes the conductive flux through the ice, and the latent, sensible, and net radiative heating of the surface. The depth of the snow layer is determined by the balance of snowfall, sublimation–evaporation, and melting. If no snow is present, then the ice layer may be reduced in thickness by melting and sublimation at the top. Growth of the ice layer occurs at the bottom, where it is assumed to thicken at a rate such that the heat of fusion will just balance the net heat flux at the bottom of the ice layer and thus keep the lower edge of the ice at the freezing point. Similarly, if the equilibrium surface temperature would exceed the freezing point, ice is assumed to melt at a rate sufficient to keep the temperature at the freezing point until all of the ice is melted.

Sea ice grows rapidly when it is thin and more slowly as thickness increases; when subjected to similar thermal forcing, ice a few centimeters thick grows nearly a hundred times faster than ice that is 2 or 3 m thick. The reason for this is because of the insulating effect of the ice itself. Consider a sheet of sea ice that is growing as a result of an imposed temperature difference between the ice surface and the seawater. The thickness of the sea ice increases at a rate necessary to balance the heat flux through the ice with the latent heat of fusion needed for freezing seawater at the base of the ice.

$$\rho_I L_f \frac{\partial h_I}{\partial t} = \frac{k_I}{h_I}(T_B - T_s) - F_w \qquad (11.5)$$

Here ρ_I is the density of ice, L_f is the latent heat of fusion for seawater, and F_w is the rate at which heat is supplied to the sea ice by ocean fluxes. It is notable that for fixed surface temperature, the growth rate of the ice thickness is inversely proportional to the thickness itself. Integrating (11.5) over time, assuming $F_w = 0$, we find that the ice thickness is proportional to the square root of the integral over time of the temperature difference, so that the ice grows more slowly as it thickens.

$$h_I^2(t) - h_I^2(t_0) = \frac{k_I}{\rho_I L_f} \int_{t_0}^{t} (T_B - T_s) dt \qquad (11.6)$$

Figure 11.4 shows solutions to (11.5) for different values of F_w, assuming that the ice thickness is initially very thin and that $T_B = -2°C$ and $T_s = -30°C$. This illustrates how thin ice initially grows very quickly, then more slowly as the ice becomes thicker. Figure 11.4 also shows that the equilibrium thickness of the ice is very sensitive to the amount of heat supplied by the ocean to the bottom of the sea ice. If the ocean heat flux to the bottom of the ice is large, then the sea ice equilibrates quickly to a thin layer. If the ocean heat flux is low, then the sea ice can grow slowly too much greater thickness.

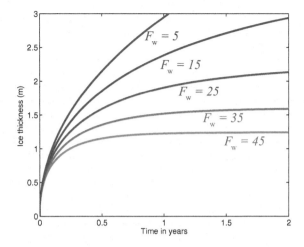

FIGURE 11.4 **Plot of sea ice thickness as a function of time from the simple model (11.5) for different values of the ocean heat flux into the bottom of the ice layer F_w in units of W m^{-2}.** Ice thickness is initially very thin, grows rapidly at first, and then reaches equilibrium thickness, which is very sensitive to the ocean flux. ($T_s = -30°C$, other parameters as described in the text.)

Sea ice models used in state-of-the-art climate models are much more complex than the simple illustration given above. Important thermodynamic processes, such as the heat capacity of the sea ice and the brine incorporated in the ice have important effects on the amount and seasonality of sea ice and its response to climate forcing (Bitz and Lipscomb, 1999). Horizontal transport of sea ice is also important in many situations and plays a significant role in heat and salt transport in the high-latitude oceans. In the Arctic Ocean, the mean ice thickness is about 3–4 m, whereas in the Antarctic the average thickness is 1–2 m. The thinner Antarctic sea ice may be related to a greater supply of heat to the sea ice by ocean fluxes in the Antarctic. Also, more mechanical deformation of the ice in the closed basin of the Arctic Ocean may contribute to the greater average thickness there compared to the Antarctic sea ice, which is less constrained by shorelines and can simply spread equatorward until it melts (Fig. 11.5). Where sea ice is driven together by winds and current, pressure ridges can be formed where the ice thickness exceeds 10 m. As part of the same process, small openings in the ice (leads) or larger areas of open water (polynyas) can be produced. In these regions of open water, enhanced heat loss and ice formation occur in the season of ice growth, and more absorption of solar radiation occurs in the melting season. Around Antarctica, the formation of leads near the shoreline by the offshore advection of sea ice enhances the freezing rate of seawater, and the associated salt rejection is an important driver of bottom water formation around Antarctica.

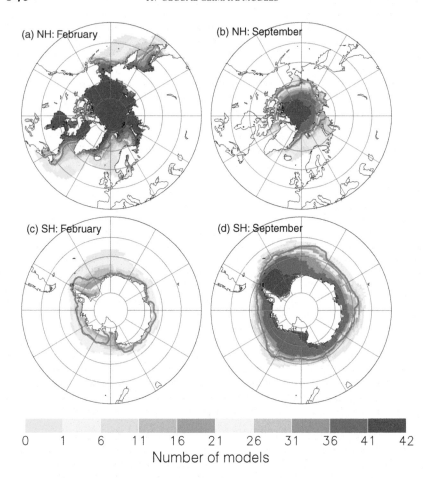

FIGURE 11.5 **Sea ice distribution (1986–2005) in the Northern Hemisphere (a,b) and the Southern Hemisphere (c,d) for February (c) and September (d).** AR5 baseline climate (1986–2005) simulated by 42 CMIP5 AOGCMs. Each model is represented with a single simulation. For each 1° × 1° longitude–latitude grid cell, the figure indicates the number of models for which at least 15% of the area is covered by sea ice. The observed 15% concentration boundaries (red lines) are based on the Hadley Centre Sea Ice and sea surface temperature (HadISST) data set (Rayner et al., 2003). *Adapted from Pavlova et al., 2011. Figure and caption reproduced exactly from IPCC WG I Report Fig. 9.43, Flato et al. (2013).*

11.7 VALIDATION OF CLIMATE MODEL SIMULATIONS

The simulations obtained with current climate models are in reasonable accord with observations of the present climate for a variety of key dynamic and thermodynamic climate variables (Flato et al., 2013). Models can produce most key features of the atmospheric general circulation discussed in Chapter 6. Current models can reproduce the observed seasonal climatology of surface temperature to fairly high accuracy, with some exceptions over high topography or in regions of ocean upwelling in the tropics (Reichler and Kim, 2008). Accurate reproduction of observed surface temperatures over the equatorial oceans where the atmosphere and ocean are strongly coupled remains a challenge for climate models. Coupled atmosphere–ocean models simulate the observed ENSO cycle with varying degrees of fidelity (Guilyardi et al., 2009). The simulation of precipitation patterns is less good than temperature, with pattern correlations of 0.82 compared to 0.99 for surface temperature. The simulation of clouds remains a significant challenge for climate models. The simulation of the observed distribution of clouds by the model ensemble still has some biases and it is recognized that cloud feedback remains one of the primary uncertainties in modeling the climate response to greenhouse gas increases.

If the observed climate forcing by greenhouse gases, aerosols, volcanoes, and total solar irradiance are specified, then climate models can simulate the observed global mean surface warming and ocean heat content increases of the twentieth century. As far as we can tell from the observed climate record, models can do a reasonable job of simulating the year-to-year variability of Northern Hemisphere and global surface temperature variations, and many models currently also simulate the natural variability associated with El Niño.

This agreement is highly encouraging, but it does not by itself mean that climate models are capable of accurately predicting the response of climate to a natural or human-induced perturbation. The reason for skepticism is that a large number of adjustable constants are introduced in the parameterizations for sub-grid–scale phenomena and processes, and these parameters often cannot be determined on the basis of fundamental principles, but rather are set to values that give the most realistic-looking simulation of the current climate. For example, ocean models are very sensitive to the specification of sub-grid–scale mixing. Real confidence in the predictive capability of climate models can be gained by testing their simulations in great detail, so that the number of observational constraints is not too small compared to the number of independently specified parameters. It is also critically important to test their response to prescribed forcings for which the response is known. Examples of prescribed forcings are the annual and diurnal cycles of solar heating, the response to an event such as a major volcanic eruption, the response of the atmosphere to an observed SST anomaly, or the response of the climate model to the boundary conditions of an ice age (see Chapter 12). In this section, we will discuss what aspects of the climate system current models can simulate well and some aspects that remain problematical.

At the present time, many organizations in many countries have developed and operate climate models that simulate the coupling of the atmosphere, ocean, cryosphere, and land surface, and thus can be used to make projections of how the climate will respond to past and future changes of the climate system that may occur as a result of natural causes or intervention in the climate system by humans. A process for conducting common experiments and sharing simulation data has been organized under the auspices of the Climate Model Intercomparison Project (CMIP). At the time of this writing the results from CMIP5 are available (Taylor et al., 2012) and were used in the Intergovernmental Panel on Climate Change (IPCC) 5th Assessment (Stocker et al., 2013). To the extent that these models are independent, comparing them gives insight into our ability to model the climate system robustly and to what extent the predictions about future climate are robust.

An example of using model ensembles to evaluate our ability to simulate climate is given in Fig. 11.5, which shows the simulation of sea ice by CMIP5 models. The simulation of sea ice is a good and important test for climate models; since it involves both the ocean circulation and atmospheric circulation, surface ice feedback is very important for climate change projections, and the fate of sea ice is one of the important questions related to global warming. The red line shows the point where the satellite-observed sea ice concentration climatology is 15%, with latitudes poleward of this point generally having greater than 15% sea ice concentration. The color shading in the plot indicates how many models have at least 15% sea ice concentration at that point. In most cases, the red line falls where the color shading is between green and blue. This means that about half the models have more than 15% concentration and half have less. This is a good result since it means that the average of the models is probably about right. On the other hand, one can clearly see that some models have far too much sea ice and others have far to little. In general, taking the average of the models, the multimodel mean, is almost always better than picking a model at random, and for most purposes also better than picking the one model that seems to be best at whatever metric of success that you consider. The error of the multimodel mean of the seasonal distribution of sea ice is less than about 10%.

11.8 FEEDBACK STRENGTH AND SENSITIVITY ESTIMATES FROM CLIMATE MODELS

Global climate models can be used to estimate how various processes interact to determine the sensitivity of climate. In this section, we review some of the key results from using GCMs to assess climate sensitivity mechanisms. The basic concepts of forcing, feedback, and sensitivity were introduced in Chapter 10. Model feedbacks and sensitivities can be assessed by attempting to simulate the observed record, or by imposing arbitrary idealized forcings on climate models and assessing their response. Climate forcings and temperature changes that have been so far observed are generally small, so a big arbitrary forcing often makes a more exciting experiment, but it is much more compelling to compare observed climate responses with observed forcings that the real climate system has actually experienced.

11.8.1 Water-Vapor Feedback

It is certain that water-vapor feedback exists and it is the strongest positive feedback, but since it is so strong, uncertainties in its magnitude are practically significant, especially since the strength of individual feedbacks magnify the effect of every other feedback. For example, if water-vapor feedback is stronger (or weaker), it is likely that it will warm more in response to a specified forcing, and the net effect of ice-albedo feedback will be greater (or less).

An observational test of water-vapor feedback in a climate model was made possible by the eruption of Mt. Pinatubo in June of 1991. Mt. Pinatubo was an explosive eruption containing large amounts of SO_2, so it made a significant aerosol cloud in the stratosphere that cooled the climate for several years. The surface cooling peaked at about 0.5 K about 18 months after the eruption. The response of the Earth radiation budget and the temperature and humidity profiles in the atmosphere to the Pinatubo eruption were well measured by global satellite observations. The absorbed solar radiation declined about 4 W m^{-2} when the stratospheric aerosol cloud was at its maximum. The specific humidity of the atmosphere decreased by about 3%, in good agreement with the Clausius–Clapeyron relation indicating 7% per °C, assuming nearly constant relative humidity.

A schematic diagram of the responses of top-of-atmospheric radiation fluxes together with tropospheric temperature and column water vapor to a volcanic eruption is shown in Fig. 11.6. A little over a year after the eruption the absorbed shortwave radiation has decreased, owing to the reflection of sunlight by stratospheric aerosols. The shortwave response is delayed about a year because of the time it takes the released SO_2 gas to convert into sulfuric acid aerosols. The troposphere starts to cool in response to the reduction in solar heating, and the column water vapor decreases with the temperature in good accord with an assumption of fixed relative humidity. The OLR decreases as the planet cools, and because the aerosol cloud also absorbs some of the OLR and emits at a lower temperature. The cooling and drying persist longer than the stratospheric aerosol cloud because of the heat capacity of the ocean.

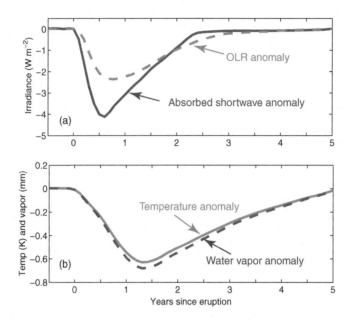

FIGURE 11.6 Schematic diagram of the observed anomalies of (a) absorbed shortwave and outgoing longwave radiation (OLR) and (b) tropospheric temperature and column water vapor to an explosive volcanic eruption that puts SO_2 into the stratosphere. *Real data for the eruption of Pinatubo in 1991 is shown in Soden et al. (2002).*

Soden et al. (2002) used a GCM to compute the response to the Pinatubo aerosols. The model produced the observed cooling and the observed reduction in tropospheric specific humidity. Using the model, they were able to show that the observed cooling is not obtained with the model unless the positive feedback associated with the specific humidity reduction is included. The experiment thus proved that the model could simulate the observed water vapor response and also that the positive water-vapor feedback of the magnitude simulated in the model was necessary to obtain the observed cooling. This and many other analyses of models and observations indicate that water-vapor feedback is strongly positive, as expected, and that GCMs can simulate it very well.

11.8.2 Planck and Water-Vapor Feedback in GCMs

Feedback processes in GCMs can be assessed using linear methods. The radiative transfer calculations performed by GCMs can be used to calculate the linear sensitivity of the top-of-atmosphere energy budget to changes of temperature and water vapor at any point in the atmosphere. Suppose that we have conducted a climate change experiment with a GCM and we want to assess the relative importance of Planck, water vapor, cloud, and surface albedo feedbacks. If we have applied some climate forcing dQ, then the temperature (T), water vapor (q), cloud (C), and surface albedo (α_s) changes must conspire to change the top of the atmosphere radiation balance by a compensating amount to achieve a new equilibrium for which $dR_{net} = dQ$. This change in radiation balance can be written,

$$dR_{net} \cong \frac{\partial R_{net}}{\partial T} dT + \frac{\partial R_{net}}{\partial q} dq + \frac{\partial R_{net}}{\partial C} dC + \frac{\partial R_{net}}{\partial \alpha_s} d\alpha_s \qquad (11.7)$$

Note that T, q, and C are functions of longitude, latitude, pressure, and time, whereas surface albedo is a function of longitude, latitude, and time, and $R_{net} = Q_{abs} - OLR$ has a solar and a terrestrial radiation component. It is also helpful to note that (11.7) can be written as,

$$dR_{net} \cong dR_{net}^T + dR_{net}^q + dR_{net}^C + dR_{net}^{\alpha_s} \qquad (11.8)$$

so that one can explicitly think of each term as contributing part of the change in the radiation balance that results from warming.

The partial derivatives can be calculated by changing T, q, C and α_s by standard amounts. Then these partial derivatives can be multiplied by the actual changes from a climate change experiment $(dT, dq, dC, d\alpha_s)$ to obtain the feedbacks. The sensitivities in (11.7) were first calculated and used by Soden et al. (2008) who termed them radiative kernels, so we introduce a kernel notation for each feedback variable, rewriting (11.7) as,

$$dR_{net} \cong K_T \, dT + K_q \, dq + K_C \, dC + K_{\alpha_s} \, d\alpha_s \qquad (11.9)$$

K_T translates the changes of temperature at every pressure, latitude, longitude, and season into a change in the top-of-atmosphere radiation balance at every latitude, longitude, and season. K_q does this for humidity changes, K_C for clouds, and K_{α_s} for surface albedo.

The feedbacks can be referenced to the global mean surface temperature change $d\bar{T}_s$, which will facilitate comparing and combining feedbacks from different models. We can thus also write (11.7) as

$$dR_{net} \cong \left(\lambda_T^{-1} + \lambda_q^{-1} + \lambda_C^{-1} + \lambda_{\alpha_s}^{-1} \right) d\bar{T}_s \qquad (11.10)$$

Here these feedbacks are defined as

$$\lambda_x^{-1} = \frac{\partial R_{net}}{\partial x} \frac{dx}{d\bar{T}_s} \qquad (11.11)$$

where $x = T$, q, α_s, and C represent the influence on feedback of temperature, humidity, surface albedo, and cloudiness.

Fig. 11.7 shows the kernels, the changes in climate parameters and the feedbacks for temperature and water vapor averaged for a suite of 28 GCMs from the Fifth Coupled Model Intercomparison Project (CMIP5, Taylor et al., 2012) averaged over season and longitude. The changes and

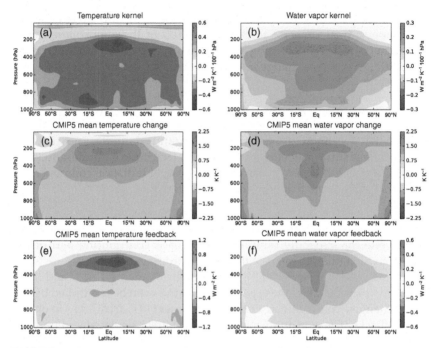

FIGURE 11.7   **Kernels, changes, and feedbacks for temperature (Planck feedback) and water vapor (water-vapor feedback) averaged over season, longitude and a suite of CMIP5 models.** Note that the abscissa in these plots is scaled by sine of latitude, so that the relative importance of latitude zones for the global mean is properly represented. The area between the equator and 30°N is the same as the area from 30°N to 90°N. *Figure courtesy M.D. Zelinka.*

feedbacks are normalized to a global mean temperature increase of 1 K to remove variations associated with the sensitivity of individual models and allow the unbiased construction of a multimodel mean. The effects of average model cloudiness are included in the computation of the kernels. Clouds partially screen the effects of water vapor, temperature, and albedo changes that occur below them. For water vapor, the increment used to compute the kernels is the change in specific humidity associated with a 1 K increase in temperature if the relative humidity is kept constant. So both the temperature and water vapor kernels are normalized for 1 K temperature increase and can be directly compared.

Figure 11.7a shows the sensitivity of R_{net} to temperature changes in 100 hPa thick layers, averaged over longitude and season. R_{net} decreases for temperature increases at all locations, since the Planck emission increases with temperature. The sensitivity is a little less very near the surface, and a little more just below the tropical tropopause and at the top of the boundary layer in the southern subtropics. These are regions where the transmission is changing with height and the OLR is more sensitive to temperature (Fig. 3.9). Figure 11.7b shows the same thing for water vapor.

R_{net} is most sensitive to water vapor changes in the upper troposphere, since this is the region where emission to space from water vapor is greatest. Increases in water vapor move the emission upward to lower temperatures and thus trap more longwave radiation, so the R_{net} sensitivity is everywhere positive. R_{net} is less sensitive to water vapor changes lower in the atmosphere, since a blanket of water vapor and clouds lies above these lower levels, and this mutes the effect of changes at lower levels on radiative fluxes at the top of the atmosphere.

Figure 11.7c shows the average temperature change, normalized so that the global mean surface warming is 1 K. Models generally warm the upper tropical troposphere and the polar regions about twice as much as the global mean. The water vapor increase shows a similar amplification in these two regions, as well as an enhancement along the equator where the models on average generate more intense convection in a warmed Earth (Fig. 11.7d). Figure 11.7e,f show the product of the upper two panels. Temperature feedback is negative everywhere in the troposphere, with stronger negative feedback in the upper tropical troposphere, not only because it warms more there but also because the kernel is larger there. The stratosphere cools in response to CO_2 increases, but the contribution of this to total feedback is fairly modest. Since the coordinates are proportional to area and mass, the total feedback can be estimated from visual averaging. The water-vapor feedback is positive and concentrated in the upper tropical troposphere (Fig. 7.11f). Note that the enhanced warming in the upper tropical troposphere causes enhanced negative feedback, but this is largely offset by increases in the positive feedback caused by the water vapor increase. This is an expression of the cancellation of lapse-rate feedback by water-vapor feedback discussed in relation to Fig. 10.6.

11.8.3 Cloud Feedback and Forcing Adjustment

Cloud feedback in climate models can also be estimated. It is of interest to know the relative contributions of cloud fractional area, cloud top height and cloud optical depth. These feedbacks can be estimated if these cloud properties are saved when the model computations are performed. It is useful to use an *'instrument simulator'*, which takes the model variables and calculates the signals that an instrument on an Earth-orbiting satellite would measure, if that instrument were orbiting above the model. These simulated cloud observations can then be used both to test the model against observations and to assess the feedbacks using *"cloud kernels"* (Zelinka et al., 2012a,b). Such experiments with a simulator of the measurements from the International Satellite Cloud Climatology Project (ISCCP, Rossow and Schiffer, 1999) were done as part of the Cloud Feedback Model Intercomparison Project (CFMIP).

When the climate is forced by the introduction of additional green-house gases, such as a doubling of CO_2, the CO_2 changes the radiative heating rates in the atmosphere, which has an almost immediate effect on the clouds, since cloud-producing convection acts to balance the radiative cooling of the atmosphere (Fig. 3.18). Because the surface temperature may not have reacted yet to the CO_2 increase, owing to the heat capacity of the ocean, the effects of these fast-response cloud changes can be isolated in models and are usually assigned to the forcing, rather than to the feedback. One way to study these effects is to double or quadruple the CO_2 instantaneously, then try to measure the direct effect of the CO_2 increase on the clouds before the climate warms up very much (Zelinka et al., 2013). Such efforts show a relatively strong fast response of clouds to CO_2 before the climate begins to warm. Cloud fractional area, cloud top height, and cloud optical depth all decrease in response to a CO_2 increase, such that the cloud adjustment produces a net increase in the radiation balance, which adds to the direct warming effect of CO_2.

One can write the change in the radiation balance due to cloud changes, dR_C as an initial adjustment F_C, plus a part that depends linearly on the global mean surface temperature change, dT_s, and thus measures the cloud feedback λ_C^{-1}.

$$dR_C = F_C + \lambda_C^{-1} dT_s \tag{11.6}$$

The cloud adjustment to forcing and the cloud feedback can be estimated with a model by applying an instantaneous forcing, such as $4 \times CO_2$ or a 2% increase in TSI and then recording the change in the radiation balance caused by cloud changes and the surface temperature as the climate warms in response to the climate forcing (Gregory and Webb, 2008). It is apparent that the adjustment will be different for greenhouse gas increases and TSI changes. For a CO_2 increase, the atmospheric cooling is suppressed and for TSI the surface heating is enhanced, giving very different initial cloud responses. It is found that, although the fast response and the adjustment to forcing are different for CO_2 and TSI forcing, the feedbacks are very similar. This similarity of the estimated feedback strength for very different forcings is another good argument for assigning the fast, prewarming, adjustments of clouds to the forcing and including only the cloud changes that respond to temperature change in the calculation of cloud feedback. A drawback of this approach is that it buries some of the uncertainty associated with clouds in the forcing, which can otherwise be calculated very precisely if the temperature, water vapor, and clouds are assumed fixed. We can call the forcing that is adjusted for rapid responses of the stratosphere and clouds the *effective forcing*. An estimate of the effective climate forcing from doubling the CO_2 in current models is $3.5 \pm 0.5 \text{ Wm}^{-2}$.

After this initial adjustment, the temperature begins to increase and cloud feedback can occur. Figure 11.8 shows the changes in cloud

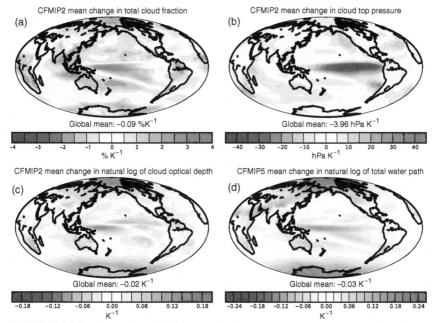

FIGURE 11.8 **Change in cloud properties in response to warming resulting from 4 × CO₂ forcing from the multimodel mean of CFMIP2 abrupt 4 × CO₂ experiment.** (a) Cloud fractional area, (b) cloud top pressure, (c) optical depth, and (d) total cloud water path. Changes are normalized for 1 K; global mean temperature increases. *Courtesy of M.D. Zelinka.*

fractional area, cloud top pressure, cloud optical depth, and cloud total water path for the CFMIP2 multimodel mean. Models show a redistribution of cloud fraction with warming, with increases in the Central Pacific and in high latitudes, and decreases elsewhere. The tropical changes are mostly related with the tendency of CMIP5 models to warm more in the eastern equatorial Pacific Ocean than elsewhere, giving an El Niño-like response. In the subtropics, enhanced entrainment of dry air and less radiative loss from cloud tops may be the cause of a general reduction in low cloud fraction. Cloud tops move upward and to lower pressures with warming (Fig. 11.8b). This is related to the increase in water vapor and increased efficiency of clear sky radiative cooling at higher altitudes in a warmed Earth, and with the greater amount of moist static energy in surface air. High clouds generally move upward with the expansion of the troposphere. Cloud optical depth and total cloud water path increase at high latitudes and in convective regions of the tropics, but decrease slightly elsewhere. This is in part related not only to the increase in water vapor with warming but also to the conversion of cloud ice to cloud water as the models warm. Cloud water persists in the atmosphere longer than cloud ice and is more reflective per unit mass, owing to the smaller particle radius of liquid water clouds.

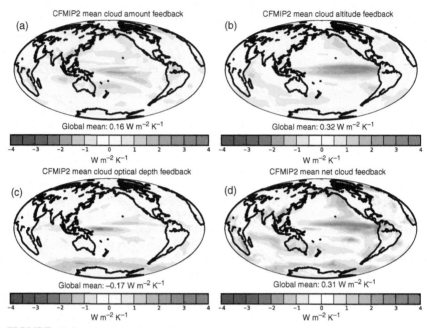

FIGURE 11.9 **Cloud feedbacks from the multimodel mean of CFMIP2 abrupt 4 × CO₂ experiment.** (a) Cloud fractional area, (b) cloud top pressure, (c) optical depth, and (d) net cloud feedback. Changes are normalized for 1 K; global mean temperature increases. *Courtesy of M.D. Zelinka.*

Figure 11.9 shows the feedbacks associated with the cloud property changes shown in Fig. 11.8. Reductions in cloud fraction give a net positive feedback in most regions because the clouds have a negative impact on the radiation balance. Much of the cloud reduction is in low clouds over the oceans, which have a particularly strong negative impact on the radiation balance (Hartmann et al., 1992). High cloud altitudes increase nearly everywhere and this gives a strong positive cloud altitude feedback. Optical depth feedback in the models is also positive over the subtropics, but strongly negative in other regions, which dominate to give a net negative feedback from cloud optical depth increases. The brightening of clouds over the Southern Ocean is particularly strong. Warming causes the microphysics in these models to convert ice clouds to water clouds, which increases the liquid water path and cloud reflection. Net cloud feedback in the models is overall fairly strongly positive, but has strong spatial gradients. It is very positive in the tropical Pacific, but quite negative over the Southern Ocean. These gradients in net feedback give rise to circulation responses. Despite the fact that the meridional gradient of surface temperature weakens, the meridional gradients of net top-of-atmosphere radiative heating and moist static energy both increase, and so does the meridional energy transport.

11.8.4 Ice–Albedo Feedback

Climate models simulate land snow cover and sea ice on a seasonal basis, and so the effect of ice–albedo feedback is included in climate models. Models designed for projecting the climate a century or two into the future generally assume that the permanent ice sheets over Greenland and Antarctica are fixed in place, although the effects of snow cover and surface melting on the albedo of ice sheets can be included. The surface albedo kernel and the magnitude of ice–albedo feedback in CMIP5 models are shown in Fig. 11.10. The impact of surface albedo changes on the energy balance would be greatest in cloud-free regions of the tropical oceans,

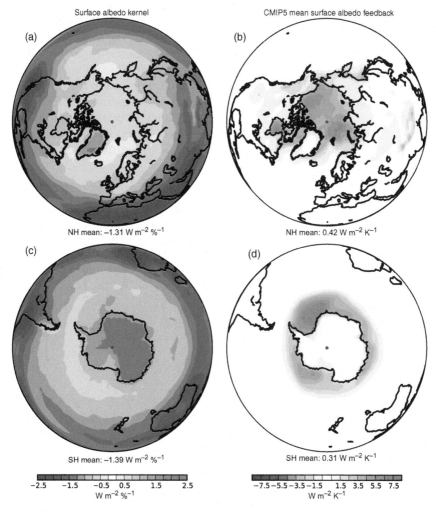

Surface albedo kernel

(a)

CMIP5 mean surface albedo feedback

(b)

NH mean: −1.31 W m⁻² %⁻¹

NH mean: 0.42 W m⁻² K⁻¹

(c)

(d)

SH mean: −1.39 W m⁻² %⁻¹

SH mean: 0.31 W m⁻² K⁻¹

-2.5   −1.5   −0.5 0.5   1.5   2.5
W m⁻² %⁻¹

−7.5 −5.5 −3.5 −1.5   1.5 3.5 5.5 7.5
W m⁻² K⁻¹

FIGURE 11.10 **Surface albedo kernel (a,c) and surface albedo feedback estimate (b,d) from CMIP5 models abrupt 4 × CO₂ experiments.** *Courtesy of M.D. Zelinka.*

since the insolation is large and the current planetary albedo is relatively low there. In high latitudes where the insolation is less and the cloud cover is greater, the sensitivity to surface albedo changes is less, but this is the region where ice and snow are present and can contract or expand with warming or cooling. The surface albedo feedback is mostly concentrated over the high latitude or high altitude land areas of the Northern Hemisphere and over the regions of sea ice in the Arctic and Antarctic Oceans.

11.8.5 Summary of Model Energetic Feedbacks

By considering an ensemble of models and adjusting the forcing for the fast response to CO_2, one can assess the magnitude of the key climate feedbacks and their uncertainties. Figure 11.11 shows feedbacks and uncertainties from a suite of CMIP5 model simulations. The uncertainties are estimated as twice the standard deviation of the model-to-model variation, which gives a general notion of which feedbacks are well quantified and which are still uncertain. A two standard deviation variation also approximates the spread of the individual model assessments of feedback from least negative to most negative. The Planck feedback is the top-of-atmosphere radiation balance change induced by a uniform increase in temperature of 1 K everywhere. This gives a strong negative feedback

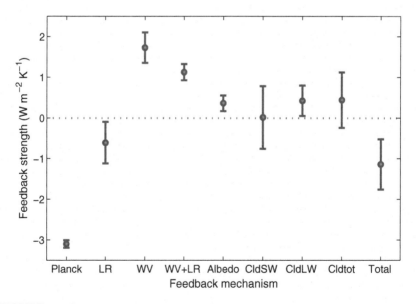

FIGURE 11.11 **Strengths and uncertainties of various feedbacks from a suite of CMIP5 models.** Uncertainties are estimated as ±2 times the standard deviation among the 28 models in the ensemble. LR, Lapse rate; WV, water vapor; Cld, Cloud; SW, shortwave; LW, longwave; Tot, LW + SW; Total is sum of all feedbacks. *Data Courtesy of Mark Zelinka.*

of about -3.1 W m^{-2} K^{-1} with little uncertainty, corresponding to the sensitivity of blackbody emission for a temperature of about 240 K.

Lapse-rate feedback is the feedback that results from the structure of the temperature change, especially the reduction of lapse rate in the tropics, which is a negative feedback. Water-vapor feedback is the change in radiation balance caused by the increase of water vapor with temperature. Note that when lapse rate and water-vapor feedback are combined, the uncertainty is reduced. This was explained in Fig. 10.6. Surface-albedo feedback is related to the melting of surface ice with warming. It is relatively small and positive. Longwave, shortwave, and total cloud feedback are highly uncertain, with total feedback slightly positive. When all the feedbacks are combined, the best estimate of the total feedback is about -1.2 ± 0.6 W m^{-2} K^{-1}, but it is uncertain by about $\pm 50\%$. Most of this uncertainty is associated with cloud feedbacks. The total feedback controls the magnitude of the warming associated with forcing such as a doubling of CO_2. Note that to obtain the global mean temperature response, the forcing is divided by the feedback (11.4). If the adjusted forcing for a doubling of CO_2 is 3.5 W m^{-2}, then the uncertainty in total feedback gives predicted equilibrium warming from doubling of CO_2 that falls in the range from 2 K to 5.8 K with a best estimate of 2.9 K. Narrowing the uncertainty in feedback is thus key to good estimates of how much impact will result from doubling or quadrupling CO_2.

11.8.6 Hydrologic Cycle Feedback

When surface temperature changes, one of the most significant responses is that the saturation water vapor increases about 7% K^{-1} of warming. This fundamental physical effect has multiple implications for climate change, including the cycling of water between the surface and atmosphere. In Section 5.5, we showed that potential evaporation should increase rapidly with warming, meaning that land areas will dry out more quickly after rain, and evaporation over the ocean will increase in a warmed earth. Since water vapor decreases exponentially with decreasing temperature, the decrease of specific humidity with altitude also becomes stronger with warming. Thus if air is pushed upward, more precipitation will result per degree of adiabatic cooling in a warmer climate. We might then expect the hydrologic cycle to get much stronger in a warmed Earth, with more evaporation and more rainfall. However, we saw in Chapter 6 that the atmospheric energy balance constrains the evaporation rate to the rate at which the atmosphere can cool radiatively, at least in the global mean (Pendergrass and Hartmann, 2014). Therefore, the atmosphere has to adjust to the fact that when air is moved upward it releases more latent heat, but that the rate at which radiative cooling of the atmosphere increases with warming is much less than the Clausius–Clapeyron rate of 7% K^{-1}, at least in part because water-vapor feedback constrains the cooling to space. One way for the atmosphere to solve this dilemma is for the circulation to slow down, with less overturning so that the upward mass flux of water vapor is reduced. Another possible solution is to confine the upward motion to a smaller area, so that cloud-free dry zones that can more efficiently radiate to space are expanded. Over the tropical oceans, the models can reduce evaporation by reducing wind speed or increasing relative humidity near the surface as the climate warms.

Let's begin by looking at how the zonal mean relative humidity in models responds to warming. Figure 11.12 shows the zonal and annual mean relative humidity for the present climate and the change in relative humidity with warming in a suite of CMIP5 models. The modeled relative humidity is generally a good depiction of the observed relative humidity shown in Fig. 10.4. With warming, the troposphere deepens, so that relative humidity

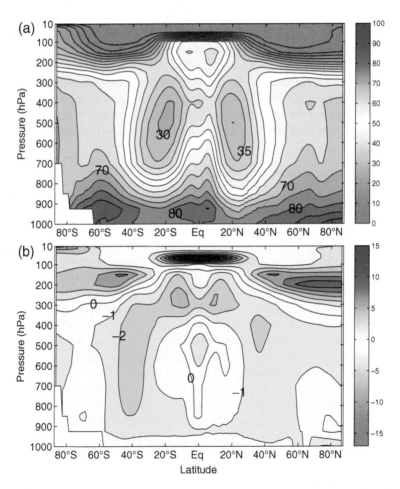

FIGURE 11.12 Zonal cross-section of percent relative humidity of (a) the historical climate of CMIP5 models and (b) the change in the relative humidity from the historical to the RCP 8.5 scenario for 2080–2100. Contour interval is 5% in (a) and 1% in (b).

increases in the vicinity of the tropopause and decreases below, except near the equator in the middle and lower troposphere where it moistens. The reduction in relative humidity is stronger around 40°N and 40°S, especially in the Southern Hemisphere. This enhanced drying in the subtropics and lower middle latitudes is associated with the poleward expansion of the dry zones that are the most prominent features of the climatology, apart from the extreme aridity of the stratosphere. These changes occur in association with an expansion of the Hadley cell and a poleward movement of the extratropical jets. The jet movement is particularly evident in the Southern Hemisphere, where the eddy-driven jet is a more dominant feature and responds fairly strongly to climate change by moving poleward (Fig. 11.13b). The troposphere expands upward to lower pressures in a warmed climate,

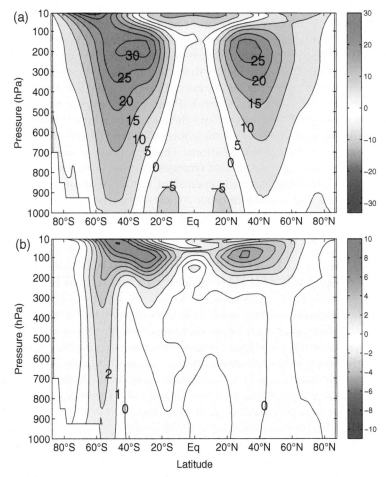

FIGURE 11.13 **Zonal cross-section of zonal mean wind of (a) the historical control simulation from CMIP5, and (b) the change from the historical simulation to the 2080–2100 RCP 8.5 simulation.** Contour interval is 5 m s^{-1} in (a) and 1 m s^{-1} in (b).

and so the winds in the upper troposphere increase as the subtropical and midlatitude jets move upward. The poleward jet movement in the Southern Hemisphere is aided by the cloud feedback, which intensifies the gradient in radiative heating by reducing the absorbed solar radiation in high latitudes and increasing it in lower latitudes (Fig. 11.9).

The net effect of the expansion of the subtropical dry zone and the wet-gets-wetter, dry-gets-drier paradigm is reflected in simulations of the changed surface hydrology in response to global warming (Fig. 11.14). Precipitation increases in the tropical Pacific and high latitudes and decreases elsewhere. Evaporation increases nearly everywhere. Relative humidity of surface air increases over the oceans, but decreases over land. The increase over the oceans is consistent with the constraint that the global evaporation cannot increase at the Clausius–Clapeyron rate like the saturation humidity, but is constrained by the rate at which the atmosphere can dispose of the latent heating by radiative emission. As previously mentioned models can reduce evaporation by increasing surface relative humidity a little, which will suppress the evaporation rate (4.32). Figure 11.14 clearly shows that most models do increase surface relative humidity over the ocean, even though the evaporation goes up with the saturation vapor pressure. E–P has a very strong zonal structure with increases in the subtropics and decreases elsewhere. This is consistent with the strengthening of the contrast in the hydrologic cycle that we expect from basic considerations. The implications of this for land areas is indicated by the modeled changes in runoff and soil moisture in Fig. 11.14. CMIP5 models fairly consistently predict drying of subtropical land areas and wetting of other regions. Regions that seem particularly likely to dry out are the American Southwest, the Mediterranean, northern South America, and southern Africa.

11.9 COUPLED ATMOSPHERE–OCEAN PROCESSES AND THE THERMOHALINE CIRCULATION

Models of the ocean in coupled AOGCMs can simulate the response of the ocean to climate change. In the case of warming, the upper ocean warms quickly, except in those regions where vertical mixing is efficient, such as the North Atlantic and the Southern Ocean surrounding Antarctica, where the surface warming is delayed. The penetration of the heating to deeper levels of the ocean proceeds more slowly, since it takes deep water formation in polar latitudes as much as a thousand years to replace the abyssal ocean water. As the heating of the surface water proceeds the ocean becomes more stable and deep water formation may slow. Also, the addition of freshwater in high latitudes owing to the increased precipitation there can slow down the meridional overturning cells. The different local responses of the ocean to warming may impart important structure to the warming. The key examples of this are the slowed warming in the North

Annual mean hydrological cycle change (RCP8.5: 2081–2100)

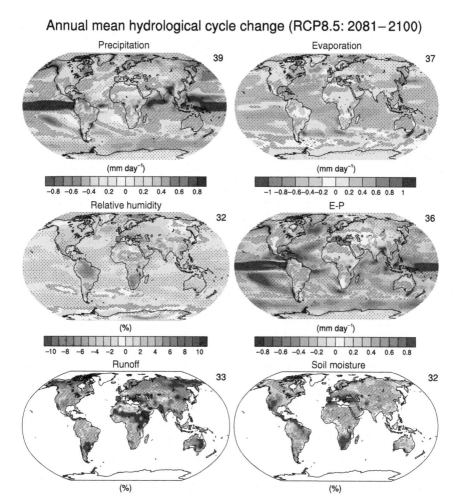

FIGURE 11.14 **Annual mean changes in precipitation (P), evaporation (E), relative humidity, E–P, runoff and soil moisture for 2081–2100 relative to 1986–2005 under the Representative Concentration Pathway 8.5 (RCP8.5).** The number of Coupled Model Intercomparison Project Phase 5 (CMIP5) models to calculate the multimodel mean is indicated in the upper right corner of each panel. Hatching indicates regions where the multimodel mean change is less than one standard deviation of internal variability. Stippling indicates regions where the multimodel mean change is greater than two standard deviations of internal variability and where 90% of models agree on the sign of change. *Figure and caption reproduced exactly from IPCC WG I Report Technical Summary Fig. TFE.1, Fig. 3, Stocker et al. (2013).*

Atlantic and the Southern Ocean associated with the deep overturning circulation, and the greater warming of the eastern equatorial Pacific compared to the tropics as a whole (Fig. 11.15). Most, but not all models have these features. Other features of the land-sea contrast of projected transient

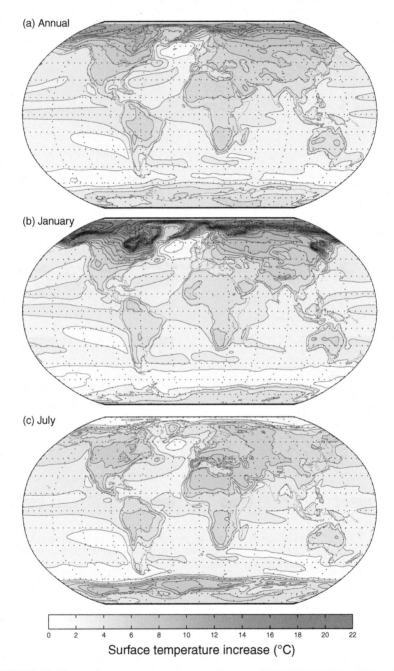

FIGURE 11.15   **Surface temperature increases between RCP8.5 2081–2100 and historical simulation from CMIP5 multimodel mean.** Contour interval is 1°C.

warming at the end of the twenty-first century are greater warming over land than ocean and greater warming in winter than summer in high latitudes. In regions with sea ice, the surface temperature cannot increase much in summer, when the ice melts a little, but still may be present, and if the ice is completely removed, then the large ocean heat capacity keeps it from warming as much in these regions. If the ice melts out in summer, then the ocean takes in much more stored heat, the sea ice is thinner, and the wintertime surface temperatures increase a lot in a warmed Earth. This is especially true in the Arctic in January, but is also true in regions of Antarctic sea ice in winter (July). In the Arctic the warming is smaller than average in summer and larger than average in winter.

The wet-gets-wetter, dry-gets drier paradigm for the hydrological cycle also has a reflection in ocean salinity. Ocean surface salinity is an integrator of the hydrological cycle, with saltier surface waters where E–P > 0 and fresher waters where E–P <0. In most models, the Atlantic gets saltier in tropical and subtropical latitudes, but fresher in the far North Atlantic as the climate warms in twenty-first century simulations. Since the north Atlantic also warms a little and the density near freezing is most sensitive to salinity, deep water formation in the North Atlantic becomes less likely, and the Atlantic meridional overturning circulation slows down in most models as the climate warms. Analysis of trends in salinity data suggest that the trend toward a saltier Atlantic and greater contrast between salty and fresh regions of the surface ocean has already begun (Durack et al., 2012).

EXERCISES

1. Suppose that the air temperature above the Arctic Ocean is –30°C and the water temperature is –2°C. Solve for the sensible heat flux to the atmosphere if the bulk aerodynamic formula for heat flux (4.26) applies with surface pressure 10^5 Pa, $C_{DH} = 10^{-3}$, and $U_r = 5$ m s^{-1}. Assume a steady state with no heat storage and ignore radiative and latent heat fluxes. Consider three cases: (a) no sea ice, (b) 1 m of sea ice, and (c) 3 m of sea ice. Solve for the temperature of the surface in those cases with sea ice. Use (11.3) with $k_l = 2$ W m^{-1} K^{-1} to solve for the flux through the ice.

2. Repeat problem 1 for the case of 1 m of sea ice with 10 cm of snow on top. Use $k_s = 0.3$ W m^{-1} K^{-1}.

3. Calculate the surface temperature and sea ice thickness in equilibrium under the following conditions: polar darkness, air temperature –40°C, surface pressure 10^5 Pa, $C_{DH} = 10^{-3}$, $U_r = 5$ m s^{-1}, and $k_l = 2$ W m^{-1} K^{-1}. The downward longwave flux is 100 W m^{-2}, and the surface has unit longwave emissivity. Consider cases where the ocean supplies heat to the base of the ice at the rate of (a) 10 W m^{-2} and (b) 20 W m^{-2}. *Hint:* Use the energy balance at the surface and at the base of the ice and linearize

the Stefan–Boltzmann emission about the air temperature. Ignore latent cooling.

4. What would be the equilibrium sea ice thickness for the two cases of problem 3 if the sea ice were covered by 20 cm of snow with $k_s = 0.3$ W m^{-1} K^{-1}?

5. Look in the tropical Pacific at the cloud changes in Fig. 11.8 and at the precipitation changes in Fig. 11.14. Explain how these are consistent with each other.

6. Compare the relative humidity changes in Fig. 11.12 with the zonal mean wind changes in Fig. 11.13. Explain how these are consistent with the general circulation of the atmosphere described in Chapter 6, and the modes of natural variability described in Fig. 8.4.

7. In Fig. 11.15, the January warming is much greater over Hudson Bay than over the surrounding land regions of North American. Can you explain this?

12

Natural Climate Change

12.1 NATURAL FORCING OF CLIMATE CHANGE

The geologic record discussed in Chapter 9 indicates that dramatic changes in climate have occurred in the past. These changes occurred in the absence of humans, for the most part, and we can call them natural climate changes. Understanding these natural climate changes is a challenging and important problem that will help us to understand and predict future natural and human-induced climate changes. Some natural forcing of climate occurs on time scales covered by the instrumental record, such as volcanic eruptions and total solar irradiance (TSI) changes. These will combine with the human impact on climate to determine the climate of the future.

In this chapter, we will evaluate several mechanisms whereby past climate changes might have been forced to occur. The distinction between a forcing mechanism and a feedback mechanism is somewhat arbitrary, and depends on what you include as part of the climate system and what is considered external to it. We will regard the atmosphere, the ocean, and the land surface as internal to the climate system, and we will consider Earth's interior and everything extraterrestrial as being external to the climate system. Therefore solar luminosity variations, variations in Earth's orbit about the Sun, and volcanic eruptions are all considered external forcing mechanisms. Another reason for making this distinction is that the solar luminosity, Earth's orbital parameters, and probably volcanic eruptions are not influenced by the climate, so the interaction is one way. Continental drift should also be considered an external forcing to climate, since continental movements are driven by convection in Earth's mantle, which is presumably not sensitive to the modest variations of Earth's skin temperature that are of interest to the climatologist. During the Holocene period when we have the most information about climate, the continental positions have been very nearly fixed, so that discussion of the effect of continental position on climate will not be expanded beyond what was presented in Chapter 9.

Global Physical Climatology. http://dx.doi.org/10.1016/B978-0-12-328531-7.00012-8

12.2 SOLAR LUMINOSITY VARIATIONS

The total energy output of the Sun is a central determinant of Earth's climate. A very direct way of forcing climate change would be to vary the luminosity of the Sun. Theories of stellar evolution suggest that the luminosity of the Sun has increased steadily by about 30% since the formation of the solar system. This increase is associated with the conversion of hydrogen to helium, leading to concomitant increases of solar density, solar core temperature, rate of fusion, and energy production. If the luminosity of the Sun were suddenly decreased by 30%, Earth would quickly become substantially cooler. Geologic evidence of sedimentary rocks 3.8 billion years old suggests that liquid water was present on Earth at this time. Other evidence seems to suggest that Earth was actually warmer early in its history, since no evidence for glaciation exists prior to 2.7 billion years ago. This combination of a less luminous early Sun and warm surface climate of the early Earth can be called the *faint young sun problem*. The most likely resolution of this paradox seems to be that the greenhouse effect was much stronger early in Earth's history, perhaps associated with elevated levels of carbon dioxide or some other biogeochemically controlled trace gas. Analysis of ancient soils suggests that ~440 million years ago atmospheric CO_2 concentration may have been ~16 times greater than its current value (Yapp and Poths, 1992). Carbon cycle models suggest that increasing solar constant and the development of land plants may have accelerated the rate of CO_2 uptake by weathering and caused the apparent decline of atmospheric CO_2 during the mid-Paleozoic (400–320 million years ago) (Berner and Kothavala, 2001). This CO_2 decline may have set the stage for the late Paleozoic glaciation. Beyond the theoretical inference that the luminosity of the Sun has increased over its history, we know little about variations in solar luminosity on climatic time scales. Although the Sun is a very dynamic entity, it is widely assumed to be a very steady source of energy.

Most of the energy received from the Sun originates in the photosphere, which has an emission temperature of about 6000 K. The dominant features seen in the photosphere are dark spots called *sunspots*, which can be seen in both visible light and in the Sun's total broadband emission. The center of a typical sunspot has an emission temperature about 1700 K colder than the average for the photosphere, so that the energy emission is only about 25% of average. The darkness of sunspots is produced by a disruption of the normal outward energy flow by the strong magnetic field disturbances that are associated with sunspots. Sunspots are transient features and range in scale from those a few hundred kilometers in diameter and lasting a day or two, to some that are tens of thousands of kilometers wide and last several months. On average sunspots persist for a week or two. The area of the visible disk of the Sun that is covered by sunspots ranges

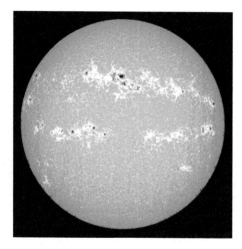

FIGURE 12.1   **Image of the Sun during an active phase, March 28, 2001.** Color scale has been altered to emphasize the faculae (white regions), which are hotter than the sunspots (black and red regions). *Image courtesy of NASA/Goddard Space Flight Center Scientific Visualization Studio. Source data courtesy of HAO & NSO PSPT project team. HAO is a division of the National Center for Atmospheric Research, which is supported by the National Science Foundation. Special thanks to Vanessa George (University of Colorado/LASP) and Randy Meisner (Michigan State University).*

from zero to about 0.1%. Dark sunspots are accompanied by bright regions called *faculae* that cover a much larger fraction of the area of the solar disk than the sunspots with which they appear to be associated (Fig. 12.1). Faculae are about 1000 K hotter than the average photosphere and emit about 15% more energy.

Sunspots are visible with the naked eye through a hazy sky at sunset, and regular reports of sunspots can be found in the Chinese literature beginning in the fourth century. Galileo observed sunspots with his small telescope in 1609. He and his contemporaries also discovered faculae and named them with the Latin word for "torches." Because sunspots are so easily observed, a long record of sunspot occurrence exists and an irregular cycle in sunspot abundance with a period of about 11 years has been documented. The number of sunspots at any given time varies from none to several hundred. The magnetic polarity of sunspot pairs alternates between sunspot maxima, so that a complete cycle of solar activity has a period of about 22 years. The sunspots result from magnetic field lines being lifted by buoyancy forces and penetrating the surface. The 22-year cycle is associated with interactions between the Sun's poloidal and toroidal magnetic fields though an interaction with the Sun's differential rotation. Reliable records of the annual mean sunspot number are available from about 1848, when the Wolf sunspot number index was defined, and the record has been extended back to 1700 (Fig. 12.2). Variations in the magnitude of the solar

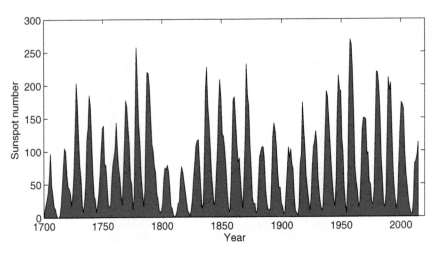

FIGURE 12.2 **Yearly sunspot number from 1700 to 2014.** *WDC-SILSO, Royal Observatory of Belgium, Brussels.*

cycle are evident, with a significant reduction in sunspot activity at the beginning of the nineteenth century. These variations are associated with an 80–90 year "Gleissberg" cycle in sunspot activity. A collection of scattered and somewhat less reliable data indicates that sunspots were virtually absent during the period 1645–1715, which is called the *Maunder Minimum* (Eddy, 1976). It is intriguing that this corresponds roughly with the period of the Little Ice Age in Europe.

Solar flares and eruptive prominences that occur with sunspots are spectacular features of the Sun's corona and cause significant increases in ultraviolet, X-ray, and gamma-ray radiation. The variation of high-energy radiation and particles associated with solar flares has a significant influence on the upper atmosphere. The effect of flares and prominences on the total energy output of the Sun is negligible, however, and their influence on the surface climate is likely to come through downward effects of the upper atmosphere on the lower atmosphere. Ozone production and chemistry are sensitive to the ultraviolet radiation variations, which strongly drive the stratosphere. The stratosphere can influence the surface, particularly through the annular modes of variability discussed in Chapter 8, and through the refraction and reflection of upward propagating Rossby waves.

Very precise measurements of the TSI can be made from Earth-orbiting satellites, and measurements from several space platforms extending over several solar cycles have been taken. The precision of TSI measurements is generally greater than their absolute accuracy, so that different estimates of the magnitude of the TSI variation over a solar cycle agree better than the estimates of the mean TSI. It has been argued that the most recent estimates have better absolute accuracy, so that previous estimates have

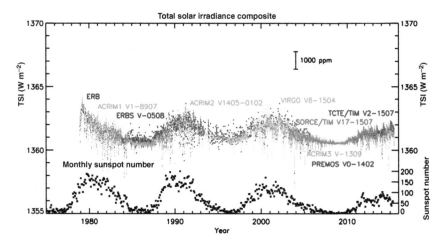

FIGURE 12.3 TSI estimates from various satellite measurements adjusted to the absolute value measured by the SOURCE-TIM instrument. Also shown are the monthly sunspot numbers. Data end in early July 2015. *Figure courtesy of Greg Kopp, University of Colorado.*

been offset to agree with the most recent set of measurements from the total irradiance monitor (TIM) instrument (Kopp et al., 2005). These measurements indicate an average solar irradiance of about 1361 W m^{-2} at the mean Earth–Sun distance. They indicate a small, less than ~0.1% or 1.5 W m^{-2}, increase in solar output from sunspot minimum to sunspot maximum (Fig. 12.3).

It is interesting that the largest solar irradiance occurs at sunspot maximum, when the most dark spots appear on the photosphere. Indices of faculae area such as Lyman-α flux have been used to show that the increased emission from faculae dominates the weakened emission from sunspots and causes the solar luminosity to increase at the maximum of solar activity. The statistical relationship between faculae and sunspot number can be used in conjunction with the observed solar irradiance variations over the period with satellite measurements to estimate the variation of the solar luminosity prior to direct measurements of TSI. Figure 12.4 shows estimates of TSI based on sunspot numbers since 1600, which includes the Maunder Minimum period. The variations have two parts, first is the solar cycle variation, which has been observed for three solar cycles and which is better constrained by observations. Also shown is a long-term secular trend from 1600 to the present. This secular trend is highly uncertain and somewhat speculative. Current best estimates from the solar physics community shown in Fig. 12.4 place the magnitude of the secular change since 1600 as about the same size as the magnitude of the variations over the most recent solar cycles, about 1 W m^{-2}.

The variations of solar irradiance associated with the 11-year sunspot cycle are of great interest, but have a relatively small influence on climate

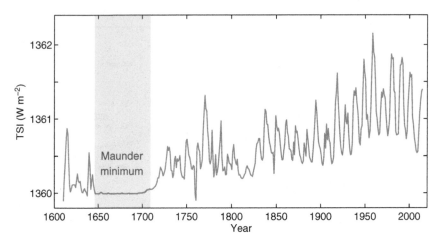

FIGURE 12.4   **TSI reconstructed from estimated sunspot numbers from about 1600 to 2015 by extrapolating the relationships between sunspots, faculae, and TSI observed directly from satellites.** *Data are based on Krivova et al. (2010), Ball et al. (2012), and Yeo et al. (2014), and updated to TIM TSI measurements as in Kopp and Lean (2011).*

(Zhou and Tung, 2013). A change in solar irradiance of 1 W m^{-2} translates into about a 0.175 W m^{-2} climate forcing, which, using a climate sensitivity of $\lambda_R = 0.5$ K/(W m^{-2}) yields an equilibrium climate response of <0.1°C. Moreover, the 11-year cycle has a time scale significantly shorter than the relaxation time scale of climate, which is determined by the large heat capacity of the oceanic mixed layer. This would mean that the magnitude of the transient response will be much less than the response of a steadily applied forcing of the same magnitude, and would be difficult to observe amid the natural variability of climate.

In summary, TSI variations associated with the 11-year solar cycle are only about one part in a thousand and should have only a minor effect on climate variations. The secular trend associated with the Maunder Minimum is of the same order of magnitude and therefore has a small effect on the energy balance compared to doubling of CO_2. To study other mechanisms of climate change, one can adopt the hypothesis that the luminosity of the Sun is effectively constant and attempt to understand past variations on that basis.

12.3 NATURAL AEROSOLS AND CLIMATE

An aerosol is a suspension of liquid or solid matter in air. According to this definition, clouds are composed of water aerosols, but we will generally use the word *aerosol* to denote all *non-cloud* aerosols. Since most aerosols contain some water, and cloud droplets generally form from growth by condensation on a tiny aerosol particle (cloud condensation nuclei or CCN), the distinction between cloud and non-cloud aerosol may seem arbitrary. A distinction is made in that cloud aerosols are at equilibrium in air with relative humidity in excess of 100%, whereas non-cloud aerosols can be in equilibrium at low relative humidity. Aerosols range in size from only slightly larger than individual air molecules to radii of several tens of micrometers. Aerosols larger than about 20 μm are removed quickly by gravitational settling to the surface. Most aerosols in the troposphere are removed within a couple of weeks by contact with cloud droplets or the surface. The smallest aerosols dominate the number of aerosols in the atmosphere, but their mass and surface area are so small that they have little direct effect on climate. The surface area of aerosols is important both for their role as cloud condensation nuclei (CCN) and also for their direct effect on radiation. The most effective aerosols for forming liquid droplets (CCN) and ice particles (ice nuclei) are different, with the former consisting of hygroscopic chemicals like liquid sulfuric acid and the latter consisting of solid substances like mineral dust, ash or solid bioaerosols such as spores, pollen, or bacteria. The largest contribution to the total surface area of atmospheric aerosols comes from aerosols with radii in the range of ~ 0.1–1 μm (Fig. 12.5). These are the aerosols that are most important for climate. Aerosol number concentrations typically range from 10^3 cm^{-3} in clean air over the oceans, to 10^6 cm^{-3} in polluted urban air.

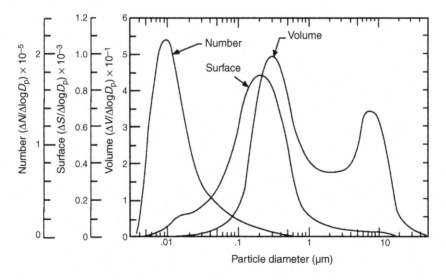

FIGURE 12.5 **Normalized plots of the number, surface area, and volume distributions for tropospheric aerosols.** *From Charlson (1988); reprinted with permission from Wiley and Sons, Inc.*

Aerosols are produced by a large number of processes and can have varied chemical compositions. Aerosols can be injected directly into the atmosphere (primary aerosols) or be formed by the condensation or chemical transformation of vapors (secondary aerosols). Except for a small contribution from meteoritic debris, the sources of aerosols are at the surface, although aerosols can also be processed in clouds. Direct emissions of aerosols come from the bursting of bubbles on the ocean surface, which produces small droplets that evaporate to leave an aerosol containing sea salt or organic material. Other direct sources of aerosols include the elevation by wind of mineral dust from dry land surfaces, injection of ash and rock by volcanoes, soot from forest fires, meteoritic debris, and biological emissions such as spores and pollen. Black carbon aerosols are produced directly by incomplete combustion. The largest contributions by mass to direct production of aerosols come from sea salt and windblown dust (Table 12.1), but a large fraction of the mass of sea salt and dust particles is contained in a relatively small number of large particles. Because they represent a comparatively small fraction of the total number of aerosol

TABLE 12.1   Global Natural and Anthropogenic Emissions of Aerosols and Aerosol Precursors

	Min.	Max.
ANTHROPOGENIC SOURCES		
Anthropogenic non-methane volatile organic compounds	98	158
Anthropogenic black carbon	4	6
Anthropogenic primary organic aerosols	6	15
Anthropogenic SO_2	43	78
Anthropogenic NH_3	35	50
Biomass burning aerosols	29	85
NATURAL SOURCES		
Sea spray	1400	6800
Mineral dust	1000	4000
Terrestrial biological primary aerosol particles	50	1000
Dimethylsulfide (DMS)	10	40
Monoterpenes	30	120
Isoprene	410	600
Secondary production from biogenic volatile organic compounds	20	380

Year 2000 emissions in Tg year^{-1} or Tg S year^{-1}.
Adapted from Boucher et al., 2013.

particles, sea salt and mineral aerosols are of less importance for climate than their large mass would suggest.

Aerosols formed by gas-to-particle conversion consist mainly of sulfates, nitrates, and hydrocarbons. The most important naturally occurring gaseous aerosol precursors come from dimethyl sulfide (DMS) emitted from the ocean and volatile organic compounds emitted by land plants. The emission of aerosol precursors by the terrestrial biosphere is complex and a subject of intense study. Human production of black carbon, sulfate, and nitrate aerosols in the atmosphere exceed natural sources, while most sea salt, primary biologic aerosols and mineral dust are of natural origin. Aerosols can have several different effects on climate; direct interaction with radiation, such as in the reflection of sunlight; indirectly by increasing the number of particles on which clouds can form; or in the case of black carbon aerosols, by absorbing radiation and heating the atmosphere and thereby causing cloud droplets to evaporate (Fig. 12.6). The effect of human production of aerosols on the reflection of solar radiation by

FIGURE 12.6   **Visible image off the coast of Oregon and California on August 29, 2002, which shows several important effects of aerosols on the energy balance.** Over the ocean, where the air is clean and thin clouds form, the emission of aerosols from diesel engines on ships can cause a brightening of the clouds to form visible ship tracks. Smoke over the land is more reflective than the underlying forest and the smoke increases the reflection of visible radiation. When this smoke is transported over bright oceanic clouds, the soot in the smoke causes a darkening of the scene as viewed from space. *NASA image based on MODIS.*

clouds is visible in ship tracks, the iconic symbol of human modification of clouds. Aerosols produced by ships can brighten up clouds by providing more CCN, which make the cloud particles smaller and thereby increasing both their efficiency in reflecting solar radiation and their lifetime in the atmosphere (Fig. 12.7).

Over the unpolluted oceans, the majority of secondary aerosol particles are produced by conversion of sulfur-bearing gases generated by ocean biology. Sea-salt aerosols account for more mass than biogenic aerosols in

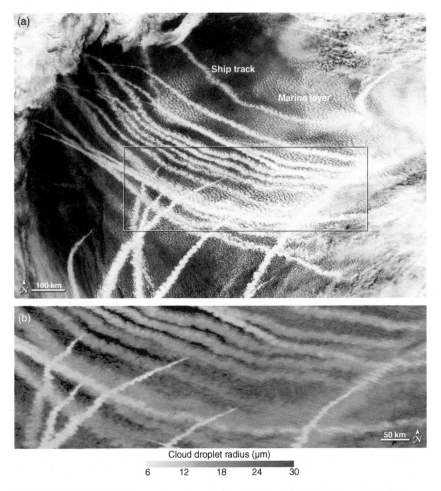

Cloud droplet radius (μm)

6   12   18   24   30

FIGURE 12.7 **Visible image showing ship tracks (a) and MODIS image of cloud droplet radius (b) from the marked box showing that the cloud particle radius is much smaller in the ship tracks, because the ship engine emissions have provided cloud condensation nuclei on which more cloud droplets can form.** *NASA image by Jesse Allen and Robert Simmon, based on data from MODIS.*

the marine boundary layer, but sea-salt aerosols are generally large, and account for only about 10% of the aerosols by number. A large fraction of the CCN over the oceans is thus biogenic sulfur aerosols or direct injection of organic material. The primary medium for carrying sulfur from the ocean to the atmosphere is DMS gas (DMS; CH_3SCH_3). DMS accounts for about 90% of the reduced volatile sulfur in seawater and is present in concentrations that are in excess of equilibrium with air, so that a DMS flux from the ocean to the atmosphere occurs. DMS flux is estimated by using measurements of DMS concentration in the ocean mixed layer and the air, together with a flux formulation not unlike the aerodynamic formulas for heat and moisture flux discussed in Section 4.5. In the atmosphere, DMS is oxidized photochemically to form sulfate (SO_4^{2-}) in aerosol form.

DMS is produced in ocean surface waters by certain species of phytoplankton, particularly algae. DMS is formed by the enzymatic cleavage of dimethylsulfonium propionate (DMSP). DMSP has many important functions for sea life, including antifreeze, antioxidant, osmolyte (salt regulation) and as a deterrent to herbivores. DMSP is released from algae when it is under stress or dies. Once in the water, DMSP is a valuable source of energy for bacteria, which break it down in a couple of ways, one of which produces DMS. DMS is a volatile gas and makes its way into the atmosphere. DMS is a significant part of the "smell of the sea". Since phytoplankton are the base of the food chain in the ocean, some sea birds follow the smell of DMS to find the larger animals that are grazing on the phytoplankton.

By regulating the release of DMS, and hence the production of oceanic CCN, phytoplankton could influence the cloud albedo over the oceans and hence regulate to some extent the climate. When more CCN are present the cloud water tends to reside in more, smaller droplets than when CCN are less abundant. As illustrated in Fig. 3.14, if the mean particle radius decreases and all else remains the same, the cloud albedo increases.

12.4 VOLCANIC ERUPTIONS AND STRATOSPHERIC AEROSOLS

Volcanoes play an important role in climate by cycling atmospheric elements such as carbon and sulfur between the lithosphere and the atmosphere. Violent eruptions can inject dust and source gases for aerosols into the stratosphere, where they remain for many months and can have a significant influence on climate. Much has been learned about the effects of volcanic eruptions in the last several decades because the major eruptions of El Chichón in Mexico (1982, 17.3°N) and Pinatubo in the Philippine Islands (1991, 15.1°N) were observed with modern satellite and *in situ* techniques. These eruptions were spectacular manifestations of explosive

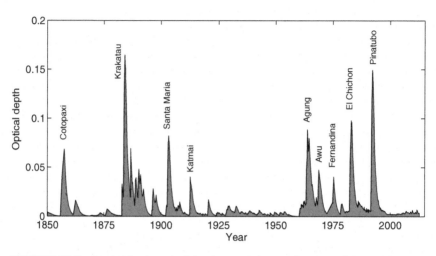

FIGURE 12.8 **Stratospheric optical depth at 0.55 µm as a function of time with major eruptions named.** *Data from Sato et al., 1993 with updates from the GISS web site.*

outgassing from Earth's interior and caused major changes in the optical properties of the stratosphere. These two eruptions resulted in peak global average stratospheric aerosol optical depths of 0.07 and 0.15, respectively. The time history of stratospheric aerosol optical depth can be measured accurately from satellites and shows these two eruptions very clearly. Their optical depth anomalies are compared with estimates for previous eruptions in Fig. 12.8.

Although ash and fine rock particles can be injected high in the stratosphere by a violent eruption, most of these aerosols are too large to remain long in the atmosphere. Most stratospheric aerosols consist of a mixture of 75% sulfuric acid (H_2SO_4) and 25% water. A nonvolcanic background aerosol level is maintained in the stratosphere through an upward flux of sulfur-bearing gases from the troposphere. These aerosols form the Junge Layer at 15–25 km altitude. A nonvolcanic source of sulfur is carbonyl sulfide (COS), which is produced in soils and has a lifetime of several years in the troposphere. COS is mixed upward into the stratosphere where it encounters ultraviolet radiation or atomic oxygen and is oxidized to form sulfur dioxide (SO_2). Sulfur dioxide is then further oxidized to sulfuric acid. Since the vapor pressure of sulfuric acid is generally above saturation in the lower stratosphere, it condenses to form aerosols. This slow emission maintains the Junge Layer even in the absence of volcanic eruptions. Sulfuric acid is ultimately removed when the aerosols become large enough to precipitate or when the air in which they reside returns to the troposphere. Sulfuric acid is very hygroscopic and collects water to add to the mass and reflectivity of the aerosol.

The enhancement of stratospheric aerosols after a volcanic eruption is caused by the direct injection of large amounts of SO_2 gas. The optical depth of the aerosol cloud peaks about 6 months after the eruption because of the time required to convert the SO_2 gas to sulfuric acid aerosols of the proper size to interact efficiently with solar and longwave radiation. The impact of the Pinatubo eruption on the energy balance is shown in Fig. 11.6. To produce a big and lasting perturbation in the aerosol optical depth of the stratosphere, volcanic eruptions must contain large amounts of sulfur in the form of SO_2, and they must be explosive enough to deliver this SO_2 to the stratosphere where the resulting aerosols can persist for long periods of time compared to their much shorter lifetime in the troposphere.

Stratospheric aerosols interact with both solar and terrestrial radiation and may both scatter and absorb radiation. Extinction is defined as the sum of scattering and absorption of the direct beam of radiation, so we may write the extinction cross section per unit mass (Section 3.6) as the sum of the scattering and absorption cross-sections.

$$k_{ext} = k_{sca} + k_{abs} \tag{12.1}$$

The extinction coefficient is the ratio of the extinction cross-section to the geometric cross-section (shadow area) of an aerosol, and can similarly be written in terms of its scattering and absorption components.

$$Q_{ext} = Q_{sca} + Q_{abs} \tag{12.2}$$

The extinction coefficient for sulfuric acid aerosols is near 1 for solar radiation and decreases to less than 0.5 for terrestrial radiation (Fig. 12.9a). The larger particles that characterize a volcanic aerosol cloud in the early phases after an eruption have a larger extinction coefficient and thus interact more efficiently with radiation, especially at longer wavelengths.

The single scattering albedo is defined as the ratio of the scattering coefficient to the extinction coefficient, and measures the ratio of extinction by scattering to total extinction during a single interaction of a photon beam with a particle.

$$\varpi = \frac{Q_{sca}}{Q_{ext}} \tag{12.3}$$

The single scattering albedo for sulfuric acid aerosols is nearly one for solar radiation, but decreases abruptly near 3 μm, so that very little absorption of solar radiation takes place, but a large fraction of the extinction of terrestrial radiation is absorption (Fig. 12.9b).

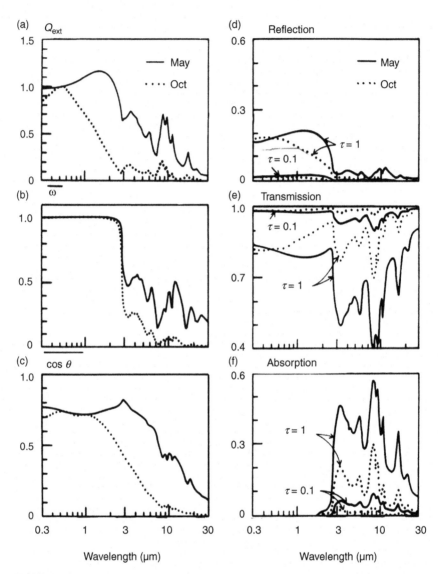

FIGURE 12.9 Single scattering and multiple scattering by 75% sulfuric acid aerosols for two size distributions characterizing the early stages 1.5 months after the eruption of El Chichón 1982 ($r_{eff} \sim 1.4$ μm, May) and the later stages when the effective particle radius was smaller ($r_{eff} \sim 0.5$ μm, October). Multiple scattering calculations are given for a plane–parallel atmosphere with visible ($\lambda = 0.55$ μm) optical depth $\tau = 0.1$ and 1.0. Single scattering parameters are explained in the text. *From Lacis et al. (1992) © American Geophysical Union.*

The probability that a scattered photon will depart in a particular direction relative to the direction of the incident beam of photons is given by the phase function, $\hat{p}$, which is normalized in the following way.

$$\frac{1}{4\pi} \int_{4\pi} \hat{P} d\omega = 1 \tag{12.4}$$

where $d\omega$ is the increment of solid angle (Section 3.3). The phase function can be characterized by the single scatter asymmetry factor $\overline{\cos\theta}$, which is defined as follows,

$$\overline{\cos\theta} = \frac{1}{4\pi} \int_{4\pi} \cos\theta \, \hat{P} \, d\omega \tag{12.5}$$

where θ is the angle between the direction of the incident beam and the scattered beam. The single scatter asymmetry factor varies between 1, which indicates forward scattering in the same direction as the incident beam, and –1, which indicates backscattering toward the direction of the source of the incident beam. Isotropic scattering, where scattering is equally probable in all directions, gives an asymmetry factor of zero. Stratospheric aerosols are strongly forward scattering for solar radiation and become increasingly isotropic scatterers for longer wavelengths (Fig. 12.9c).

As a result of the radiative properties of stratospheric sulfuric acid aerosols, the modest optical depths associated with recent large volcanic eruptions ($\tau \sim 0.1$) give rise to a modest reflection of solar radiation and a modest absorption of terrestrial radiation (Figs. 12.9d,f). Because of their relatively efficient absorption of infrared radiation, volcanic aerosols generally heat the layer in which they occur (20–25 km). Their effect on the tropospheric climate is more complex, as they both reflect solar radiation and provide a small greenhouse effect. The balance of these two opposing effects is sensitive to the size of the aerosol particles. Smaller particles have a substantial influence on solar radiation, but not on infrared radiation, so that a reduction of net radiation of 3–4 W m^{-2} results for particles smaller than ~ 1 μm (Fig. 12.10). The greenhouse effect increases approximately linearly with particle radius, whereas the solar reflection becomes independent of particle radius when the radius is larger than the effective wavelength, so that a net warming results for particles larger than about 2 μm. This result is relatively insensitive to particle size dispersion (Fig. 12.10b). A 1-μm spherical sulfuric acid particle at 20 km falls about 2 km per month, so that particles larger than this are rare. Therefore, except for the brief period immediately after the eruption, when large dust and ash particles may be abundant, we expect that volcanic aerosol clouds in the stratosphere will cause a significant reduction in net energy input to the climate system, with a large eruption such as Mt. Pinatubo

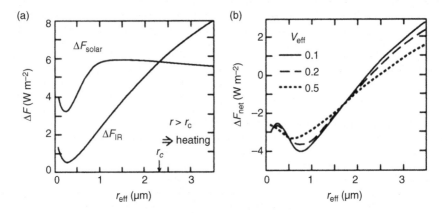

FIGURE 12.10 Decrease of incoming solar and outgoing infrared fluxes at the tropopause caused by adding a stratospheric aerosol layer (20–25 km) with visible optical depth
$\tau = 0.1$ in a one-dimensional radiative–convective model with fixed surface temperature.
(a) Plotted as a function of effective radius, r_{eff}, with an effective variance of particle radius
$v_{eff} = 0.2$. (b) Change of net radiation at the tropopause for three values of effective variance
of particle radius. *From Lacis et al. (1992), © American Geophysical Union.*

producing a 0.15 optical depth cloud and a -4 W m^{-2} climate forcing during its peak (Fig. 11.6). Because the aerosol cloud enhancement resulting
from a single eruption is flushed out of the stratosphere in a year or two,
and the thermal capacity of the ocean mixed layer gives it a thermal-
response time scale of a decade, the response of the climate system to a
single volcanic eruption is rather modest, unless it is an exceptionally
large eruption with a high sulfur content. However, a sequence of major
volcanic eruptions each spaced about a year apart could cause a significant climate signal.

The effect of a volcanic eruption on climate depends on the amount of
long-lasting stratospheric aerosol produced, which depends both on the
explosiveness of the eruption and the abundance of sulfur in the gases
released. Evidence suggests that large eruptions in the past have had a
significant effect on short-term climate anomalies. One of the largest eruptions we know of was the 1815 eruption of Mt. Tambora on Sumbawa
Island, Indonesia. Approximately 150 km^3 of ash and pumice was spread
over a large area and the eruptive plume was estimated to have reached
50-km altitude. Within a few months, the volcanic cloud had spread
around the world and its optical effects were observed in Europe. The Sun
and stars were dimmed noticeably by the stratospheric aerosol cloud for
nearly 2 years after the eruption. The following year (1816) was known as
"the year without a summer", since it was much colder than normal both
in Europe and North America. Unusual spring and summer frosts and
a 6-in. snowfall in the second week of June caused crop failures in New

England. It is unlikely that these climate anomalies can be completely attributed to the Tambora eruption, since the entire decade (1810–1820) was somewhat colder than normal.

12.5 THE ORBITAL PARAMETER THEORY OF ICE AGES

12.5.1 Historical Introduction

The idea that Earth has experienced ice ages was first suggested in Europe in the eighteenth century. Continental ice sheets were offered as an explanation for the appearance of large boulders far from their source of origin and the presence of deep grooves on exposed bedrock. At the time, many people believed that the erratic boulders were transported by a great flood, and the idea of great ice sheets was slow to be accepted. Credit for convincing the scientific community that the European Alps were once covered by a great ice sheet is usually given to the Swiss-born geologist Louis Agassiz, who argued the case forcefully and published a monograph on the subject in 1840, at about which time the argument was beginning to turn in his favor.

Among the initial explanations offered for glacial ages was an astronomical theory by the French mathematician Joseph Alphonse Adhémar published in 1842. Adhémar clearly stated the hypothesis that the primary cause of the ice age succession was secular variation in Earth's orbit. His theory was based on the precession of the equinoxes, which in 1754 d'Alembert had shown completes a cycle every 22,000 years. Adhémar based his ice age timing on the number of days of daylight per year. Because Earth travels faster when it is closer to the Sun (perihelion) than when it is far away (aphelion), and Southern Hemisphere summer currently occurs near perihelion, the Southern Hemisphere receives fewer hours of daylight than the Northern Hemisphere. He concluded that ice ages alternated between the hemispheres every 11,000 years, and that the Southern Hemisphere is currently in an ice age. Adhémar apparently did not understand that energy input and not hours of daylight is the critical consideration, and that annual mean insolation in each hemisphere is independent of the precession cycle.

In 1864, the Scotsman James Croll extended the work of Adhémar to include the modulation by eccentricity variations of the effect of precession of the equinoxes. Eccentricity variations had been established by the work of LeVerrier in 1843, who also showed that the annual insolation is only weakly affected by eccentricity variations. Croll deduced that if eccentricity variations were to have an effect on ice ages, it must arise from the larger variations in seasonal insolation that accompany eccentricity changes. He adopted the idea that colder winters lead to growing glaciers. His theory predicts that an ice age will occur in one hemisphere or the other whenever two conditions occur simultaneously: a markedly eccentric orbit and a winter solstice near aphelion. According to his theory, the last ice age occurred 80,000 years ago. He also hypothesized that ice ages would be more likely when the tilt of Earth's axis of rotation (obliquity) was less, because at these times the polar regions would receive less insolation. The timing of the obliquity variations was not known, however, so he could not incorporate this factor in his theory for past glaciations. In developing his theory, Croll introduced the idea of ice albedo feedback in order to amplify the effect of the small insolation changes associated with changes in Earth's orbital parameters. He also included a feedback mechanism involving the tradewinds, ocean currents, and the geometry of the Atlantic that would amplify cooling in one hemisphere by diverting oceanic heat fluxes to the other hemisphere. Croll's theory met with a great deal of interest and approval, since by this time it was accepted that not one but many ice ages had occurred in the past, and the astronomical theory predicts cyclic ice ages associated with the cycles in Earth's orbital parameters. Although geologic dating techniques were very primitive at that time, what evidence could be gathered suggested that the last glaciation was much more recent than 80,000 years ago as Croll predicted. In addition, evidence came in from the Southern Hemisphere suggesting that glacial ages there occurred simultaneously with those in the Northern Hemisphere, which is inconsistent with the predictions of Adhémar and Croll that glacial ages would alternate between the hemispheres. By the end of the nineteenth century Croll's theory had been largely dismissed.

The orbital parameter theory of climate change was next taken up by the Serbian applied mathematician Milutin Milankovitch. Milankovitch began his work in 1911 with the benefit of Pilgrim's determination of all of the orbital parameters necessary to compute insolation as a function of time and latitude. His work on the orbital parameter theory extended over 30 years and was interrupted by three wars. In 1920, his work attracted the attention of the German climatologist Vladimir Köppen, who encouraged Milankovitch and engaged in a correspondence with him about the physical basis of climate. Milankovitch had developed the mathematical tools for calculating the insolation at any point and time as a function of the orbital parameters, but he was unsure at what latitude and season the ice sheets would be most sensitive to the insolation. Köppen indicated that a reduction of the summertime insolation in the Arctic would be the critical factor in producing growth of ice sheets. He reasoned that the temperatures were always cold enough for snow to form in the winter, but that a reduction in the summertime melting of glaciers would allow the ice sheets to expand. This was a critical departure from the work of Adhémar and Croll, who had focused on the coldness of winter as the critical parameter. With this advice, Milankovitch set about to calculate the summertime insolation at 55°, 60°, and 65° N for the last 650,000 years. After 100 days of calculation, he sent his insolation curves to Köppen, who replied that they matched fairly well with what was known about the timing of the last several glacial epochs. The astronomical climate theory of Milankovitch was accepted to varying degrees until the 1950s when radiocarbon dating showed that the ages of many important glacial drift features on the land surface did not correspond to the times predicted by the radiation curves.

By the mid-1960s, the early paleoclimatic evidence that had been used to support the Milankovitch theory had been dismissed and the theory itself was increasingly questioned. However, new evidence began to emerge that supported the theory. The sources of new information included new dating techniques that extended the 40,000-year time reach of radiocarbon dating. Coral terraces that could be associated with rising sea level were dated at 122,000, 103,000, and 80,000 years ago and these dates corresponded to times when Milankovitch had calculated higher than normal summertime insolation in northern mid-latitudes, and thus that ice sheets would be reduced, the climate would be warmer, and sea level would be rising.

The discovery of a reversal of Earth's magnetic field whose date could be fixed at 700,000 years ago was used to set the relationship between depth and time for ocean sediment cores by assuming a constant sedimentation rate. The oxygen $^{18}O/^{16}O$ ratio in these cores was found to be a good measure of the global ice volume. The continuous nature of the global ice volume record contained in ocean sediment cores allowed the use of time series analysis techniques. Using these techniques it was shown that the oxygen isotope record in ocean cores contains periodicities that correspond to variations in Earth's orbital parameters at periods of 100,000, 41,000, 23,000, and 19,000 years (Fig. 9.4). It is somewhat puzzling that the record of the last 700,000 years is dominated by a 100,000-year periodicity, since the direct forcing of seasonal insolation by eccentricity variations at this long time scale is much less than the forcing associated with the obliquity and longitude of perihelion variations at 41,000 and about 20,000 years, respectively. Prior to 700,000 years ago, the available global ice volume records constructed from ocean cores appear to be dominated by the higher frequencies as the Milankovitch hypothesis would suggest (Figs. 9.3 and 9.4). The reason for the apparent amplification of the 100,000-year periodicity in the more recent era is still not completely understood, but seems to be related to increased total amount of ice during recent glacial maxima.

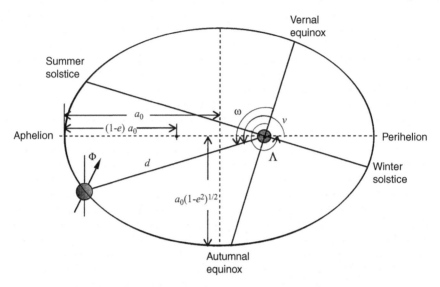

FIGURE 12.11 **Schematic diagram of Earth's elliptical orbit about the Sun showing the critical parameters of eccentricity (e), obliquity (ϕ), and longitude of perihelion (Λ) defined relative to the vernal equinox.** The size of the orbit is defined by the greatest distance between the ellipse and its center point, which is called the semimajor axis length, a_0. The Earth–Sun distance at any time (d), the angle between the position of Earth and perihelion that we call the true anomaly (v), and the angle between the position of Earth and the vernal equinox (ω) are also shown.

12.5.2 The Orbital Parameters and Their Influence

Early in the seventeenth century, Johannes Kepler discovered his first law, that the orbits of the planets are ellipses with the Sun at one focus (Fig. 12.11). The degree to which the orbit departs from a circle is measured by the eccentricity of the ellipse. The point where the planet is closest to the Sun is called *perihelion*, and the point on the orbit most distant from the Sun is called *aphelion*. If the Earth–Sun distance is d_p at perihelion and d_a at aphelion, then the eccentricity is defined as the ratio of the difference to the sum of these two distances.

$$e = \frac{d_a - d_p}{d_a + d_p} \tag{12.6}$$

The maximum distance from the center of the ellipse to the orbit, a_0, occurs at perihelion and aphelion and is called the semimajor axis. From geometric considerations it can be shown that the minimum distance from the center of the ellipse to the orbit is given by $a_0\sqrt{(1-e^2)}$. The Earth–Sun distance d as a function of the true anomaly v is given by

$$d(v) = \frac{a_0(1 - e^2)}{1 + e \cos v} \qquad (12.7)$$

The variation of the Earth–Sun distance as Earth moves along its orbit contributes to the annual cycle of insolation and is one of the key factors in the orbital parameter theory.

Of even greater importance for the annual cycle of climate is the obliquity. The obliquity is the angle between the axis of Earth's rotation and the normal to the plane of Earth's orbit about the Sun. The magnitude of the seasonal variation of isolation in high latitudes and the annual mean insolation in high latitudes both increase with increasing obliquity. This can be seen most easily by considering a perfectly circular orbit with zero eccentricity. The daily insolation at any latitude and season, Q, can then be written as

$$Q = \frac{S_0}{4} \tilde{s}(\Phi, x, t) \qquad (12.8)$$

where S_0 is the TSI at the mean Earth–Sun distance $\bar{d}$, and $\tilde{s}(\Phi, x, t)$ is the distribution function for a circular orbit and a particular obliquity Φ, sine of latitude x, and time of year t. This distribution function is normalized such that its area average is 1.

$$\frac{1}{2} \int_{-1}^{1} \tilde{s}(\Phi, x, t) dx = 1 \qquad (12.9)$$

The distribution function for a circular orbit with an obliquity of $23.5°$ is shown in Fig. 12.12a. Except for the effect of eccentricity, this is just the normalized form of Fig. 2.6. The sensitivity of this distribution to the obliquity can be illustrated by plotting the difference between $\tilde{s}$ for $\phi = 24.5°$ and $\phi = 22.5°$ (Fig. 12.12b). Increasing the obliquity by $2°$ increases the summertime insolation in high latitudes by more than 12% and decreases the wintertime insolation in mid-latitudes by more than 4%. The magnitude of the change of insolation during summer caused by an increase of obliquity from the minimum to the maximum experienced by Earth is also about 12% or more than 50 W m^{-2}. The annual average insolation at the pole is also sensitive to the obliquity. If the obliquity is zero, the annual mean insolation at the pole is zero. For a circular orbit the annual mean value of $\tilde{s}$ at the pole varies as $4 / \pi (\sin \Phi)$ (Held, 1982).

The longitude of perihelion is defined as the angle λ, between the line from Earth to the Sun at vernal equinox and the line from the Sun to perihelion. It defines the direction of Earth's orbital tilt relative to the orientation of the orbit, and thereby determines the season at which Earth passes through perihelion.

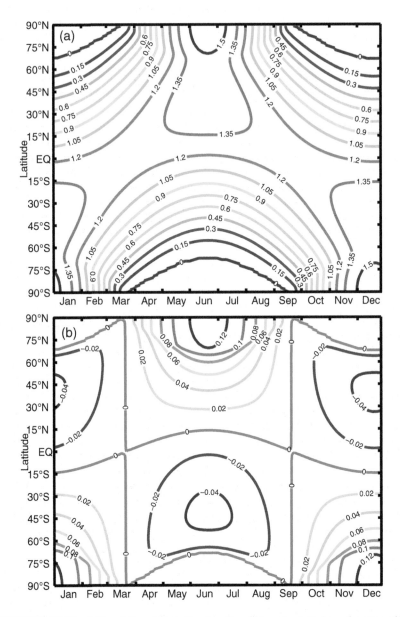

FIGURE 12.12   (a) Insolation distribution function $\tilde{s}(\Phi, x, t)$ plotted as a function of latitude and season for an obliquity of $\Phi = 23.5°$. (b) Sensitivity of the insolation distribution function to obliquity, $\Delta\Phi(\partial\tilde{s} / \partial\Phi)_{\Phi=23.5}$, evaluated for $\Delta\Phi = 2°$.

We are now ready to write the insolation as a function of the solar irradiance at some mean distance $\bar{d}$, $S_0/4$, and a distribution function, s, that depends on the eccentricity (e), obliquity (ϕ), longitude of perihelion (Λ), sine of latitude (x), and position in the orbit defined in terms of the true anomaly (v). This distribution function can be written as the product of the distribution function $\tilde{s}$ that contains the obliquity dependence, and the inverse square dependence on Earth–Sun distance, d.

$$\frac{S_0}{4} s(e, \Lambda, \Phi, x, v) = \frac{S_0}{4} \tilde{s}(\Phi, x, v) \frac{\bar{d}^2}{d(v)^2} \tag{12.10}$$

To make use of (12.10), we must relate the time to the true anomaly v. From Kepler's second law, we know that the Earth–Sun vector sweeps out equal area in equal time, and that the velocity of Earth in its orbit is greater at perihelion than aphelion. From the conservation of angular momentum it follows that

$$\frac{dv}{dt} = \frac{M_e}{d^2} \tag{12.11}$$

where M_e is the constant angular momentum per unit of Earth mass. If we define the unit of time, t', such that the period of the orbit

$$P_0 = \frac{2\pi}{M_e} \bar{d}^2 \sqrt{1 - e^2} \tag{12.12}$$

corresponds to 2π units of time, then (12.11) becomes

$$\frac{dv}{dt'} = \frac{\bar{d}^2}{d^2} \sqrt{1 - e^2} \tag{12.13}$$

We may now average the insolation over an annual cycle by integrating over t'

$$\frac{1}{2\pi} \int_0^{2\pi} \frac{S_0}{4} \tilde{s}(\Phi, x, v) \frac{\bar{d}^2}{d(v)^2} dt' = \int_0^{2\pi} \frac{S_0}{4} \tilde{s}(\Phi, x, v) \frac{\bar{d}^2}{d(v)^2} \frac{dt'}{dv} dv$$
$$= (1 - e^2)^{-1/2} \int_0^{2\pi} \frac{S_0}{4} \tilde{s}(\Phi, x, v) dv \tag{12.14}$$

From (12.14) we deduce that the annual-mean insolation at any point is independent of the longitude of perihelion. The annual mean insolation at every latitude is proportional to $(1 - e^2)^{-1/2}$, which for small eccentricity is approximately $(1 + 0.5e^2)$. This dependence of annual insolation on eccentricity is very weak, since in going from a circular orbit to the most eccentric orbit Earth experiences ($e \sim 0.06$) causes a change of annual isolation of only $\sim 0.18\%$.

The variations of annual mean insolation associated with eccentricity changes are too small to produce a significant effect, if our estimates of climate sensitivity are approximately correct. It is for this reason that Milankovitch sought an explanation for the ice ages in the changes of the seasonal and latitudinal distribution that are caused by variations in the orbital parameters. Summertime insolation in high latitudes will be larger when the obliquity is larger, and summertime insolation in the Northern Hemisphere will be greater when the northern summer solstice occurs at perihelion. The effect of longitude of perihelion variations on insolation increases with the eccentricity of the orbit. To see this quantitatively, we can insert (12.7) into (12.10), and use Fig. 12.11 to deduce that $v = \omega - \Lambda$, so that we can write an expression for the insolation that shows its dependence on eccentricity and longitude of perihelion.

$$\frac{S_0}{4} s(e, \Lambda, \Phi, x, v) = \frac{S_0}{4} \tilde{s}(\Phi, x, v) \frac{(1 + e\cos(\omega - \Lambda))^2}{(1 - e^2)^2} \quad (12.15)$$

At northern summer solstice $\omega = \pi/2$, so that (12.15) becomes

$$\frac{S_0}{4} \tilde{s}(\Phi, x, v) \frac{(1 + e\sin\Lambda)^2}{(1 - e^2)^2} \approx \frac{S_0}{4} \tilde{s}(\Phi, x, v)(1 + 2e\sin\Lambda) \quad (12.16)$$

where the approximation holds for small values of e. The critical parameter that describes the influence of eccentricity and longitude of perihelion on northern summer insolation is $e\sin\Lambda$, which we will call the precession parameter.

A summary of the effects of the orbital parameters on insolation reveals their importance for the seasonal and latitudinal distribution of insolation.

1. The annual mean insolation varies only as a result of eccentricity in proportion to $(1 - e^2)^{-1/2}$, which, for the observed eccentricities $e < 0.06$, produces changes that seem too small to be important.
2. Obliquity controls the annual mean equator-to-pole gradient of insolation and, together with the eccentricity and longitude of perihelion, the amplitude of the seasonal variation of insolation at a point. Obliquity variations within the observed range of 22.0–24.5° can produce ~10% variations in summer insolation in high latitudes. The annual mean insolation at the poles increases with the obliquity, according to $4/\pi(\sin\Phi)$.
3. The precessional parameter, $e\sin\Lambda$, controls the modulation in seasonal insolation due to eccentricity e and longitude of perihelion Λ variations. The combined effects of eccentricity and longitude of perihelion can result in ~15% changes in high-latitude summer insolation.
4. The combined effects of the three orbital parameters can cause variations in seasonal insolation as large as ~30% in high latitudes.

The time histories of Earth's orbital parameters can be constructed very accurately from celestial mechanics calculations assuming that Earth and other planets orbit the Sun without perturbations from any large bodies passing through the solar system. The time variation of the eccentricity of Earth's orbit is shown in Fig. 12.13a for the period from 1 million years ago to 500,000 years into the future. The dominant periodicities of about 100 and 400 thousand years can be clearly seen. The present eccentricity of about 0.015 is relatively small compared to maximum values of about 0.05 attained about 200, 600, and 950 thousand years ago. The obliquity and the precession parameter vary with periods of about 40 and 20 thousand years, respectively (Fig. 12.13b,c). The precession parameter is strongly modulated by the eccentricity variations at longer time scales, so that it also has 100 and 400 thousand year periods embedded in the envelope of the much faster longitude of perihelion variations.

The obliquity and the precession parameter are shown again in Fig. 12.14 for the period from 200 thousand years ago to 50 thousand years in the future. The obliquity has a dominant periodicity of about 41,000 years and is currently near the middle of its range between about 22.5° and 24.5°. The longitude of perihelion cycle has a period around 22,000 years, but its effect on the precession parameter is modulated by the longer period variation of eccentricity. The precession effect will be small for the next 50,000 years because of the small eccentricity during this period, and the obliquity will remain near the center of its range. Also shown in Fig. 12.14 is the insolation at 65°N at summer solstice, which Milankovitch computed as a predictor of ice ages. According to the Milankovitch theory, a relatively ice-free period will occur as a result of large Northern Hemisphere summertime insolation, which occurs when the obliquity is large and the precession parameter is positive and large. Also shown in Fig. 12.14b is the $\delta^{18}O$ from ocean sediment cores, which when low indicates a relative minimum in land ice. It is plotted inverted so that when it peaks it indicates a relatively warm and ice-free period. You can see that about 12 thousand years ago the summer insolation at 65°N was at a relative maximum and this was a time when the ice volume was decreasing rapidly leading up to the current interglacial period. A similar transition from full glacial to interglacial conditions occurred about 130 thousand years ago when the summer insolation at 65°N was even higher. The glacial maximum preceding each of these transitions is associated with a relative minimum in summer insolation at 65°N. By comparing the 65°N insolation with the isotope record in Fig. 12.14b one can see, however, that the relationship between them, while apparently synchronized, is not simply linear. The ice volume seems to grow slowly to a maximum value before it makes a quick transition to an ice minimum, suggesting that the time scale for land ice accumulation is slower than the time scale for ice removal.

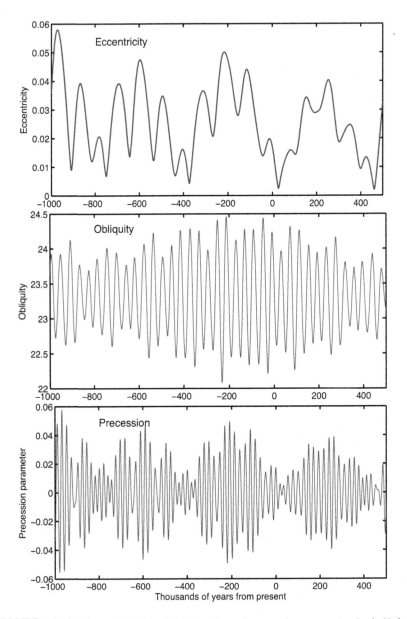

FIGURE 12.13 Eccentricity (*e*), obliquity (Φ), and precession parameter (*e* sinΛ) for Earth's orbit as a function of time from 1 million years before present to 500,000 years into the future. *Calculated using the algorithms of Berger and Loutre (1991) and Laskar et al. (2004).*

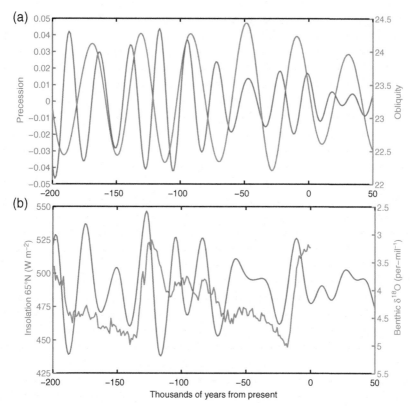

FIGURE 12.14   **(a) Obliquity (Φ) and precession parameter ($e \sin\Lambda$) from 200,000 years before present to 50,000 years in the future. (b) Insolation at 65°N at summer solstice and benthic $\delta^{18}O$.** *From Lisiecki and Raymo (2005).*

The combined effect of the orbital parameters can be seen by computing the insolation as a function of latitude for various times in the past and future. The temporal variation of daily averaged insolation anomalies at the solstices is shown in Fig. 12.15. Anomalies from the time average over the period from 200,000 years ago to 50,000 years in the future are shown. Figure 12.15a shows the insolation variations at Northern Hemisphere summer solstice, which have amplitudes as large as 60 W m^{-2} near the pole. Large positive anomalies at 130,000 years ago and 10,000 years ago correspond fairly well with the times of the last two interglacial periods. The last glacial maximum about 20,000 years ago was preceded by a relative minimum in northern summertime insolation. Insolation anomalies in the future will not stray far from the average over the 250 thousand year period shown in Fig. 12.15. Anomalies at the equinoxes are centered on the equator, can be more than 30 W m^{-2}, depending on the eccentricity, and are opposite in the spring and fall seasons (not shown). These variations

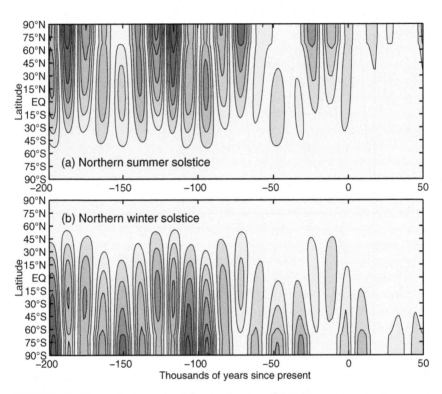

FIGURE 12.15 **Insolation anomalies as a function of latitude and time for the period from 200 thousand years ago until 50 thousand years in the future for (a) Northern summer solstice and (b) Northern winter solstice.** The average for the entire period is subtracted from each individual year to make anomalies. The contour interval is 10 W m^{-2}, with positive values shaded red and negative values shaded blue, and the zero contour is not shown.

drive changes in the equator-to-pole gradient of insolation, which may also be important.

12.5.3 Testing the Milankovitch Theory

Because we know Earth's orbital parameters for several million years into the past and future, and because we can deduce global ice volume for the past five million years or so from ocean cores, the Milankovitch theory is a testable hypothesis. It should be apparent from the discussion of the previous sections that the orbital parameter mechanism for producing secular changes in climate must depend heavily on the physics of the annual cycle of climate, and perhaps particularly in high latitudes during summer where insolation variations are large and ice–albedo feedback is important. The mechanism usually envisioned has to do with spring, summer, and fall snowmelt and the effect of insolation on that process.

The idea is that if the insolation during summer is lower than normal, then the snowmelt will be slower. Some regions that would otherwise be snow free by fall will remain snow covered and the perennial ice cover will expand. Because of the near symmetry of the perihelion cycle, when summer insolation is less than normal winter insolation will be greater than normal (Fig. 12.15). This means that when insolation changes act to make the summer temperatures colder than average, they also make the winter temperatures warmer than average. Because warmer air can contain more moisture, and wintertime temperatures in high latitudes are normally well below freezing, warmer wintertime temperatures may allow greater snow accumulation during the cold season. The combination of increased winter accumulation, decreased summer ablation of ice, and ice–albedo feedback can produce ice volume growth. When summer insolation is enhanced, all the arguments are reversed so that reduced ice volume and warming would result. These arguments all apply to the Northern Hemisphere, where most of the land has been for the past 10 million years.

The preceding arguments all sound plausible, but are very difficult to model quantitatively in a convincing manner, particularly for the time scales required for ice sheet growth. Rather than jump directly into physical modeling, one may first consider an empirical approach as Imbrie and Imbrie (1980) did. Suppose we assume that the variations of global ice volume over the past 500,000 years are due entirely to the orbital parameter variations. We can then develop a simple empirical model with a few adjustable parameters to see how well we can fit the data and whether the relationship between the variables in the fit is consistent with the Milankovitch theory.

Assume that the equilibrium response of the global ice volume to orbital variations is of the form

$$I_{eq} = I_0 - b_\Phi \frac{(\Phi - \Phi_0)}{\sigma(\Phi)} - b_e \frac{e}{\sigma(e)} \cos(\Lambda - \varphi_0) \qquad (12.17)$$

Here I_0 is the ice volume at zero eccentricity and some standard value of the obliquity Φ_0. The parameters b_Φ and b_e set the magnitude of the equilibrium response to obliquity variations and the perihelion cycle, φ_0 determines the phase of the perihelion cycle at which the ice is minimum, and $\sigma(\Phi)$ and $\sigma(e)$ are the standard deviations of obliquity and eccentricity over the past 500,000 years. Any linear response of ice volume to the orbital parameters would take a form such as (12.17).

Further assume that the ice volume relaxes toward its equilibrium value with a time scale that depends on whether the ice volume is increasing or decreasing

$$\frac{\partial I}{\partial t} = \frac{(I_{eq} - I)}{\tau_I} \qquad (12.18)$$

where

$$\tau_I = \begin{cases} \tau_c & for \ I < I_{eq} \\ \tau_w & for \ I > I_{eq} \end{cases} \qquad (12.19)$$

Thus τ_c is the accumulation time scale and τ_w is the ablation time scale. The global ice volume record suggests an asymmetry between slow growth and rapid decay, and this is allowed for by using separate time scales for growth and decay of ice sheets.

A good fit to the ice volume record is obtained with $\tau_c = 42,000$ years, $\tau_w = 10,600$ years, $b_\Phi > 0$, $b_e/b_\Phi = 2$, and $\varphi_0 = 125°$. These optimal parameters, if taken at face value, imply that if the ice volume had time to adjust, the smallest ice volume would occur for large obliquity and for perihelion between northern summer solstice and autumnal equinox ($\varphi_0 = 90°$ would indicate minimum equilibrium ice volume when perihelion occurs at summer solstice), and with the obliquity and perihelion cycles being of comparable importance. These empirical determinations are in general accord with the original hypothesis of Milankovitch.

The empirical fit between the orbital parameter forcing and the global ice volume record indicates that the time scale for ice accumulation is 42,000 years. It is interesting to consider what processes in the climate system could account for this long time scale. Several candidates suggest themselves.

1. *The deep-ocean circulation.* As discussed in Chapter 7, the time scale for the deep ocean may be thousands of years because of its large heat capacity and slow overturning time. A liberal upper limit on the time scale would be a few thousand years, however, which is too short to explain the 10,000-year time scales required.

2. *Glacial and ice sheet dynamics.* The current precipitation in Green Bay, Wisconsin is about 1 in. of water for each of the months of December, January, and February. If we assume that all of the winter precipitation (3 in.) survives the summer melt season, it would take 39,000 years to grow a 3-km thick ice sheet. Glaciers and ice sheets also move around slowly and spread under their own weight, which would probably add to the time required.

3. *Isostatic adjustment.* The continents sink slowly under glacial loading and bounce back slowly when ice sheets melt and the water flows into the ocean. The time scale for the response of the bedrock ranges from 5,000 to 20,000 years, depending on the size of the ice sheet and the character of the rock. The sinking of the continents under the ice sheets may allow the oceans to intrude over what would otherwise be continent and contribute to the rapid decay of the ice sheet via calving of icebergs from the ice sheets and heat and ice transport by the ocean

currents. This has been offered as one mechanism to explain the more rapid decay than growth of the ice sheets. This asymmetry may also be because the precipitation rate limits the growth rate but no similar constraint applies to the melting rate.

4. *Geochemistry.* The carbon dioxide content of the atmosphere varies significantly between glacial and interglacial conditions, and the radiative effect of reduced atmospheric CO_2 contributes significantly to the cooling during an ice age. It can be argued that the CO_2 change must be the result of a change in ocean chemistry, since the CO_2 changes persisted for thousands of years and on this time scale the atmospheric CO_2 concentration is slave to the ocean. The ocean contains 60 times more carbon than the atmosphere. Ocean geochemistry and the budgets of certain key nutrients have time scales that are quite long and could contribute to the low-frequency response of the climate system.

12.6 MODELING OF ICE AGE CLIMATES

Global climate models (GCMs) can be used to understand how past glacial climates were maintained and estimate the relative importance of the known feedback mechanisms. Paleoclimatic data can be used to estimate the extent and thickness of ice sheets, land vegetation, sea-surface temperature, and atmospheric composition as a function of time since the last glacial maximum, approximately 20,000 years ago. Each of these factors can be specified individually in a climate model to determine its effect on the simulated climate, or they can be predicted independently to gauge the ability of the GCM to faithfully simulate the observed relationships among them. The orbital parameters can also be varied to determine how they affect the climate simulation. Currently, climate models of the most realistic type cannot be integrated for a long enough period to simulate ice sheets growth or changes in vegetation patterns, so these parameters have been specified in most simulations conducted to date. The quality of the simulation and the effect of these changes on ice age climate are best judged by comparing the simulated SST and air temperatures with estimates from paleoclimatic data.

The Paleoclimate Model Intercomparison Project (PMIP) has conducted a suite of experiments focused on the climate of: the last glacial maximum 21 kBP (LGM), the Mid-Holocene 6000 years before present (6 kBP) and the preindustrial climate (Braconnot et al., 2007a,b). A reconstruction of surface conditions during the LGM from paleoclimatic data was made by the CLIMAP Project (CLIMAP Project Members, 1976), which can be used to force an atmosphere model or to compare with the simulation from a climate model with a coupled ocean. The insolation distribution during the LGM was similar to the preindustrial one, but at 6 kBP the insolation was greater in the Northern Hemisphere during the summer solstice season and the ice was modern, so that the effect of the changed insolation pattern can be tested. Thus by comparing LGM with the preindustrial the effect of the ice sheets and CO_2 reductions can be studied, and by comparing 6 kBP with the preindustrial the effect of changed orbital parameters can be studied.

12.6.1 Simulation of the Last Glacial Maximum

The land ice, CO_2 and vegetation changes inferred from paleoclimatic data can be imposed on a coupled climate model to see if the model produces a climate consistent with the inferred temperature and land hydrology changes. The reduction in CO_2 to 185 ppm provides an estimated -2.8 W m^{-2} forcing for the LGM climate. The direct effect of the increased ice sheet cover is estimated to be about -2.5 W m^{-2}. The effects of these climate forcings are amplified by water vapor and surface albedo feedback. Increased dust could provide significant additional negative climate forcing, but this was not considered in the PMIP experiments. Land ice is increased and sea level is lowered consistent with estimated conditions for the LGM (Fig. 9.7).

When these conditions are imposed on a suite of atmosphere-ocean climate models, the global mean surface temperature is reduced by 4–6°C, the global mean precipitation decreases and the land areas become drier, in agreement with inferences from paleoclimate data. The drying occurs mostly because of the reduced amount of water vapor in the cooler atmosphere, so that the flux of water from ocean to land is decreased. This is particularly significant in the tropics, where the insolation is similar despite a cooler Earth.

12.6.2 Simulation of the Mid-Holocene

The primary difference between the Mid-Holocene (6 kBP) and preindustrial conditions is the orbital parameters. The difference in daily insolation between 6 kBP and present is shown in Fig. 12.16b. Shown for comparison are the daily insolation anomalies for 12 kBP, when the anomalies at the summer solstice in the north were most positive and were helping to melt the ice sheets (Fig. 12.16a). During the LGM at 21 kBP the insolation anomalies were a modest -10 W m^{-2} during July in the North and during November in the South (not shown). By 6 kBP, the ice sheets were retracted to modern levels, but a fairly strong difference in the northern summer insolation was still present, with a northward gradient of the anomaly across the equator during July. This insolation gradient is thought to be a strong driver of anomalous monsoon circulations during this time. Recall from Fig. 6.24 and the discussion in Section 6.5 that the

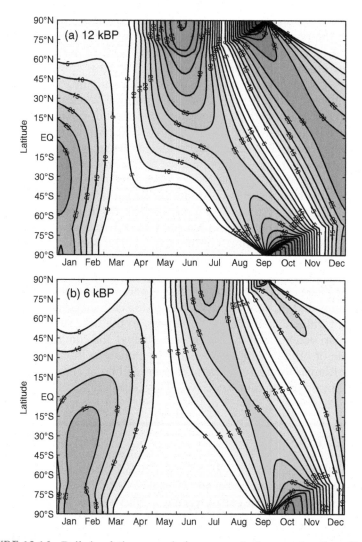

FIGURE 12.16 **Daily insolation anomaly from present values as a function of latitude and season for (a) 12 kBP (e = 0.020, Φ = 24.2°, Λ = 81°) and (b) 6 kBP (e = 0.019, Φ = 24.1°, Λ = 180°).** Contour interval is 5 W m^{-2} and the zero contour is not shown. Positive value shaded red.

surface low-pressure belt and low-level wind convergence move northward and southward more or less following the Sun over Africa.

The major response of the model climate to the Mid-Holocene insolation anomalies is a larger seasonal variation in the Northern Hemisphere, especially over land. This leads to a stronger thermal low over the land during summer, which drives convergence and enhanced precipitation. The monsoons of the tropics are strengthened, especially in North Africa, and the Sahara region shows increased precipitation, which is consistent with the paleoclimatic evidence that it was wetter then. The greater Sahel and Sahara precipitation is associated with a northward shift of the intertropical convergence zone (ITCZ) during summer, which is driven by the insolation gradient, a resulting shift in SST, and vegetation feedbacks. When it rains more, vegetation is more robust, and this lowers the surface albedo and amplifies the effect of enhanced insolation to produce a stronger thermal low and more moisture convergence and moisture recycling. For Africa, the modeled changes over land and over the Atlantic Ocean conspire to enhance the northward shift of the ITCZ to moisten the Sahara. In the Indian Ocean, the ocean changes tend to counteract the effect of enhanced land heating on the summer monsoon in models. As the enhanced heating over the land drives stronger ascent over the Indian subcontinent, enhanced winds drive upwelling and cooling in the Indian Ocean north of the equator. The ocean cooling tends to damp the effect of the land heating.

The evidence is overwhelming that orbital parameter changes have a profound effect on climate. It is equally apparent that many other forcing mechanisms and the internal dynamics of the climate system also have a significant influence on climate variations. In the near future, it is likely that human actions will be the primary driver for climate change. The time scale of orbital parameter forcing of climate is thousands of years, whereas human forcing of climate change will be of similar magnitude, but introduced over a century or two. This is the subject of Chapter 13.

EXERCISES

1. Show that if the planetary albedo is 0.3, a change in total solar irradiance of 1 W m^{-2} results in a global mean climate forcing of 0.175 W m^{-2}.

2. The current magnitude of the greenhouse effect is measured by the difference between the emission from the surface and the emission from the top of the atmosphere, $G = \sigma T_s^4 - \sigma T_e^4 \approx 150$ W m^{-2}. What would be the required magnitude of the greenhouse effect to maintain the surface temperature at 288 K if the solar constant were reduced by 30%? By what distance would the effective emission level of the atmosphere need to rise, if the lapse rate is approximately 6 K km^{-1}?

3. Calculate the percent change in annual mean insolation at a planet if the eccentricity is changed from 0 to 0.2. How does this compare with the change in insolation at summer solstice if summer solstice occurs at perihelion?

4. Calculate the difference of insolation at 60°N at the summer solstice for conditions with perihelion at northern winter solstice and an obliquity of 22.5° and conditions with perihelion at northern summer solstice and

an obliquity of 24.5°. Assume a total solar irradiance of 1361 W m^{-2} at the mean Earth–Sun distance and an eccentricity of 0.04. What are the relative contributions of precession and obliquity variations to this difference? *Hint*: Use (12.16) and Fig. 12.12.

5. What do the paleoclimatic data and their use in modeling studies such as that described in Section 12.6 imply about the energy balance albedo feedback models of the ice caps described in Section 10.4?

6. Discuss what adjustments to the Milankovitch theory described in Section 12.5 might be required by the results of the numerical experiments discussed in Section 12.6.

13

Anthropogenic Climate Change

13.1 THE WINGS OF DAEDALUS[1]

The greenhouse effect that warms the surface of Earth above its emission temperature results because a few minor constituents absorb thermal infrared radiation very efficiently. As a result of human activities, the atmospheric concentrations of some of these natural greenhouse gases are increasing, and entirely new man-made greenhouse gases have been introduced into the atmosphere. The human-produced increase in the atmospheric greenhouse effect is warming the surface of Earth. When the effects of feedback processes internal to the climate system are taken into account, it becomes clear that human activities are leading to a global climate change that may produce a mean surface temperature on Earth as warm as any for more than a million years. It is one of the great challenges of global climatology to predict future climate changes with adequate detail and sufficiently far in advance to allow humanity to adjust its behavior in time to avert the worst consequences of such a global climate change.

This challenge consists of several interlocking parts. We must first understand and predict the human-induced changes to the environment that are most important for climate, which may include atmospheric gaseous composition, aerosol amount and type, and the condition of the land surface. Then we must predict the climate change that will result from these changed climate parameters. These two steps are not independent, since climate change itself may feed back on the surface conditions and atmospheric composition through physical, chemical, and biological processes.

---

[1]In Greek mythology Daedalus was an Athenian inventor, architect, and engineer. He was imprisoned with his son Icarus in a tower, but managed to escape by constructing wings from feathers and wax. Icarus was young and imprudent and, despite warnings from his father, flew too close to the Sun, where his wings melted and he fell into the sea and drowned.

Global Physical Climatology. http://dx.doi.org/10.1016/B978-0-12-328531-7.00013-X

13.2 HUMANS AND THE GREENHOUSE EFFECT

Long-lived greenhouse gases that appear to be increasing as a direct result of human activities include carbon dioxide (CO_2), methane (CH_4), nitrous oxide (N_2O), and halocarbons (e.g. CFC-11 and CFC-12). Figure 13.1 shows records of CO_2 and CH_4 from year 1 to 2014, and combines ice core measurements from the Law Dome, East Antarctica, and atmospheric measurements from Cape Grim, Tasmania. The concentrations of these two key greenhouse gases remained approximately constant from year 1 until about 1800, after which they began a rapid increase driven by human activities. This increase accelerated after 1950. The insert shows more detail since the end of preindustrial times, which is generally defined as 1750. In 2014, CO_2 continues on an exponentially increasing trend, but CH_4 began to level off in the 1990s. Methane has increased by a factor of 2.5 since preindustrial times, and it appears that its sinks have begun to balance its sources. Methane has a much shorter lifetime in the atmosphere than carbon dioxide.

Most of the key greenhouse gases that humans are changing have very long lifetimes in the atmosphere so that the amounts we release into the atmosphere today will remain in the atmosphere for up to two centuries, depending on the gas in question (Table 13.1). For each gas in the table (except CO_2), the "lifetime" is defined here as the ratio of the atmospheric content to the total rate of removal. This time scale also characterizes the rate of adjustment of the atmospheric concentrations if the emission rates are changed abruptly. Carbon dioxide is a special case since it has no real

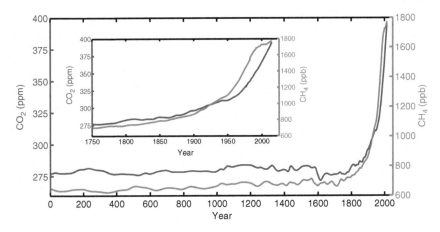

FIGURE 13.1 Carbon dioxide and methane concentrations from ice core measurements at Law Dome, Antarctica, and *in situ* and flask atmospheric measurements from Cape Grim, Tasmania combined together to shown annual means from year 1 to 2014 CE. Insert shows detail since 1750. In situ *data are courtesy of CSIRO and the Australian Bureau of Meteorology. Ice core data are from Etheridge et al. (1996) and MacFarling Meure et al. (2006).*

TABLE 13.1 Characteristics of Some Key Greenhouse Gases That are Influenced by Human Activities

Parameter	CO_2	CH_4	CFC-11	CFC-12	N_2O
Preindustrial atmospheric concentration (1750)	278 ppm	715 ppb	0	0	270 ppb
Current atmospheric concentration (2011)	390 ppm	1803 ppb	238 ppt	527 ppt	324 ppb
Current annual rate of atmospheric accumulation	2.0 ppm (0.5%)	4.8 ppb (0.27%)	−2.25 ppt (−0.93%)	−2.25 ppt (−0.42%)	0.8 ppb (0.25%)
Atmospheric lifetime (years)	(50–200)	10	65	130	150

ppm, parts per million by volume; ppb, parts per billion by volume; ppt, parts per trillion by volume. *Hartmann et al., 2013, Section 2.2.*

sinks, but is merely circulated between various reservoirs (atmosphere, ocean, biota). The "lifetime" of CO_2 given in the table is a rough indication of the time it would take for the CO_2 concentration to adjust to changes in the emissions. Simulations of the carbon cycle coupled with climate suggest, however, that the CO_2 increases that we are introducing now will take many centuries to be removed by natural processes and are essentially permanent.

Most of the important greenhouse gases are naturally occurring, but many of the halocarbons are industrially created and have no sources in nature. Although water vapor is the most important greenhouse gas, we exclude it from this discussion because its atmospheric abundance is not under direct human control, but responds freely to the prevailing climate conditions and helps to determine them. The turnover time for atmospheric water vapor is only about 9 days. We consider the changes in long-lived trace species to provide a forcing to the climate system, and we regard the changes in water-vapor abundance that result from this warming to be a feedback process. Nonetheless, we must keep in mind that any temperature change associated with human activity will be composed of a large contribution associated with water-vapor feedback and other feedbacks internal to the climate system. The net effect of all of these feedbacks is not known with great precision (Chapter 10).

13.3 CARBON DIOXIDE

Carbon dioxide is a naturally occurring atmospheric constituent that is cycled between reservoirs in the ocean, the atmosphere, and the land. Figure 13.2 shows how the carbon cycle has been changed by the actions

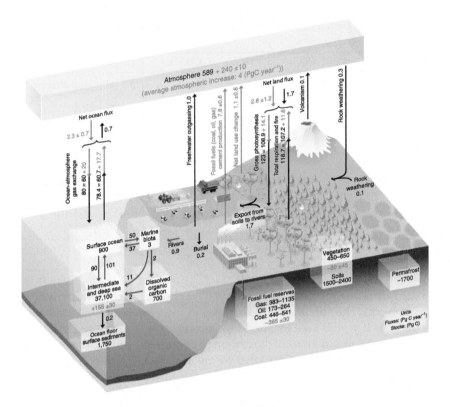

FIGURE 13.2   **Simplified schematic of the global carbon cycle.** Numbers represent reservoir mass, also called "carbon stocks" in PgC (1 PgC = 10^{15} gC) and annual carbon exchange fluxes (in PgC year^{-1}). Black numbers and arrows indicate reservoir mass and exchange fluxes estimated for the time prior to the Industrial Era, about 1750. Red arrows and numbers indicate annual "anthropogenic" fluxes averaged over the 2000–2009 time period. These fluxes are a perturbation of the carbon cycle during Industrial Era post-1750. These fluxes (red arrows) are: fossil fuel and cement emissions of CO_2, net land use change, and the average atmospheric increase of CO_2 in the atmosphere, also called "CO_2 growth rate." The uptake of anthropogenic CO_2 by the ocean and by terrestrial ecosystems, often called "carbon sinks," are the red arrows part of net land flux and net ocean flux. Red numbers in the reservoirs denote cumulative changes of anthropogenic carbon over the Industrial Period 1750–2011. By convention, a positive cumulative change means that a reservoir has gained carbon since 1750. The cumulative change of anthropogenic carbon in the terrestrial reservoir is the sum of carbon cumulatively lost through land use change and carbon accumulated since 1750 in other ecosystems. Note that the mass balance of the two ocean carbon stocks, surface ocean, and intermediate and deep ocean, includes a yearly accumulation of anthropogenic carbon (not shown). Uncertainties are reported as 90% confidence intervals. To achieve an overall balance, the values of the more uncertain gross fluxes have been adjusted so that their difference matches the net land flux and net ocean flux estimates. Fluxes from volcanic eruptions, rock weathering (silicates and carbonates weathering reactions resulting into a small uptake of atmospheric CO_2), export of carbon from soils to rivers, burial of carbon in freshwater lakes and reservoirs, and transport of carbon by rivers to the ocean are all assumed to be preindustrial fluxes, that is, unchanged during 1750–2011. *Figure reproduced from IPCC WG I Report (Fig. 6.1), Ciais et al. (2013), more details and references can be found in Chapter 6. Caption shortened with permission.*

of humans. Fossil fuel reserves are sufficient to continue the current rate of human production of CO_2 for a century or more into the future. CO_2 increased by about 40% between 1750 and 2014 as a result of human activities, about two-thirds of this from fossil fuel combustion and cement manufacture and about one-third from land use change related to changing forest cover (Ciais et al., 2013). About half of the emissions from humans stayed in the atmosphere and about half has been taken up in equal parts by the ocean and the land biosphere. Human emissions of CO_2 have increased in the 2000s and the increase in the first decade of the twenty-first century averaged 2 ppm year^{-1}, more rapidly than in any previous decade.

Carbon exchanges rapidly between the atmosphere and the ocean, so that the particular CO_2 molecules in the atmosphere are changed in about 4 years. We may call this the *turnover time*. However, the time required for atmospheric CO_2 to achieve a new equilibrium in response to a perturbation such as fossil fuel burning is much longer, because of the slow rate at which carbon is exchanged between the surface waters and deep ocean. It requires 50–200 years for the atmospheric CO_2 concentration to achieve a new steady state. CO_2 estimates from ice cores indicate that the preindustrial atmosphere maintained a relatively constant CO_2 of about 278 ppmv for centuries. The recent rapid increase in atmospheric CO_2 concentration parallels very closely the known increase in fossil fuel combustion. The fossil fuel origin of the recent CO_2 increase is further confirmed by changes in the isotopic abundance of ^{13}C and ^{14}C. Radiocarbon, the ^{14}C isotope, is created in the upper atmosphere by cosmic rays, and has a half-life of 5570 years. Therefore, since fossil carbon has been buried in the earth for millions of years, it has no radiocarbon. The decrease of about 2% in atmospheric ^{14}C from 1800 to 1950 is consistent with a fossil carbon source. Fossil carbon is also very deficient in ^{13}C because plants strongly favor the lighter isotope, and the decreasing fraction of ^{13}C in CO_2 indicates that the increased CO_2 is coming from organic carbon rather than from a geochemical exchange with the ocean or volcanoes. The decreasing amount of O_2 in the atmosphere is also definitive evidence for fossil fuel combustion as the source of increasing CO_2 in the atmosphere, since O_2 is needed to burn carbon to make CO_2. If fossil carbon fuel consumption increases at current rates, atmospheric CO_2 concentration will be double its preindustrial value by around 2060 and continue to increase after that.

13.4 METHANE

Methane (CH_4) is produced in a wide variety of anaerobic environments, including natural wetlands, rice paddies, and the guts of animals. It is also released during oil and gas drilling and coal mining. At the present time, it is believed that the human sources of methane from agriculture and waste is about equal to the emissions from natural wetlands, and that anthropogenic emissions are 50–65% of total emissions. The primary removal mechanism is oxidation by hydroxyl (OH) in the atmosphere. Methane oxidation by OH is the dominant source of water vapor in the stratosphere. Water in the stratosphere can be an important greenhouse gas, and both water vapor and ice can influence the photochemistry of ozone.

13.5 HALOCARBONS

Halogens such as chlorine, bromine, and iodine have a wide variety of industrial applications, and in compounds with carbon they produce a number of useful gases. Although present in the atmosphere in very small amounts, industrially produced gases such as CFC-11 (CCl_3F) and CFC-12 (CCl_2F_2) have a substantial influence on the greenhouse effect of Earth. The reason for the strong greenhouse effect of these gases is that they have strong absorption lines in the 8–12-μm region of the longwave spectrum where the surface emission is large. In this wavelength interval, naturally occurring gases do not absorb strongly, so the natural atmosphere is relatively transparent. These gases and other halocarbons are manufactured for use as the working fluid in refrigeration units, as foaming agents, as solvents, and in many other applications. Fully halogenated compounds are extremely unreactive and have very long lifetimes in the atmosphere. They are photodissociated by ultraviolet radiation in the stratosphere, where the chlorine and bromine thus released participate in the catalytic destruction of ozone. Concentrations of these ozone-depleting substances have been decreasing owing to successful international regulation through the Montreal Protocol. Alternatives to fully halogenated compounds are now being used. In their manufacture, one of the halogens is replaced with a hydrogen atom, making the molecules more reactive, giving them shorter lifetimes in the atmosphere, and producing less free chlorine and bromine when they dissociate. The most abundant of these is currently HCFC-22 ($CHClF_2$ ~200 ppt).

13.6 NITROUS OXIDE

Nitrous oxide (N_2O) is produced by biological sources in soils and water. Its primary sinks are in the stratosphere, where it is removed by photolysis and by reaction with electronically excited oxygen atoms. Anthropogenic sources include the use of artificial fertilizers in cultivated soils, biomass burning, and a number of industrial activities.

13.7 OZONE

Ozone (O_3) is increasing in the troposphere and has been depleted in the stratosphere, both owing to human actions. The tropospheric increase is related to industrial and automobile pollution that leads to the photochemical production of ozone near the ground. Ozone in the stratosphere has been depleted by human production of industrial gases such as CFCs, mostly through the action of bromine and chlorine released from them. Ozone near the surface is an environmental hazard because of its health effects on humans and plants. Ozone in the troposphere and lower stratosphere is an effective greenhouse gas, primarily because of the position of its 9.6 μm absorption band in the middle of the water vapor window. When stratospheric ozone recovers in the mid- to late twenty-first century, it will add a little to the greenhouse effect, but not as much as the reduction of CFC-11 and CFC-12 will reduce the greenhouse effect.

13.8 ANTHROPOGENIC AEROSOLS AND CLIMATE

Tropospheric aerosols have direct and indirect effects on the radiation balance, as discussed in Chapter 12. Aerosols directly influence solar and longwave radiative transmission. They also serve as cloud condensation nuclei (CCN), and the abundance of CCN influences the number, size, and atmospheric lifetime of cloud droplets or particles. Cloud abundance and radiative properties have a substantial influence on the net radiation balance of Earth. As discussed in Section 12.3, a big fraction of tropospheric aerosols is produced by conversion of SO_2 gas to sulfate aerosol, and more than half of the total sulfate aerosol production is anthropogenic and mostly related to fossil fuel combustion. These sulfur emissions come mostly from the Northern Hemisphere and have led to a serious problem with acid rain, which motivated successful emission reductions through a cap and trade system in North America that was initiated in 1991, along with the Acid Rain Treaty between Canada and the USA.

It is generally about 1 week between the release of an aerosol precursor gas such as SO_2, its subsequent conversion to sulfate aerosol, and final precipitation in solution within a raindrop. Therefore, the aerosol burden of the troposphere is based on emissions from at most the previous 2 weeks, and if human production of aerosols were to cease, the aerosol burden of the atmosphere would return to its natural level within a similar period. Because of this short lifetime, the aerosol loading of the troposphere is highly variable, and tends to be highest near the sources of aerosols or their precursor gases. Recent trends in aerosol burden have been downward in Europe and North America because of efforts to clean up industrial emissions, and upward in Asia in association with rapid growth in emissions from coal-powered power plants, factories, and vehicles. Over the long term, the cooling effect of aerosols will be overwhelmed by the warming effect of greenhouse gases. In the near term, though, the difficulty of quantifying the anthropogenic aerosol effect, especially through modification of clouds, imposes a large uncertainty on efforts to use observations of warming to constrain the sensitivity of climate.

13.9 CHANGING SURFACE CONDITIONS

Deforestation in mid-latitudes and the tropics, expansion and contraction of deserts, and urbanization of the landscape can have significant effects on the local surface conditions and local climate. One of the most direct effects is that caused by surface albedo changes, which have a strong influence on the energy balance. When the relatively small fraction of land area on Earth and the effect of cloud cover in screening surface albedo changes are taken into account, however, land surface albedo changes directly caused by humans seem to have a relatively small effect on the global average energy balance, about 0.15 W m^{-2}.

Although land albedo changes may not be of first-order importance for global climate, most people live on continents and the regional climate and ecological changes associated with deforestation and urbanization can be quite important for human populations. It is estimated, for example, that complete removal of the Amazon rain forest would have a substantial influence on local surface temperature and hydrology. About half of the rainfall over the Amazon basin is derived from evapotranspiration from the forest. Removing the forest will change the ratio of runoff to evapotranspiration, and the hydrological balance of the region may be seriously altered. The raised surface albedo also lessens the solar heating that ultimately drives upward motion and low-level moisture convergence. Although model simulations of complex interactions between vegetation and climate are at an early stage of development, some simulations suggest that removing the Amazon rain forest would result in a reduction of both moisture convergence and evapotranspiration, so that runoff would decrease significantly.

13.10 CLIMATE FORCING BY HUMANS

An important contribution of science to the policy options regarding human-induced climate change is to quantify the magnitude of the effect of various human actions in producing global warming. To do this we assume the paradigm that global temperature changes can be divided into a part produced by a forcing and a part contributed by a feedback that depends on the global temperature change. To compute the forcing, for example, one could take a climate model, leave the temperature, humidity, and clouds unchanged, introduce a change such as increasing the CO_2 from the preindustrial value to today's value, and then compute the change in the top-of-atmosphere (TOA) radiation balance. The change in the radiation balance, thus computed, is then the climate forcing that humans have so far produced by increasing the atmospheric CO_2. Early on, however, it was noted that the stratosphere cools rapidly when the CO_2 is increased, before any surface temperature change has occurred. This change in stratospheric temperature changes the TOA radiation balance and changes the effective forcing of increasing CO_2. To maintain the forcing–feedback paradigm intact, the forcing is now calculated as the change in TOA radiation balance after the stratosphere has come into a new equilibrium with the increased CO_2, but before the surface temperature has changed. A little later, it was discovered that the clouds change rather quickly in response to the CO_2 increase also, as we would expect, since we know that convection is a response to the radiative cooling of the atmosphere, and CO_2 affects the radiative cooling of the atmosphere (Chapter 3). The rapid response of the clouds to CO_2 produces an even larger fast response of the TOA radiation balance and requires another adjustment to the estimate of forcing. To incorporate the fast response of clouds into the forcing from CO_2, one procedure is to introduce the CO_2 change, while keeping the SST fixed in climate model, then run the model long enough so that the clouds have equilibrated to the CO_2 and a new estimate of the effective forcing can be computed. To get stable statistics the model may need to be run for 20 years. The atmospheric temperature and the land temperatures are allowed to respond to the CO_2. The global mean precipitation decreases and shifts a little from the ocean to the land, since the land is allowed to warm and the oceans are not. So the forcing–feedback paradigm has become a little strained in this process, but the goal is to provide the best numbers for use in policymaking. The most recent IPCC report (Stocker et al., 2013) provided a more traditional stratosphere-adjusted radiative forcing (RF), and a cloud-adjusted effective radiative forcing (ERF).

Figure 13.3 shows the temporal evolution of natural and anthropogenic climate forcing from 1750 to 2011 as estimated by the IPCC WG I Fifth Assessment. Values for the major elements of forcing are shown in a bar chart with uncertainty bars on the right. Forcing by CO_2 and other well-mixed greenhouse gases (WMGHG) are large, positive, and relatively certain. Forcings from aerosols and aerosol–cloud interactions are large, negative, and very uncertain, but are thought to offset a significant fraction of the greenhouse gas forcing at the present time. However, aerosol forcing is not changing much at the present time. Total anthropogenic forcing began to increase more rapidly after about 1970 and continues to increase. Forcing from solar variability is small and regular, and volcanic forcing is often large, but is very episodic (Fig. 13.3).

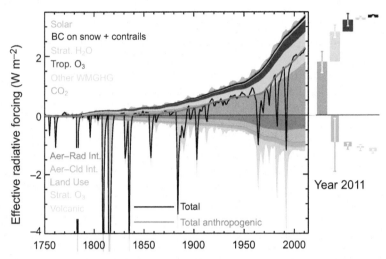

FIGURE 13.3   **Time evolution of forcing for anthropogenic and natural forcing mechanisms.** Bars with the forcing and uncertainty ranges (5–95% confidence range) at present are given in the right part of the figure. For aerosol the ERF due to aerosol–radiation interaction and total aerosol ERF are shown. The uncertainty ranges are for present (2011 versus 1750). For aerosols, only the uncertainty in the total aerosol ERF is given. For several of the forcing agents, the relative uncertainty may be larger for certain time periods compared to present. The total anthropogenic forcing was 0.57 (0.29–0.85) W m^{-2} in 1950, 1.25 (0.64–1.86) W m^{-2} in 1980, and 2.29 (1.13–3.33) W m^{-2} in 2011. *Figure and caption reproduced from IPCC WG I Report Chapter 8 Fig. 8.18, Myhre et al. (2013) where more details are provided.*

13.11 GLOBAL WARMING POTENTIAL

For planning and regulation purposes it is useful to know the climate impact of a particular human action. For example, "How much warming should we expect from the emission of 1 kg of CO_2?" The answer to this question depends on the climate forcing, as discussed in the previous section, which depends on the efficiency with which one kilogram of CO_2 can reduce the emission of longwave radiation. In addition to this, we need to know how long that extra kilogram of CO_2 will stay in the atmosphere after we have emitted it. This depends on the atmospheric lifetime of the climate forcing in question. These two factors can be combined into the global warming potential, but one must decide over what time frame to integrate the climate forcing. Table 13.2 shows lifetimes, radiative efficiencies, and global warming potentials (GWP) for some long-lived greenhouse gases for time horizons of 20, 100, and 500 years. A good example is methane, which has a much higher radiative efficiency than CO_2, and although its lifetime is much shorter than CO_2, its GWP is higher than that of CO_2, especially on relatively short time scales. The reason we are more

TABLE 13.2   Lifetimes, Radiative Efficiencies, and Global Warming Potentials (GWP) Relative to CO_2

Name	Chemical formula	Lifetime (years)	Radiative efficiency $(W\,m^{-2}\,ppb^{-1})$	20-year	100-year	500-year
Carbon dioxide	CO_2	50–200	1.4×10^{-5}	1	1	1
Methane	CH_4	12	3.7×10^{-4}	72	25	7.6
Nitrous oxide	N_2O	114	3.03×10^{-3}	289	298	153
CFC-11	CCl_3F	45	0.25	6,730	4,750	1,620
CFC-12	CCl_2F_2	100	0.32	11,000	10,900	5,200
HCFC-22	$CHClF_2$	12	0.2	5,160	1,810	549

CO_2 lifetime is not easily characterized with a single number, and so is shown as a range. The GWP for CO_2 was calculated using a decay time from carbon cycle model.
Numbers were taken from Solomon et al. (2007), Section 2.10.2.

concerned about CO_2 is that there is much more of it in the atmosphere and it is projected to increase much faster, so that its net effect on surface temperature will dominate in the future. Although the 100-year GWP of methane is 25 times that of CO_2, the annual increase of CO_2 is 400 times that of methane (Table 13.1). Similarly HCFC-22 has a massive GWP compared to CO_2, but its concentration in the atmosphere is tiny, so its effect on climate is currently small and will likely remain so until after CO_2 has greatly warmed the surface. Nonetheless, the warming effect of methane is currently more than a third of that of CO_2 and it has a relatively short lifetime in the atmosphere. So if you wanted to produce a quick reduction in greenhouse gas forcing, one way might be to reduce methane emissions, since the methane would begin to decline quickly and the reduction in greenhouse gas forcing would be significant.

13.12   EQUILIBRIUM CLIMATE CHANGES

A standard experiment with a global climate model is to calculate the equilibrium climate for present and for doubled atmospheric carbon dioxide concentrations and study the differences between the two climate states. In such studies, the transient nature of the carbon dioxide increase and the climate response are ignored and only the state of the climate system when it is in balance with a particular concentration of CO_2 is considered. The response of the equilibrium climate to doubled CO_2 is easier to compute than the transient response and gives useful information about the nature of the transient climate response to be expected. Doubled CO_2 is used as a surrogate for an equivalent climate forcing that may consist of contributions from many greenhouse gases.

13.12.1 One-Dimensional Model Results

A simple estimation of the response to changed greenhouse gases or other radiative forcings can be obtained from one-dimensional radiative–convective equilibrium models (Section 3.10). The vertical temperature profiles from such calculations for changed CO_2 concentration and changed total solar irradiance (TSI) are shown in Fig. 13.4. These calculations were performed for average cloudiness, fixed relative humidity, and a moist adiabatic lapse rate. Notice that for a CO_2 increase and a TSI increase the surface warming is similar, but that the temperature change in the stratosphere is opposite, warming in the case of increased TSI and cooling in the case of CO_2 increase. Surface temperature increases with CO_2 concentration because of its enhancement of the total greenhouse effect, but the stratospheric temperatures decline with increasing CO_2. The reasons for this can be easily understood by remembering that the approximate stratospheric energy balance is between heating through absorption of solar radiation by ozone and cooling by emission from CO_2 (Fig. 3.18).

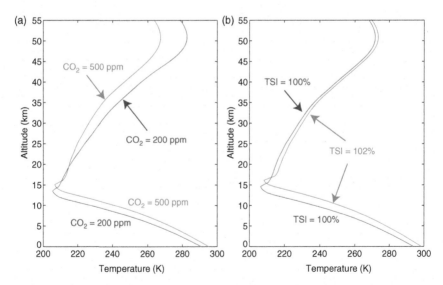

FIGURE 13.4 **Radiative–convective equilibrium temperature profiles computed with a one-dimensional model with a moist adiabatic lapse rate.** (a) CO_2 concentrations of 200 ppm and 500 ppm and (b) total solar irradiance of 100% and 102% of the current value.

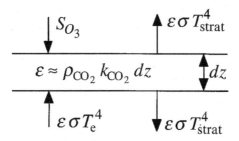

FIGURE 13.5 Diagram showing the energy fluxes for a thin layer of stratospheric air in which ozone absorbs an amount of solar radiation S_{O_3}, and that absorbs and emits terrestrial radiation with an emissivity of $\varepsilon \ll 1.0$.

We can illustrate the dependence of stratospheric temperature on its longwave emissivity through a very simple model (Fig. 13.5). Suppose that we consider a thin layer of the stratosphere with a longwave emissivity that is much less than 1 ($\varepsilon \ll 1.0$). This layer will absorb a fraction of the outgoing longwave radiation, and it will emit up and down according to its assumed emissivity. In addition, it will receive solar heating through the absorption by ozone S_{O_3}. The absorption by this layer of solar and outgoing longwave radiation (σT_e^4) must be balanced by its longwave emission.

$$S_{O_3} + \varepsilon \sigma T_e^4 = 2\varepsilon \sigma T_{strat}^4 \tag{13.1}$$

Solving for the temperature of the stratospheric layer shows the relationship of the stratospheric temperature to the emissivity of the layer.

$$T_{strat} = \sqrt[4]{\frac{\varepsilon^{-1} S_{O_3} + \sigma T_e^4}{2\sigma}} \tag{13.2}$$

Since CO_2 is the principal longwave absorber in the middle and upper stratosphere, and from (3.29) $\varepsilon \approx \rho_{CO_2} k_{co_2} dz$, we know that the emissivity would increase when the CO_2 concentration is doubled. If we assume that the ozone amount and the planetary albedo remain unchanged when the CO_2 concentration is increased, then the absorbed solar radiation and the outgoing longwave radiation will be the same for a doubled CO_2 climate in equilibrium. Therefore, nothing within the radical in (13.2) will change except the emissivity of the stratospheric layer, and the stratospheric temperature will decline with increasing CO_2.

Figure 13.4 shows that the vertical structure of the warming response for a CO_2 increase and a TSI increase are different. This is also true for the latitude structure of the response, since a TSI change will produce a stronger radiative forcing in the Tropics, whereas the radiative forcing from a CO_2 increase may actually be stronger in high latitudes, since the fractional increase in the greenhouse effect is larger in high latitudes where much less water vapor is present to compete with CO_2 in absorbing longwave radiation. The effective radiative forcing adjustment for CO_2 and TSI are also different, since CO_2 warms the atmosphere and suppresses clouds, whereas a TSI increase heats the surface more than the atmosphere and enhances clouds. So the rapid adjustments to CO_2 forcing and TSI forcing are also different.

13.13 DETECTION AND ATTRIBUTION

Climate science has been able to answer two very key questions regarding climate change. These are, *detection*: "Is the surface of Earth warming?" and *attribution*: "Are humans causing the surface of Earth to warm"? To answer the detection question it must be shown that the warming that has been observed over the past century cannot be explained as change that has resulted from natural internal variability of the climate system, or from natural forcing such as volcanoes or solar variability. To answer the attribution question, it must be shown that the observed warming could not have occurred without the climate forcing that humans have provided. These questions are answered in the affirmative with multiple lines of observational evidence and with many experiments with global climate models.

Independent instrumental temperature records all indicate warming. Surface air from land stations and ocean observations, upper air from balloons and satellites, sea surface temperature from *in situ*, and satellite measurements and ocean heat content are all warming. In addition, other instrumental records are also showing change that is consistent with warming. Atmospheric specific humidity is increasing, Arctic sea ice is declining, mountain glaciers are shrinking, and sea level is rising. Careful work with data and climate models has indicated that most of these changes are unlikely to have occurred without the warming associated with the production of greenhouse gases by humans.

Figure 13.6 illustrates the method by which climate models are used to confirm detection of global warming and attribution of warming to humans. On the left are the observed global mean temperatures in black, along with the global mean temperature from a large suite of model simulations from CMIP3 and CMIP5, the model integrations done for the IPCC 4th and 5th assessments, respectively (Taylor et al., 2012). These integrations were done separately with the best estimate of only the natural climate forcings and with the best estimate of the natural plus anthropogenic forcings. Fig. 13.6a shows that none of the model experiments can simulate the warming after 1980 if only the natural forcings are included, whereas if the human forcing is included the models can simulate the warming after

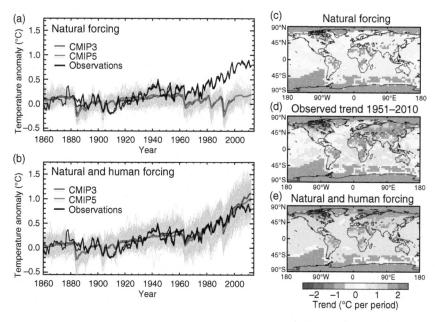

FIGURE 13.6 (Left) Time series of global and annual averaged surface temperature change from 1860 to 2010. (a) The results from two ensemble of climate models driven with just natural forcings, shown as thin blue and yellow lines; ensemble average temperature changes are thick blue and red lines. Three different observed estimates are shown as black lines. (b) The simulations by the same models, but driven with both natural forcing and human-induced changes in greenhouse gases and aerosols. (Right) Spatial patterns of local surface temperature trends from 1951 to 2010. (c) The pattern of trends from a large ensemble of Coupled Model Intercomparison Project Phase 5 (CMIP5) simulations driven with just natural forcings. (d) The pattern of observed trends from the Hadley Centre/Climatic Research Unit gridded surface temperature data set 4 (HadCRUT4) during this period. (e) The trends from a corresponding ensemble of simulations driven with natural + human forcings. *Figure and caption reproduced exactly from IPCC WG I Report Chapter 10 Fig. FAQ 10.1 Figure 1, Bindoff et al. (2013).*

1980 very well. The conclusion is that the warming after 1980 would not have happened without humans. Some of the earlier warming also seems to have been caused by humans, but it was not yet big enough to exceed the natural year-to-year variability of global annual mean temperature. On the right in Fig. 13.6 are maps of the surface temperature trends from 1951 to 2010 as observed, as modeled with only natural forcing and as modeled with human forcing included. The models suggest that without humans the climate would have cooled slightly from 1951 to 2010, mostly as a result of volcanic eruptions, but that with human forcing included the magnitude and pattern of warming produced by the models is similar to the observed warming. Such analyses have also been used to attribute other changes to human influences, including warming on all continents except Antarctica, the cooling of the stratosphere, the warming of the upper oceans, sea level rise, and Arctic sea ice decline (Bindoff et al., 2013).

13.14 TIME-DEPENDENT CLIMATE CHANGES

The changes in climate that result from human activities occur gradually in response to steadily increasing climate forcing by greenhouse gas increases, aerosol production, and land surface modification. The transient response of the climate system to the anthropogenic shift in climate forcing may be very different from the equilibrium response, and equilibrium may never be achieved since it is unlikely that the anthropogenic forcing of climate will remain constant for the many centuries required for a steady state to be established. Moreover, the immediate need is to understand the history of climate that we are likely to experience over the next 100 years or so given a particular climate-forcing scenario. The response of the climate system to changed thermal forcing will be delayed by the large heat capacity of the ocean. In addition, the ocean currents and their slow response to this thermal forcing may yield geographic variations in the temperature change that are different from those of an equilibrium calculation. The ocean model is thus critical to the prediction of the transient and equilibrium response to climate forcing.

Much of the initial lag of the warming behind the forcing is associated with storage of heat in the ocean mixed layer. The response of the mixed-layer ocean models can be understood using very simple conceptual models. In these models, the ocean is represented by a wet surface with the heat capacity of the mixed layer of the ocean. We may represent the global mean transient temperature response T' to an imposed climate forcing, Q, with a first-order differential equation.

$$c\frac{dT'}{dt} = -\lambda_R^{-1}T' + Q \tag{13.3}$$

The time tendency term on the left represents the storage of energy in the ocean, c is the heat capacity of the ocean, and λ_R is the climate sensitivity parameter defined in (10.2). The solution to (13.3) for the case in which the temperature perturbation is initially zero is easily obtained.

$$T' = e^{-t/\tau_R} \int_0^t c^{-1} Q e^{t'/\tau_R} dt' \tag{13.4}$$

The response time, $\tau_R = c\lambda_R$, is proportional to the product of the heat capacity of the system times the sensitivity parameter. This makes it difficult to estimate the equilibrium climate change from the initial history of an induced warming, since both the sensitivity and the effective heat capacity are uncertain. Both the magnitude of the equilibrium response and the time required to achieve it are proportional to λ_R. The system (13.3) with formal solution (13.4) has only two independent parameters. So if you have a solution for an initial set of parameters c_1, Q_1 and λ_{R_1}, the solution will be exactly the same for a different sensitivity, λ_{R_2}, so long as you choose the forcing and heat capacity such that,

$$Q_2 = Q_1 \frac{\lambda_{R_1}}{\lambda_{R_2}} \text{ and } c_2 = c_1 \frac{\lambda_{R_1}}{\lambda_{R_2}} \tag{13.5}$$

So one temperature history can be produced by an infinite number of combinations of c, Q and λ_R.

If the perturbation temperature is forced by an instantaneous switch-on of steady forcing

$$Q = \begin{cases} 0, & t \le 0 \\ Q_0, & t > 0 \end{cases} \tag{13.6}$$

where Q_0 is a constant, the solution (13.4) becomes

$$T' = \lambda_R Q_0 (1 - e^{-t/\tau_R}) \tag{13.7}$$

so that the equilibrium temperature perturbation $T' = \lambda_R Q_0$ is approached exponentially with an e-folding time scale of τ_R.

For CO_2 increasing exponentially with time, the change of climate forcing is approximately linear with time, because the radiative forcing scales approximately as the logarithm of CO_2 concentration. If we apply a forcing that increases linearly in time

$$Q = \begin{cases} 0, & t \le 0 \\ Q_t t, & t > 0 \end{cases} \tag{13.8}$$

and insert it into the solution (13.4) for (13.3), we obtain

$$T' = \lambda_R Q_t \{t + \tau_R (e^{-t/\tau_R} - 1)\} \tag{13.9}$$

The exponential term within the brackets represents an initial transient, which for $t \gg \tau_R$ is small compared to the other two terms, after which time the solution becomes approximately

$$T' \approx \lambda_R Q_t \{t - \tau_R\} \tag{13.10}$$

The forcing at any time is $Q_t t$ to which the equilibrium response would be $\lambda_R Q_t t$. From (13.10), we infer that the transient response to linearly increasing forcing is just the equilibrium response, delayed by the response time τ_R. Choosing the thermal capacity to be that of 150 m of water [$c = 7.3 \times 10^8$ J K^{-1}m^{-2} from (4.5)] and choosing the sensitivity parameter to be such that doubling CO_2 leads to an equilibrium response of 4°C ($\lambda_R = 1$ K (W m^{-2})$^{-1}$), yields a response time of $\tau_R = 20$ years. The response of this simple model to linearly increasing forcing is similar to the global mean response of a GCM with a mixed-layer ocean. The transient response is very much like the equilibrium response, except delayed by about 20 years. The temperature responds sooner over the continental interiors where the effect of the ocean's heat capacity is less directly felt.

The simple model (13.3) has three independent parameters that determine the structure of the transient climate response: the forcing, the heat capacity, and the climate sensitivity. The real-world value of each of these parameters is uncertain, and we can use this model to illustrate why the global mean temperature record does not help to constrain these parameters very well. Figure 13.7a shows example solutions for a forcing that increases linearly at a rate of 4 W m^{-2} per 70 years, approximately the rate of increase of human forcing by greenhouse gas emissions. Solutions are shown with climate sensitivities that differ by a factor of two and mixed layer heat capacities that vary inversely with the sensitivity to keep the time scale $\tau_R = c\lambda_R$ constant. For the first 50 years, these two simulations are virtually indistinguishable, certainly within the observational precision, but they diverge greatly in the second half of the century, when the more sensitive climate warms much more. Therefore, even if the forcing was precisely known, which it is not, the temperature record would not allow us to constrain the sensitivity to within a factor of two, if the rate at which the ocean will take up heat is also uncertain. Figure 13.7b shows the

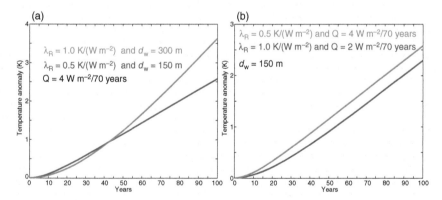

FIGURE 13.7 **Response of a simple climate model to linearly increasing forcing from equation** (13.9). (a) Forcing increasing 4 W m^{-2} per 70 years, comparing doubled sensitivity and doubled mixed-layer depth (red) with baseline values (blue). (b) Mixed-layer depth of 150 m, but comparing doubled sensitivity and halved forcing (red) with baseline values (blue).

case where we assume we know the heat capacity very well, but the forcing and the sensitivity are both uncertain. In this case, the more sensitive model jumps out to an early lead, but because the equilibrium climate sensitivity is the same, the two models maintain the same rate of warming once the initial transient has died away. If the sensitivity and the forcing errors compensate, then the rate of warming is the same, despite the time scale for adjustment being different. So the uncertainty in each of heat capacity, sensitivity, and forcing precludes any of them being constrained very precisely by the observational record, even though we can establish detection and attribution with a high degree of confidence.

In atmosphere–ocean climate model simulations with dynamically active oceans, the timing and spatial structure of the response are more complex. When ocean currents are included in the simulation, the warming is much reduced around Antarctica and over the north Atlantic (Fig. 11.15), where mixing with the deep ocean occurs. Ocean models have two time scales, a rapid adjustment of the mixed layer, and a much longer time scale of hundreds of years associated with the deep circulation of the ocean (Held et al., 2010). Also, it is reasonable that the sensitivity of climate is not constant, as sensitivity is a property of the mean climate. A simple example is surface ice. When less surface ice is present, the surface ice feedback is reduced.

13.15 PROJECTIONS OF FUTURE CLIMATE

It is very important to predict how the climate will change in the future. Robust projections of future climate can be used to plan actions that will reduce the impact of climate change on the natural and built environment, either by reducing positive climate forcing, or by adapting to the changes that will occur. To make useful projections for more than several decades into the future, it is necessary to specify the actions that humans will take that affect the climate. For the IPCC Fifth assessment, four Representative Concentration Pathways (RCPs) were defined in order to cover the likely spread of possible future human emissions of greenhouse gases (van Vuuren et al., 2011). They are named after the approximate anthropogenic human forcing in the year 2100, which are $(+2.6 \ W \ m^{-2}, +4.5 \ W \ m^{-2}, +6.0 \ W \ m^{-2}, +8.5 \ W \ m^{-2})$. The resulting CO_2 concentrations are shown in Fig. 13.8. The lowest emission scenario is RCP 2.6, which represents a reduction in greenhouse gas emissions sufficient to keep global mean warming under about 2°C. It stabilizes CO_2 at about 450 ppm by about 2075. The highest emission scenario, RCP 8.5, represents the path we are currently on, sometimes called the "business as usual" scenario. For RCP 8.5, the CO_2 reaches 1000 ppm shortly after 2100. Other greenhouse gases besides CO_2 are expected to contribute to

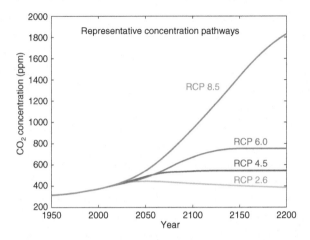

FIGURE 13.8 Past and projected CO_2 concentrations for the four representative concentration pathways utilized by the IPCC Fifth Assessment to project possible future climates.

future warming. When the effect of these other greenhouse gases are incorporated and translated into an equivalent CO_2 concentration, the CO_2 equivalent exceeds 1300 ppm when the actual CO_2 reaches 1000 ppm. To move from the RCP 8.5 scenario we are currently following to any of the lower emissions scenarios requires a rapid reduction in CO_2 emissions, most especially from fossil fuel combustion. This could be achieved through a combination of increased efficiency, shifting to energy sources such as solar, wind and nuclear, and carbon capture and storage at fossil fuel power plants.

The response of the global mean surface temperature to these climate forcing scenarios is uncertain, especially because of uncertainty in the climate feedbacks described in Chapters 10 and 11. Figure 13.9a shows projected changes in global mean surface temperature for the RCP 2.6 and RCP 8.5 scenarios, plus the uncertainty in these projections, which is large because the forcing, feedback, and ocean storage rate are all somewhat uncertain. The best estimates for 2200 are about 2°C for RCP 2.6 and 6°C for RCP 8.5. Some of the expected impacts of RCP 8.5 were described in Chapter 11, especially changes in the hydrologic cycle in Fig. 11.14, and the spatial distribution of warming in Fig. 11.15. Changes in the hydrologic cycle will be very significant, with the intensity of both floods and droughts increasing as the spatial and seasonal contrasts in the hydrologic cycle are increased. Global average precipitation will increase at 1–3% per Kelvin of warming, but not as fast as the specific humidity, which will go up at about the Clausius–Clapeyron rate 6–7% K^{-1}. In regions of moisture convergence, the precipitation may go up at faster than the

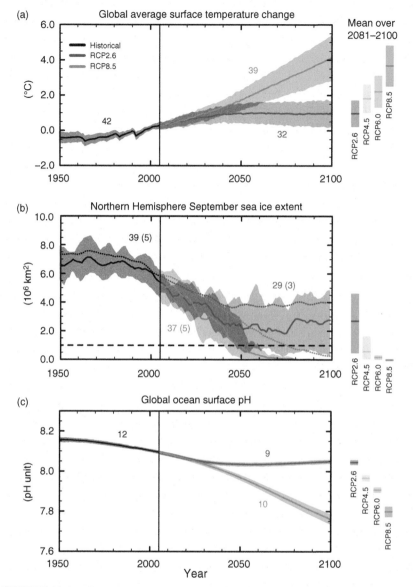

FIGURE 13.9 **CMIP5 multimodel simulated time series from 1950 to 2100 for (a) change in global annual mean surface temperature relative to 1986–2005, (b) Northern Hemisphere September sea ice extent (5-year running mean), and (c) global mean ocean surface pH.** Time series of projections and a measure of uncertainty (shading) are shown for scenarios RCP 2.6 (blue) and RCP8.5 (red). Black (gray shading) is the modeled historical evolution using historical reconstructed forcings. The mean and associated uncertainties averaged over 2081–2100 are given for all RCP scenarios as colored vertical bars. The numbers of CMIP5 models used to calculate the multi-model mean is indicated. For sea ice extent (b), the projected mean and uncertainty (minimum–maximum range) of the subset of models that most closely reproduce the climatological mean state and 1979–2012 trend of the Arctic sea ice is given (number of models given in brackets). For completeness, the CMIP5 multimodel mean is also indicated with dotted lines. The dashed line represents nearly ice-free conditions (i.e., when sea ice extent is less than 10⁶ km² for at least 5 consecutive years). *Figure and caption reproduced exactly from IPCC WG I Report Summary for Policymakers Fig. SPM.7, Stocker et al. (2013).*

Clausius–Clapeyron rate, but in other regions it will need to go up much more slowly or even decrease to satisfy the atmospheric energy balance.

Models and observations indicate that warming will be greater in high latitudes and will have a very visible impact on the landscape through the reduction in snow cover, sea ice and the melting of permafrost. The reduction in Arctic Sea ice has been dramatic over the period of satellite observation. Current models vary greatly in their projections of sea ice, and the future of sea ice is regarded as uncertain, but models that produce the most realistic current sea ice conditions predict that the Arctic will become ice free by about 2050 under the RCP 8.5 scenario (Fig. 13.9b). Under RCP 2.6, a minimum of 2 million square kilometers of sea ice in September is maintained through 2100. Another important impact of CO_2 increase is acidification of the ocean. CO_2 dissolved in water forms carbonic acid, the sharp taste of carbonated beverages, which lowers the pH of seawater. Sea life is adapted to basic water, particularly organisms that form calcite shells. Increased CO_2 in the atmosphere is thus a threat to organisms that form the base of the food chain in the ocean. The RCP 8.5 scenario produces a stronger reduction in ocean pH and so a more acidic ocean (Fig. 13.9c), providing another motivation to reduce emissions of CO_2.

Sea ice concentrations have declined during the period of satellite observations more rapidly than most models predicted, but it is uncertain whether the sea ice was not sensitive enough to warming in the current generation of models or whether natural variability has added onto the recent decline of Arctic Sea ice forced by anthropogenic warming. The year-to-year variability of Arctic sea ice extent is fairly large. Figure 13.10 shows the satellite-measured Arctic sea ice area and sea ice extent during

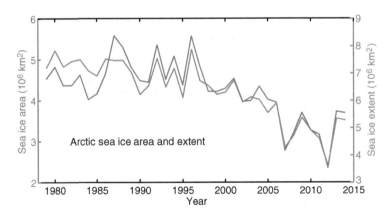

FIGURE 13.10 Arctic sea ice area and extent during September from NASA passive microwave satellite records from 1979 through 2014.

September from 1979 to 2014. Satellite microwave measurements are used to make an estimate of the fraction of the ocean surface that is sea ice in an area. If the concentration is over the threshold of 15%, then sea ice is defined to be present. Sea ice extent is the total area where ice is present, and September sea ice extent averaged about 7 million square kilometers in the Arctic from 1979 to 1996. The sea ice area is the product of the area of a geographic grid box and the sea ice concentration in that box, summed over the Arctic or Antarctic. The extent is thus always greater than the area, but it also seems to be a little more stable from year-to-year, especially early in the record. Both area and extent show a period of little trend before about 1996, and a significant downward trend after that (Cavalieri and Parkinson, 2012). The year with the least sea ice extent in this period was 2012, when the extent was about 3.5 million square kilometers, compared to 7 in the early record, or a decline of 50%, which is far outside the range of variability earlier in the record.

Figure 13.11 shows the daily sea ice extent for each year from 1979 to 2014. The first 7 years (1979–1985) are plotted in blue and the last 7 years (2007–2014) are plotted in red. In the Arctic the sea ice extent appears to have been reduced in every season, but most notably in September, the month of minimum Arctic sea ice extent. In the Antarctic the sea ice trends have been smaller, but in all seasons the sea ice extent appears to have increased a little over this period, contrary to the predictions of climate models (Parkinson and Cavalieri, 2012). In the long run, sea ice extent is expected to decline with global warming, so it is likely that the Antarctic sea ice increase during this period is a product of natural variability, perhaps related to the trend toward more La Niña-like conditions that was discussed in Chapter 8. It has also been hypothesized that melting of glacier ice has freshened the surface water around Antarctica, allowing

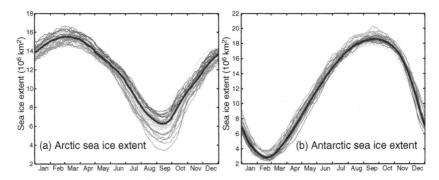

FIGURE 13.11 **Annual cycles of daily (a) Arctic and (b) Antarctic sea ice extent.** 2007–2014 are shown in red, while 1979–1985 are shown in blue. All other years are shown in gray. *NASA satellite data.*

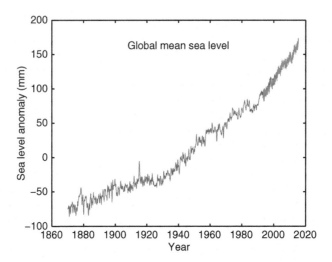

FIGURE 13.12 **Global mean sea level anomalies from gauge data (blue, 1870–2001) and from satellite altimetry measurements (red, 1999–2014).** Satellite data have been adjusted to have the same mean as the gauge data during the period of overlap. *Gauge data are from Church and White (2011) and satellite data from Nerem et al. (2010).*

sea ice to form farther from the coast (Hansen et al., 2015), so that the Antarctic sea ice expansion is a transient response to human-induced changes in the climate system.

One of the more significant impacts of human-induced global warming is sea level rise. Estimates from tide gauges suggest that the global sea level has risen by about 19 cm from 1901 to 2010 (Fig. 13.12). Between 1993 and 2010, the rate of sea level rise was about 3.2 mm year^{-1}, which is faster than the average rise over the past century. This rise in sea level parallels the rise in temperature over this period. Evidence from paleo sea level evidence for the past 3 million years suggests that sea level was 5 m higher in the past when the temperature was about 2 K warmer than present.

The major contributors to sea-level rise over the period of satellite observation (1993–2010) are: thermal expansion of the ocean ($\sim$ 40%), melting of mountain glaciers and small ice caps ($\sim$ 30%), and melting of the Greenland and Antarctic Ice Sheets ($\sim$ 25%) (Church et al., 2013). Recent evidence suggests an acceleration of Greenland Ice Sheet melting, and observations and models also suggest that some ice sheets in West Antarctica that are grounded below sea level are unstable and will eventually flow into the ocean, adding to sea level rise (Joughin et al., 2014; Rignot et al., 2014). Many glaciologists believe it is plausible that the Greenland ice sheet will melt away if the human-induced warming exceeds 2°C, but it is likely that it will take centuries to completely melt. If this happens the sea level rise will be about seven meters.

The future behavior of the major ice sheets is more uncertain than the thermal expansion of the ocean in response to warming. Projections of sea level rise over the next century are 26 to 53 cm for the RCP 2.6 scenario and 52 to 98 cm for RCP 8.5. For RCP 8.5 the rate of seal level rise in 2090 would be 8 to 16 mm year^{-1}, about 4–8 times the average rate during the twentieth century (1.7 mm year^{-1}) (Church et al., 2013).

13.16 OUTLOOK FOR THE FUTURE

It is very clear that human activities have altered the atmosphere and that if we continue to release CO_2 and other greenhouse gases we can produce a climate warmer than any for the past million years. The magnitude and timing of this global climate change are both uncertain and depend on economic and social decisions that society has yet to make. The key question is whether we will continue to burn fossil carbon energy sources and release the combustion products into the atmosphere. Figure 13.13 shows the projected warming of global mean surface temperature since 1870 as a function of the total CO_2 emissions since 1870. All 4 of the RCP scenarios are shown, and all fall along the same line. This means that for the global mean surface temperature increase it matters little at what rate we release CO_2, but only the total amount of CO_2 that is released. Also of note is that in 2010 we were about half way to the amount of CO_2 emission that will lead to a 2°C warming. Scientific evidence and some treaty agreements under the United Nations Framework Convention on Climate Change both indicate 2°C as the level of warming that becomes "dangerous" with respect to its potential impacts, nonlinearities and uncertainties. Under the RCP 8.5 scenario, the one we are currently following, we will reach this dangerous level by 2040, just 25 years from the time of this writing. Under the RCP 2.6 scenario, dangerous warming is averted.

Energy production from renewables like solar and wind have been increasing rapidly, but are still small compared to fossil fuel energy, which made up 82% of world energy production in 2012. Fossil fuel reserves are large and energy production from fossil fuels has increased very rapidly in the early twenty-first century. Global mean surface temperature has increased about 0.8°C in the past century, mostly because of fossil fuel burning, and another 0.5°C will occur if no further greenhouse gas additions are made, since the warming lags behind the climate forcing. So while it is technically possible to reduce emissions through a variety of means and achieve a stabilization of climate as envisioned in the RCP 2.6 scenario (e.g. Pacala and Socolow, 2004), the world has continued on the path of rapid fossil fuel exploitation, with every indication that efforts to mitigate global warming will be too little and too late when they finally occur.

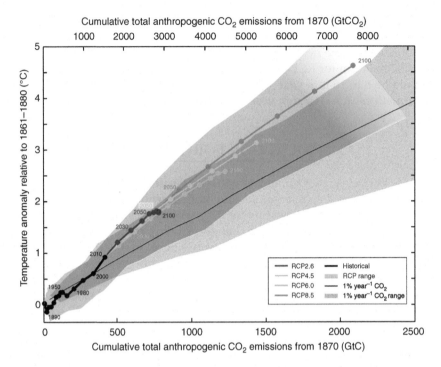

Cumulative total anthropogenic CO_2 emissions from 1870 (GtCO$_2$)

FIGURE 13.13   **Global mean surface temperature increase as a function of cumulative total global CO_2 emissions from various lines of evidence.** Multimodel results from a hierarchy of climate-carbon cycle models for each RCP until 2100 are shown with colored lines and decadal means (dots). Some decadal means are labeled for clarity (e.g., 2050 indicating the decade 2040–2049). Model results over the historical period (1860 to 2010) are indicated in black. The colored plume illustrates the multi-model spread over the four RCP scenarios and fades with the decreasing number of available models in RCP8.5. The multi-model mean and range simulated by CMIP5 models, forced by a CO_2 increase of 1% per year (1% year^{-1} CO_2 simulations), is given by the thin black line and gray area. For a specific amount of cumulative CO_2 emissions, the 1% per year CO_2 simulations exhibit lower warming than those driven by RCPs, which include additional non-CO_2 forcings. Temperature values are given relative to the 1861–1880 base period, emissions relative to 1870. Decadal averages are connected by straight lines. *Figure and caption reproduced exactly from IPCC WG I Report Summary for Policymakers Fig. SPM.10, Stocker et al. (2013).*

13.17 CLIMATE INTERVENTION: GEOENGINEERING THE CLIMATE OF EARTH

It is technically feasible to mitigate global warming by reducing the emissions of greenhouse gases while still maintaining growth in global energy consumption, and this is likely the best option. Since moving to a low CO_2 emission pathway has proved difficult for political, economic and social reasons, it has been proposed that we consider interventions to mitigate the warming produced by our introduction of greenhouse gases into the atmosphere. These actions fall into two broad categories: increasing the albedo of Earth, and removing CO_2 from the atmosphere after it has been introduced.

We have already seen several examples of how nature or humans have increased the albedo of Earth. One way is to inject aerosol precursor gases like SO_2 into the stratosphere to simulate the cooling that a major volcanic eruption produces. This is technically and economically feasible. Another way is to increase the albedo of clouds in the marine boundary layer by introducing cloud condensation nuclei. So far, this seems less technically feasible. A major problem with these approaches is that they would have to be maintained in perpetuity, since if they were ever stopped a large and rapid warming would occur. Also, enhanced CO_2 in the atmosphere is making the ocean more acidic, with potentially serious consequences for the ocean food chain. Cooling the Earth by reflecting more solar radiation would not stop the acidification of the ocean. In addition, increasing the albedo of Earth while increasing the CO_2 of the atmosphere would change the driving of convection and could lead to shifts in the amount or distribution of precipitation and evaporation (e.g. Fig. 13.4). Albedo modification strategies do not address the root cause of human-induced warming, which is greenhouse gas increases, and they may raise additional risks, but they are relatively cost-effective and could produce significant cooling within a few years.

CO_2 removal strategies focus on increasing the rate of carbon uptake by the ocean or land biosphere, or direct removal from the air and subsequent storage. Removal of CO_2 from the air is much less efficient than removing it from smokestacks at power plants. For permanent burial the right kind of underground geology is required. Increasing carbon uptake by the land can have additional benefits that increase its economic feasibility. Growing plants for biofuel and then capturing and burying the CO_2 when the fuel is burned produces energy and net reduction of CO_2. Agricultural practices and soil amendments to enhance carbon content of soils also benefit agricultural efficiency. Fertilizing the ocean with trace minerals can increase ocean productivity, but experiments so far indicate that the CO_2 thus taken up is not stored very long in the ocean, and thus is not effective for mitigation of warming. It is unclear whether these approaches can be scaled up enough to be effective in offsetting the increased production of CO_2 by burning of coal, oil, and natural gas. CO_2 removal strategies address the root cause of human-induced warming, and introduce fewer additional risks, but they are expensive, slow and may produce fairly modest results in the near term.

If we continue to alter the atmosphere in the future at an accelerating rate, and the climate is as sensitive as some models suggest, then within a century we could produce a climate on Earth that would be warmer than any in more than a million years. Moreover, the rate of temperature increase would be very large by natural standards and would make it difficult for plants, animals, and humans to adapt. Because of the long time scale of perturbations in the carbon cycle, once a CO_2-induced warming is initiated, it is essentially permanent from a human perspective. Also, the full warming from a given CO_2 increase does not appear for decades after the CO_2 injection, so that one cannot wait until a "dangerous" 2°C or greater warming has occurred to begin actions to mitigate further warming. Climate science thus has the important challenge of trying to predict the global and regional consequences of human-induced warming as far into the future and with as much detail as possible, both to illustrate the benefits of mitigation and to inform strategies for adaptation to future warming, some of which is already inevitable.

EXERCISES

1. Suppose that a volcanic eruption gives rise to a stratospheric aerosol cloud that changes the instantaneous Earth energy balance by 4 W m^{-2}. The cloud persists unchanged for 2 years. If the heat capacity of the climate system on this time scale is equivalent to a layer of water 50 m deep and the sensitivity of the climate is characterized by a sensitivity parameter $\lambda_R = 1$ K (W m^{-2})$^{-1}$, what is the surface temperature anomaly produced by the aerosol cloud at the end of the 2-year period? What would the temperature response be if the aerosol cloud remained forever? What fraction of this long-term equilibrium response is achieved after 2 years?

2. Consider a situation as in problem 1, except assume that the forcing from the aerosol cloud begins at 4 W m^{-2} but then decreases exponentially with an e-folding time of 2 years. Forcing = 4 W m^{-2} exp $(-t/2$ years). What is the temperature response after 2 years?

3. If the tropical SST is raised by 2°C from 299 to 301 K, and the lapse rate follows the moist adiabatic profile, estimate by how much the temperature at 300 mb will increase?

4. A 1-km-thick layer of the atmosphere at about 55-km altitude is heated by absorption of solar radiation by ozone at a rate that would warm the layer by 10°C day^{-1}. In equilibrium, this solar heating is balanced by longwave cooling. The air density in the layer is about 4×10^{-4} kg m^{-3}. The outgoing longwave radiation is about 240 W m^{-2} on which the stratosphere at this altitude is assumed to have very little effect.

   a. What is the broadband longwave emissivity of this stratospheric layer if it has a radiative equilibrium temperature of 280 K? The emissivity is the ratio of the actual emission from the layer to that of a black body at the same temperature, $\varepsilon = $ (emission/σT^4). Assume local thermodynamic equilibrium so that the longwave emissivity of the layer is equal to its absorptivity.

b. The emissivity of the layer depends primarily on the CO_2 concentration, which was initially 300 ppmv. The CO_2 is doubled to 600 ppmv and the climate finds a new equilibrium. The albedo of the planet is assumed to remain constant during this climate change, so that the emission temperature of the planet is unchanged as a result of doubling CO_2. If the emissivity of the air increases approximately as the natural logarithm of the CO_2 concentration so that the new emissivity is 112% of the original one, what will be the new radiative equilibrium temperature of the stratosphere in the doubled CO_2 climate?

5. Suppose we can assume that the climate system has an effective heat capacity equivalent to that of 100 m of ocean, but the climate sensitivity can be either $\lambda_R = 0.5$ or 1.0 K/(W m^{-2}), and we do not know which. A climate forcing is applied that increases linearly with time as $Q = Q_t t$, where $Q_t = 4$ W m^{-2} in 50 years. If the measurement uncertainty for global mean temperature is 0.5°C, how many years will it be before we could distinguish whether the climate sensitivity is characterized by $\lambda_R = 0.5$ or 1.0 K/(W m^{-2})? That is, when would the temperature for the more sensitive climate exceed that of the less sensitive climate by 0.5°C? What are the temperature perturbations of the global climate for the two sensitivities when they differ by 0.5°C? If the climate forcing is held constant at its value when the two responses become distinguishable, what will be the equilibrium response for each sensitivity? *Hint:* Start by assuming that the time in question is significantly longer than the response time τ_R, so that the exponential transient in the solution (13.9) can be ignored. Determine a posteriori whether this assumption is justified.

Calculation of Insolation Under Current Conditions

A.1 SOLAR ZENITH ANGLE

Consider a unit circle representing the Earth, as pictured in Fig. A.1. We are interested in calculating the solar zenith angle θs and the solar azimuth angle ξ at the point X on the surface of the sphere, located at latitude ϕ. We draw a radial from the center of the sphere through the point X to define the local zenith direction, Z. Another radial from the center of Earth to the Sun crosses through the surface of the sphere at the subsolar point, ss. The point ss occurs at a latitude of δ, which is equal to the declination angle. At the point X we draw another line to the Sun. Since the Sun is many Earth radii from Earth, the lines drawn to the Sun from the center of the Earth and point X are parallel. The arc length on the unit sphere between X and ss is equal to the desired solar zenith angle, θs. We can now apply the law of cosines to the oblique spherical triangle defined by the points at the pole, P, the subsolar point, ss, and the point, X, where we want the solar zenith angle. For this triangle, we know the arc length of two sides and one interior angle, and we wish to know the length of the third side. The law of cosines requires that,

$$\cos\theta_s = \cos(90-\phi)\cos(90-\delta) + \sin(90-\phi)\sin(90-\delta)\cos h \quad (A.1)$$

or

$$\cos\theta_s = \sin\phi\sin\delta + \cos\phi\cos\delta\cos h \quad (A.2)$$

The solar azimuth angle ξ, the angle between the subsolar point and due south (or equatorward) can be obtained from the law of sines for an oblique spherical triangle.

$$\frac{\sin(180-\xi)}{\sin(90-\delta)} = \frac{\sin h}{\sin\theta_s} \quad (A.3)$$

Global Physical Climatology. http://dx.doi.org/10.1016/B978-0-12-328531-7.00014-1

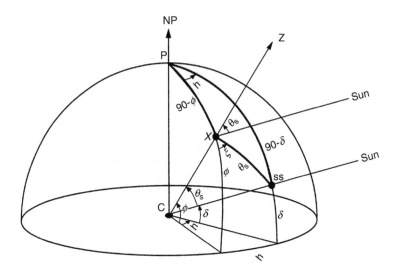

FIGURE A.1   Spherical geometry for solar zenith angle calculation.

or

$$\sin \xi = \frac{\cos \delta \sin h}{\sin \theta_s} \qquad (A.4)$$

A.2 DECLINATION ANGLE

The annual variation of the declination angle for current conditions can be defined to a good approximation in terms of a truncated Fourier series in the time of year. If the time of year is expressed in radians according to the following formula, where d_n is the day number, which ranges from 0 on January 1 to 364 on December 31, then

$$\theta_d = \frac{2\pi d_n}{365}. \qquad (A.5)$$

The declination angle is given to a good approximation by the Fourier series,

$$\delta = \sum_{n=0}^{3} a_n \cos(n\theta_d) + b_n \sin(n\theta_d) \qquad (A.6)$$

where the coefficients are as given in the following table.

n	a_n	b_n
0	0.006918	
1	−0.399912	0.070257
2	−0.006758	0.000907
3	−0.002697	0.001480

Because of the eccentricity of the Earth's orbit, the Earth–Sun distance varies according to the time of year.

A Fourier series formula for the squared ratio of the mean Earth–Sun distance to the actual distance has also been derived by Spencer (1971).

$$\left(\frac{\overline{d}}{d}\right)^2 = \sum_{n=0}^{2} a_n \cos(n\theta_d) + b_n \sin(n\theta_d) \qquad (A.7)$$

where the coefficients are given by

n	a_n	b_n
0	1.000110	
1	0.034221	0.001280
2	0.000719	0.000077

Symbol Definitions

Many symbols are required to represent the quantities used in physical climatology. The letters of the alphabet are less than the number of variables, even if both the English and Greek alphabets are fully utilized. Modern climatology crosses the boundaries between subdisciplines within earth sciences, and the traditional symbol for a quantity in one discipline may be the same as the traditional symbol for some other quantity in another discipline. In developing the symbols used in this book, an attempt was made to balance simplicity and tradition, on one hand, against the clarity of having a unique symbol for every variable, on the other. Some symbols are used to represent more than one variable in order to maintain the traditional usage. It is hoped that the meaning will be obvious from the context. A table containing the symbols used, their meaning, and the equation, section, or figure where they first appear is provided here to assist the student.

ENGLISH SYMBOLS

a	The mean radius of Earth $= 6.37 \times 10^6$ m	(2.23), (6.4)
a_n	A series of expansion coefficients with index n	(A6)
a_o	Semi-major axis of the Earth's orbit	Fig. 12.8, (13.7)
a_p	Planetary absorptivity $= 1 - \alpha_p$	(10.12)
A	Area	Fig. 2.5
A_c	Total fractional area coverage by clouds	(10.43)
A_b	Fractional area coverage by black daisies	(10.45)
A_g	Fractional area coverage by ground in which daisies can grow	(10.48)
A_w	Fractional area coverage by white daisies	(10.44)
A_v	Absorption of radiation at frequency v	(3.26)
b_e, b_Φ	Regression coefficients for eccentricity and obliquity	(12.17)

Global Physical Climatology. http://dx.doi.org/10.1016/B978-0-12-328531-7.00015-3

b_n	A series of expansion coefficients with index n	(A6)
B	A coefficient relating OLR to surface temperature	(10.16), (10.21)
B_e	Equilibrium Bowen ratio	(4.34)
$B_v(T)$	Planck's black body emission at frequency v and temperature T	(3.7)
B_o	Bowen ratio = SH/LE	(4.34)
c	Heat capacity per unit area	(13.3)
c^*	Speed of light = 3×10^8 m s^{-1}	(3.1)
c_p	Specific heat of air at constant pressure	(1.15), (3.24)
c_v	Specific heat of air at constant volume	(1.12)
c_s	Specific heat of soil	(4.8)
c_c	Specific heat of organic matter in soil	(4.8)
c_w	Specific heat of water	(4.5)
$\bar{C}_a$	Total effective heat capacity for the atmosphere per unit area	(4.4)
$\bar{C}_o$	Total effective heat capacity for the ocean per unit area	(4.5)
$\bar{C}_{eo}$	Total effective heat capacity for the surface per unit area	(4.3)
C_D	Aerodynamic transfer coefficient for momentum, or drag coefficient	(4.18)
C_{DE}	Aerodynamic transfer coefficient for vapor	(4.27)
C_{DH}	Aerodynamic transfer coefficient for heat	(4.26)
C_s	Volumetric heat capacity of soil	(4.8)
^{14}C	The isotope of carbon with atomic weight of approximately 14	Sec. 13.3
^{13}C	The isotope of carbon with atomic weight of approximately 13	Sec. 13.3
d	Distance of Earth from the Sun	(2.4), (12.7)
d	Total derivative prefix	(3.3), (10.1)
d_a	Earth–Sun distance at aphelion	(12.6)
d_e	A depth scale for baroclinic disturbances	(10.30)
d_n	Day of the year with index n	(A5)
d_p	Earth–Sun distance at perihelion	(12.6)
d_s	An increment of optical path length	(3.12)

d_w	A depth of water	(4.5)
D	Surface condensation (dewfall plus frost)	(5.1)
D_o	Depth of ocean	(7.15)
DJF	December, January and February	
D_T	Thermal diffusivity	(4.9)
$\dfrac{D}{Dt}$	Material derivative following motion	(6.4), (11.2)
e	Eccentricity of an orbit	(12.6)
e_s	Saturation partial pressure of water vapor	(1.9), (10.8)
e^x	2.71828 raised to the power x	(1.6), (3.17), (7.4)
E	Evapotranspiration or evaporation	(5.1)
E_a	Energy content of the atmosphere per unit area	(6.1)
E_{ao}	Energy content of the atmosphere-ocean climate system	(2.19)
E_{air}	Component of evaporation associated with dryness of air	(5.13)
E_{BB}	Radiative energy emission from a black body	(2.5)
E_{en}	Component of evaporation associated with energy supply to the surface	(5.10)
E_R	Radiative energy emission	(2.6)
E_s	Energy content of the surface	(4.1)
E_{sub}	Mass rate of sublimation of snow	(5.15)
E_ν	Emission of radiation at frequency ν	(3.26)
f	Coriolis parameter $= 2\Omega \sin\phi$	(7.1)
F	Energy flux	(3.12)
$F^\uparrow(z)$	Upward longwave flux at altitude z	(3.33), (4.11)
$F^\downarrow(z)$	Downward longwave flux at altitude z	(3.34), (4.11)
$F_{net}^\downarrow(z)$	$F^\downarrow(z) - F^\uparrow(z)$ net downward longwave flux at altitude z	(10.35)
$\vec{F}_a$	The vector of horizontal of energy flux in the atmosphere	(6.1)
$\vec{F}_{ao}$	Horizontal energy flux in the atmosphere plus ocean	(2.19), (7.14)
F_{eo}	Horizontal energy flux below the surface in earth or ocean	(4.1)
F_I	Vertical flux of heat at the base of sea ice	(11.2)

$\vec{F}_o$	The horizontal flux of energy in the ocean	(7.15)
FRH	A subscript indicating a process that takes place with fixed relative humidity	(10.10)
F_s	The vertical flux of energy through soil or snow	(4.6), (11.3)
F_v	Energy flux in some infinitesimal range of frequency centered on v	(3.3)
F_ϕ	Meridional flux of energy in the atmosphere–ocean	(2.23)
F_∞	Downward radiation flux at the top of the atmosphere	(3.17)
g	Acceleration of gravity, 9.81 m s^{-2}	(1.2)
g_w	Storage of water at and below the surface	(5.1)
g_{wa}	Storage of water in the atmosphere	(5.3)
G	Energy storage in the surface	(4.1)
h	Hour angle	(2.15), (A1)
h	Planck's constant = 6.625×10^{-34} J s	(3.2)
h_c	Water equivalent depth of soil water capacity	(5.14)
h_I	Depth of sea ice	(11.3)
h_o	Hour angle at sunset and sunrise $(-h_o)$	(2.16)
h_s	Water equivalent depth of snow	(5.19)
h_T	Penetration depth of temperature perturbations in soil	(4.10)
h_v	Water equivalent depth of soil moisture below which plants transpire at less than the potential rate	(5.22)
h_w	Equivalent depth of soil moisture	(5.18)
H	Scale height = RT/g	(1.5), (3.18)
$\vec{i}$	Unit vector in eastward direction	(7.11)
I	OLR scaled by global insolation	(10.22)
I	Global ice volume	(12.18)
I_v	Intensity of radiation at frequency v	(3.3), (3.26)
ITCZ	Intertropical Convergence Zone	
$\vec{j}$	Unit vector in northward direction	(7.11)
JJA	June, July, and August	
k	Boltzmann's constant = 1.37×10^{-23} J K^{-1}	(3.7)
$\vec{k}$	Vertical unit vector	(7.11), (10.2)
k_{abs}	Absorption cross-section	(3.12), (12.1)

k_{ext}	Extinction cross-section	(12.1)
k_e	Extinction cross-section	(3.54)
k_I	Thermal conductivity for sea ice	(11.1)
k_s	Thermal conductivity for snow	(11.2)
k_{sca}	Scattering cross-section	(12.1)
k_ν	Absorption cross-section at frequency ν	(3.27)
K_H	A coefficient for horizontal heat transport	(10.19), (10.26)
L_e	A horizontal mixing length scale for atmospheric disturbances	(10.27)
L	Latent heat of vaporization for water $= 2.5 \times 10^6$ J kg^{-1}	(4.25)
L_o	Luminosity of the Sun	(2.2)
LAI	Leaf area index	Sec. 5.4
LE	Surface cooling by evaporation	(4.1), (6.1)
L_f	Latent heat of fusion for water $= 3.34 \times 10^5$ J kg^{-1}	(11.5)
L_R	Rossby radius of deformation	(10.28)
LWC	Liquid water content	(3.56)
m_a	Molecular weight of dry air	App. E
m_w	Molecular weight of water	App. E
M	Angular momentum	(6.16)
M_a	Mass mixing ratio of absorber in air	(3.19), (3.25)
M_e	Orbital angular momentum of Earth about the Sun	(12.11)
MAM	March, April, and May	
M_s	Mass rate of snow melting per unit area	(5.19)
N	The buoyancy frequency	(10.28)
N	Total number density of cloud particles	(3.55)
$n(r)$	Number density of cloud particles of radius r	(3.55)
^{18}O	The isotope of oxygen with atomic weight of approximately 18	(9.1)
OLR	Outgoing longwave radiation $= F^\uparrow(\infty)$	Sec. 3.11
p	Pressure	(1.2)
p_o	Reference pressure, usually 1000 mb	(1.20)
p_s	Surface pressure	(1.6), (3.19), (10.1)
P	Precipitation by rain and snow	(5.1)

PE	Potential evaporation rate	(5.14)
P_o	Period of Earth's orbit about the Sun	(12.12)
$\hat{P}$	Scattering phase function	(12.4)
P_r	Mass rate of rainfall per unit area	(5.18)
P_s	Mass rate of snowfall per unit area	(5.19)
q	Specific humidity-mass mixing ratio of water vapor in air	(1.10)
q^*	Saturation specific humidity	(4.28)
q_a	Specific humidity at anemometer level	(4.27)
q_s	Specific humidity at the surface	(4.27)
q_s^*	Saturation specific humidity at the surface	(4.32), (10.37)
Q	Heating or energy input	(2.1)
Q_{ABS}	Absorbed solar radiation	(10.11)
Q_{abs}	Absorption coefficient	(12.2)
Q_{ext}	Extinction coefficient	(12.2)
Q_{sca}	Scattering coefficient	(12.2)
r_{eff}	Effective particle radius	Fig. 12.9
r_p	Radius of a planet	(2.7)
r_{photo}	Radius of the photosphere of the Sun	(2.2)
R	Gas constant for air	(1.3)
R^*	Universal gas constant	Appendix D
R_a	Net radiative heating of the atmosphere per unit area	(6.1)
RH	Relative humidity $= q/q^*$	(4.30)
Ri	Richardson number	(4.20)
R_s	Net radiative energy input at the surface	(4.1), (4.11)
R_{TOA}	Net radiative energy input at the top of the atmosphere	(2.19), (3.59)
R_v	Gas constant for water vapor $= 461 \text{ J K}^{-1}\text{ kg}^{-1}$	(1.10), (10.9)
s	Distribution function for solar radiation	(10.12), (12.10)
$\tilde{s}$	Distribution function for solar radiation for a circular orbit	(12.8)
S_o	Total solar irradiance at the mean Earth–Sun distance $(1360.8 \pm 0.5 \text{ W m}^{-2})$	(2.7)

S_d	Solar irradiance in W m^{-2} at some distance d from the Sun	(2.4)
SH	Sensible cooling of the surface	(4.1)
SON	September, October, and November	
S_{O_3}	Heating from solar radiation absorption by ozone	(13.1)
SPCZ	South Pacific convergence zone	
$S^{\uparrow}(z)$	Upward solar flux at altitude z	(4.11)
$S^{\downarrow}(z)$	Downward solar flux at altitude z	(4.11)
T	Temperature	(1.1)
T^*	Deviation of temperature from its zonal average	(6.13)
T_A	An atmospheric temperature	(2.11), (3.39)
T_B	Temperature at the bottom of a layer of sea ice	(11.3)
T_a	Temperature of air at anemometer level	(4.26)
T_e	Emission temperature	(2.8), (3.47), (10.7)
T_i	Local emission temperature at point i	(10.46)
T_o	A reference temperature	(4.20)
T_s	Surface temperature	(2.12)
T_{eo}	Effective temperature of the land or ocean surface for heat storage	(4.3)
T_{photo}	Emission temperature of the photosphere of the Sun	(2.2)
T_{SA}	Temperature of the near-surface air	(3.53)
$T_{z_{cb}}$	Temperature of the air at the altitude of cloud base	(3.45)
$T_{Z_{ct}}$	Temperature of the air at the altitude of cloud top	(3.46)
t	Time	(3.24), (13.3)
t_o	Initial time	(10.13)
U	Internal energy	(2.1)
U	Wind speed	(4.16)
U_E	Eastward transport velocity in the oceanic Ekman layer	(7.6)
U_r	A reference wind speed	(4.18)
u	Eastward component of velocity relative to surface	(6.4)
u_E	Eastward component of current velocity in the oceanic Ekman layer	(7.6)

u_{earth}	Eastward velocity of the surface associated with Earth's rotation	(6.15)
u_ϕ	Eastward velocity air will have at any altitude ϕ, if it is zero at the equator	(6.17)
u_*	Friction velocity	(4.15)
u^*	The deviation of u from its zonal average	(6.13)
v	Northward component of velocity	(6.4)
v^*	The deviation of v from its zonal average	(6.6)
v_E	Northward component of current velocity in the oceanic Ekman layer	(7.5)
V	A horizontal velocity scale	(10.27)
$\vec{V}$	Horizontal velocity vector	(7.10)
V_{eff}	Variance of particle radius	Fig. 12.9
V_E	Northward transport velocity in the oceanic Ekman layer	(7.8)
V_I	Northward current velocity in the ocean interior	(7.14)
w	Vertical component of velocity	(4.22), (6.5)
w_E	Vertical velocity at the base of the Ekman layer	(7.10)
w_w	Soil water mass per unit area	(5.18)
w_s	Surface snow mass per unit area	(5.19)
W	A vertical velocity scale	(10.33)
W	Work	(2.1)
x	Eastward spatial coordinate and distance	(7.5)
x	Sine of latitude	(10.11)
x	A symbol for an arbitrary variable	(6.12)
x_i	Sine of latitude poleward of which perennial ice cover exists	(10.21)
y	Northward spatial coordinate and distance	(7.5), (10.26)
z	Vertical spatial coordinate and distance	(1.1)
z_E	Ekman depth	(7.6)
z_o	Roughness height	(4.17)
z_r	A reference height	(4.19)
z_s	Elevation of the surface, often zero	(3.35)
z_{cb}	Altitude of cloud base	(3.45)
z_{ct}	Altitude of cloud top	(3.46)

GREEK SYMBOLS

α	Specific volume	(1.12)
α_p	Planetary albedo	(2.7)
α_g	Albedo of bare ground	(10.48)
α_w	Albedo of white daisies	(10.48)
α_b	Albedo of black daisies	(10.48)
α_s	Surface albedo	(4.12)
β	Meridional derivative of the Coriolis parameter	(7.13)
β	Birth rate of daisies	(10.44)
β_E	Ratio of evapotranspiration to potential evaporation	(5.21)
∂	Partial differential symbol	(1.1)
δ	Solar declination angle	(2.15), (A1)
δ	γ/B ratio of horizontal transport to longwave cooling coefficients	(10.23)
$\delta^{18}O$	Normalized deviation of $^{18}O/^{16}O$ fraction from normal in ‰	(9.1)
$\bar{\nabla} \bullet$	Divergence of vector operator	(10.19)
Δ	A prefix to signify a difference	(3.58)
Δ	A prefix to signify a divergence operator	(2.21)
Δf	Runoff or divergence of horizontal water flux at and below the surface	(5.1)
Δf_a	Divergence of horizontal water flux in the atmosphere	(5.3)
ΔF_{ao}	The divergence of the horizontal transport of energy by the atmosphere and ocean	(2.21), (10.17)
ΔF_a	Divergence of horizontal heat flux in the atmosphere	(6.1)
ΔF_{eo}	Divergence of the horizontal heat flux below the surface $\approx \Delta F_o$	(4.1)
ΔF_o	Divergence of the horizontal heat flux in the ocean	(4.1)
γ	A heat exchange coefficient	(10.17)
Γ	Lapse rate	(1.1)
Γ_d	Dry adiabatic lapse rate	(1.24)
Γ_s	Saturated adiabatic lapse rate	(1.25)
$\bar{\nabla}$	Vector gradient operator	(7.11), (10.19)

ε	Emissivity	(2.6)
ε_ν	Emissivity at frequency ν	(3.28)
φ_o	A phase angle	(12.17)
φ	An azimuth angle	Fig. 3.2, (3.4)
ϕ	Latitude Earth coordinate	(2.15), Fig. 6.3
Φ	Geopotential height $= gz$	(10.2)
Φ	Obliquity angle	(12.8), Fig. 12.11
κ	von Kármán constant	(4.16)
η	Horizontal transport efficiency parameter	(10.51)
λ	Wavelength of radiation	(3.1)
λ	Longitude Earth coordinate	(6.4), Fig. 6.3
λ_R	A gross sensitivity parameter for climate	(10.2), (10.6)
Λ	Longitude of perihelion	(12.10), Fig. 12.11
μ	$\cos\theta$	(3.21)
ν	Diffusivity	(7.3)
ν	Frequency of radiation	(3.1)
ν	True anomaly angle of the Earth's orbit	(12.10)
ν	Northward component of velocity on a sphere	(6.4), Fig. 6.3, (7.2)
π	Pi = 3.141592654, ratio of the circumference to diameter of a circle	(2.2)
ρ	Density of air	(1.2)
ρ_a	Density of absorber in air	(3.12)
ρ_{as}	Density of absorber in air at the surface	(3.18)
ρ_c	Density of organic matter in soil	(4.8)
ρ_o	Mean density of seawater	(7.5)
ρ_s	Density of soil material	(4.8)
ρ_I	Density of sea ice	(11.5)
ρ_w	Density of water	(4.5), (5.19)
σ	Stefan–Boltzmann constant, 5.67×10^{-8} W m^{-2} K^{-4}	(2.5)
$\sigma(x)$	The standard deviation of x	(12.17)
ρ_t	Potential density of seawater	Fig. 7.1
θ	A general zenith angle	(3.4), (3.13)
θ_s	Solar zenith angle	(2.15), (Fig. A.1)

Θ	Potential temperature	(1.20)
Θ_e	Equivalent potential temperature	(1.26)
Θ_v	Virtual potential temperature	Fig. 4.6
τ_I	A time scale for global ice volume	(12.19)
τ	Optical depth	(3.15)
τ	A time scale	(4.10)
$\vec{\tau}$	$= \vec{i}\,\tau_x + \vec{j}\,\tau_y$, Wind stress vector	(7.14)
τ_o	Surface stress	(4.15)
τ_x	Eastward component of wind stress	(7.5)
τ_y	Northward component of wind stress	(7.5)
τ_ν	Optical depth at a particular frequency ν	(3.31)
τ_R	Response time scale	(13.4)
ϖ	Single scattering albedo	(12.3)
ψ_M	Meridional mass stream function	(6.9)
ω	Angle between the Earth, the Sun and Earth's position at vernal equinox	Fig. 12.11
ω	Rate of change of pressure following an air parcel	(6.5)
ω	Solid angle	(3.4), (12.4)
Ω	The rotation rate of Earth $= 7.292 \times 10^{-5}\ \mathrm{s}^{-1}$	(6.15), (7.12)
χ	Death rate of daisies	(10.44)
ξ	Azimuth angle between south and the position of the Sun	(A3)
ζ_a	Vertical component of absolute vorticity $= f + \zeta_r$	(7.12)
ζ_r	Vertical component of relative vorticity	(7.12)

MISCELLANEOUS SYMBOLS

$\mathcal{T}\{z, z'\}$	Broadband slab transmissivity between altitudes z and z'	(3.35)
$[x]$	The average of x around a latitude circle, the zonal average of x	(6.7)
$\lvert x \rvert$	Absolute value of x	
$\tilde{T}$	The global average of T	(10.18)
$\bar{x}$	The time average of x	(6.8)

$\vec{\nabla} \times \vec{x}$	Vector curl of $\vec{x}$	(7.11)
%	Percent	
‰	Per thousand	
°C	Degrees Celsius temperature unit	
K	Kelvin temperature unit	
°E	Degrees east Earth coordinate	
°F	Degrees Fahrenheit temperature unit	
°N	Degrees north Earth coordinate	
°W	Degrees west Earth coordinate	
°S	Degrees south Earth coordinate	

Système International (SI) Units

In most fields of science, the SI system of units is the accepted standard. The SI base units are consistent with the metric system of measurement and are given in Table C.1. In addition to the base units, derived units are used in many applications. A selection of SI-derived units frequently used in physical climatology are given in Table C.2.

Supplementary dimensionless units for angles and some derived units using these are given in Table C.3. Prefixes to indicate decimal multiples of units are given in Table C.4. Some non-SI units that are in common use or appear in older references are given in Table C.5 with their SI equivalents.

TABLE C.1 SI Base Units

Quantity	Name	Symbol
Length	Meter	m
Mass	Kilogram	kg
Time	Second	s
Electric current	Ampere	A
Thermodynamic temperature	Kelvin	K
Amount of substance	Mole	mol
Luminous intensity	Candela	cd

TABLE C.2 Examples of SI-Derived Units

Quantity	Name	Symbol In terms of SI base units	Symbol For special name	Symbol In terms of other units
Area	Square meter	m^2	–	–
Volume	Cubic meter	m^3	–	–
Speed, velocity	Meter per second	$m\ s^{-1}$	–	–

(Continued)

Global Physical Climatology. http://dx.doi.org/10.1016/B978-0-12-328531-7.00016-5

TABLE C.2   Examples of SI-Derived Units (*cont.*)

Quantity	Name	Symbol In terms of SI base units	For special name	In terms of other units
Acceleration	Meter per second squared	$m\,s^{-2}$	–	–
Divergence	Per second	s^{-1}	–	–
Vorticity	Per second	s^{-1}	–	–
Wave number	1 per meter	m^{-1}	–	–
Geopotential; dynamic height	Meter squared per second squared	$m^2\,s^{-2}$	–	–
Density	Kilogram per cubic meter	$kg\,m^{-3}$	–	–
Specific volume	Cubic meter per kilogram	$m^3\,kg^{-1}$	–	–
Luminance	Candela per square meter	$cd\,m^{-2}$	–	–
Frequency	Hertz	s^{-1}	Hz	–
Force	Newton	$m\,kg\,s^{-2}$	N	–
Pressure	Pascal	$m^{-1}\,kg\,s^{-2}$	Pa	$N\,m^{-2}$
Energy	Joule	$m^2\,kg\,s^{-2}$	J	$N\,m$
Power	Watt	$m^2\,kg\,s^{-3}$	W	$J\,s^{-1}$
Electric charge	Coulomb	$s\,A$	C	$A\,s$
Electric potential	Volt	$m^2\,kg\,s^{-3}\,A^{-1}$	V	$W\,A^{-1}$
Capacitance	Farad	$m^{-2}\,kg^{-1}\,s^4\,A^2$	F	$C\,V^{-1}$
Electric resistance	Ohm	$m^2\,kg\,s^{-3}\,A^{-2}$	Ω	$V\,A^{-1}$
Conductance	Siemens	$m^{-2}\,kg^{-1}\,s^3\,A^2$	S	$A\,V^{-1}$
Magnetic flux	Weber	$m^2\,kg\,s^{-2}\,A^{-1}$	Wb	$V\,s$
Magnetic flux density	Tesla	$kg\,s^{-2}\,A^{-1}$	T	$Wb\,m^{-2}$
Inductance	Henry	$m^2\,kg\,s^{-2}\,A^{-1}$	H	$Wb\,A^{-1}$
Luminous flux	Lumen	$cd\,sr$	lm	–
Illuminance	Lux	$m^{-2}\,cd\,sr$	lx	–
Dynamic viscosity	Pascal second	$m^{-1}\,kg\,s^{-1}$	–	$Pa\,s$
Moment of force	Meter Newton	$m^2\,kg\,s^{-2}$	–	$N\,m$

TABLE C.2 Examples of SI-Derived Units (*cont.*)

Quantity	Name	In terms of SI base units	For special name	In terms of other units
Surface tension	Newton per meter	$kg\,s^{-2}$	–	$N\,m^{-1}$
Irradiance	Watt per square meter	$kg\,s^{-3}$	–	$W\,m^{-2}$
Entropy	Joule per kelvin	$m^2\,kg\,s^{-2}\,K^{-1}$	–	$J\,K^{-1}$
Gas constant, universal	Joule per kelvin	$m^2\,kg\,s^{-2}\,K^{-1}$	–	$J\,K^{-1}$
Specific heat capacity	Joule per kilogram kelvin	$m^2\,s^{-2}\,K^{-1}$	–	$J\,kg^{-1}\,K^{-1}$
Specific energy	Joule per kilogram	$m^2\,s^{-2}$	–	$J\,kg^{-1}$
Thermal conductivity	Watt per meter kelvin	$m\,kg\,s^{-3}\,K^{-1}$	–	$W\,m^{-1}\,K^{-1}$
Energy density	Joule per cubic meter	$m^{-1}\,kg\,s^{-2}$	–	$J\,m^{-3}$

TABLE C.3 SI Supplementary Units and Derived Units Formed Using Supplemental Units

Quantity	Name	Symbol
Plane angle	Radian	rad
Solid angle	Steradian	sr
Angular velocity	Radian per second	$rad\,s^{-1}$
Angular acceleration	Radian per second squared	$rad\,s^{-2}$
Radiant intensity	Watt per steradian	$W\,sr^{-1}$
Radiance	Watt per square meter per steradian	$W\,m^{-2}\,sr^{-1}$

TABLE C.4 Prefixes for Decimal Multiples and Submultiples of SI Units

Multiple	Prefix	Symbol	Submultiple	Prefix	Symbol
10^{18}	Exa	E	10^{-1}	Deci	d
10^{15}	Peta	P	10^{-2}	Centi	c
10^{12}	Tera	T	10^{-3}	Milli	m
10^{9}	Giga	G	10^{-6}	Micro	μ
10^{6}	Mega	M	10^{-9}	Nano	n
10^{3}	Kilo	k	10^{-12}	Pico	p
10^{2}	Hecto	h	10^{-15}	Femto	f
10^{1}	Deka	da	10^{-18}	Atto	a

TABLE C.5   Non-SI Units Commonly Used in Current or Past Literature
With Conversion Factors

	Name	Symbol	Value in SI unit
Time	Minute	min	1 min = 60 s
	Hour	h	1 h = 60 min = 3600 s
	Day	d	1 d = 24 h = 86,400 s
Distance	Nautical mile	n mi	1 n mi = 1852 m
	Knot	kt	1 kt = 1 n mi h^{-1} = (1852/3600) m s^{-1}
	Ångström	Å	1 Å = 0.1 nm = 10^{-10} m
	Mile (USA, statute)	mi	1 mi = 1609.3 m
	Foot	ft	1 ft. = 0.3048 m
Mass	Pound	lb	1 lb = 0.4336 kg
	Metric ton	t	1 t = 10^3 kg
Area	Are	a	1 a = 1 dam^2 = 10^2 m^2
	Acre	acre	1 acre = 4046.8 m^2
	Hectare	ha	1 ha = 1 hm^2 = 10^4 m^2
Volume	Liter	l	1 l = 1 dm^3 = 10^{-3} m^3
	Gal	gal	1 gal = 3.785 × 10^{-3} m^3
Angle	Degree	°	1° = (π/180) rad
	Minute	′	1′ = (1/60)° = (π/10,800) rad
	Second	″	1″ = (1/60)′ = (π/648,000) rad
Force	Dyne	dyn	1dyn = 10^{-5} N
Energy	Calorie	cal	1 cal = 4.186 J
	British thermal unit	Btu	1 Btu =1054.6 J
	Erg	erg	1 erg = 10^{-7} J
Pressure	Bar	b	1 b = 0.1 MPa = 10^5 Pa
	Millibar	mb	1 hectoPascal (hPa)
	Standard atmosphere	atm	1 atm = 101,325 Pa
Water Flow	Sverdrup	Sv	1 Sv = 10^6 m^3 seawater s^{-1}

Useful Numerical Values

FUNDAMENTAL CONSTANTS

Universal gas constant (R^*)	$8.3143 \text{ J K}^{-1} \text{ mol}^{-1}$
Boltzmann's constant (k)	$1.38 \times 10^{-23} \text{ J K}^{-1}$
Stefan–Boltzmann constant (σ)	$5.67 \times 10^{-8} \text{ W m}^{-2} \text{ K}^{-4}$
Planck's constant (h)	$6.63 \times 10^{-34} \text{ J s}$
Speed of light (c^*)	$2.998 \times 10^{8} \text{ m s}^{-1}$
Gravitational constant	$6.67 \times 10^{-11} \text{ N m}^2 \text{ kg}^{-2}$

SUN

Luminosity	$3.92 \times 10^{26} \text{ W}$
Mass	$1.99 \times 10^{30} \text{ kg}$
Radius	$6.96 \times 10^{8} \text{ m}$

EARTH

Average radius (a)	$6.37 \times 10^{6} \text{ m}$
Equatorial radius	$6.378 \times 10^{6} \text{ m}$
Polar radius	$6.357 \times 10^{6} \text{ m}$
Standard gravity (g)	9.80665 m s^{-2}
Mass of Earth	$5.983 \times 10^{24} \text{ kg}$
Mass of ocean	$1.4 \times 10^{21} \text{ kg}$
Mass of atmosphere	$5.3 \times 10^{18} \text{ kg}$
Mean angular rotation rate (Ω)	$7.292 \times 10^{-5} \text{ rad s}^{-1}$
Total solar irradiance (S_o)	$1360.8 \pm 0.5 \text{ W m}^{-2}$
Mean distance from Sun (d)	$1.496 \times 10^{11} \text{ m}$

Global Physical Climatology. http://dx.doi.org/10.1016/B978-0-12-328531-7.00017-7

DRY AIR

Average molecular weight (m_a)	28.97 g mol^{-1}
Gas constant (R) – R^*/m_a	287 J K^{-1} kg^{-1}
Density at 0°C and 101325 Pa	1.293 kg m^{-3}
Specific heat at constant pressure (c_p)	1004 J K^{-1} kg^{-1}
Specific heat at constant volume (c_v)	717 J K^{-1} kg^{-1}

WATER

Molecular weight (m_w)	18.016 g mol^{-1}
Gas constant for vapor ($R_v = R^*/m_w$)	461 J K^{-1} kg^{-1}
Density of pure water at 0°C	1000 kg m^{-3}
Density of ice at 0°C	917 kg m^{-3}
Specific heat of vapor at constant pressure	1952 J K^{-1} kg^{-1}
Specific heat of vapor at constant volume	1463 J K^{-1} kg^{-1}
Specific heat of liquid water at 0°C	4218 J K^{-1} kg^{-1}
Specific heat of ice at 0°C	2106 J K^{-1} kg^{-1}
Latent heat of vaporization at 0°C	2.5×10^6 J kg^{-1}
Latent heat of vaporization at 100°C	2.25×10^6 J kg^{-1}
Latent heat of fusion at 0°C	3.34×10^5 J kg^{-1}

APPENDIX

E

Answers to Selected Exercises

CHAPTER 1

2. For a mean temperature of 260 K, 69% is below you.
3. If $T_s = 288$K, $\bar{T} = 260$ K, $p_s = 1013.25$ hPa, $\Gamma = 6.5$ K km^{-1}, then $T = 223$ K and $p = 272$ hPa.
4. If H increases from 7.6 km to 7.76 km, then p increases by ~6.9 hPa.
5. 6.11 hPa at 273 K for both. At 30°C = 303 K. 13.4 hPa for (1.10) and 43.7 hPa for (1.11). The linearization is not so good for such a big temperature change.

CHAPTER 2

1. For Jupiter, $S_0 = 1361$ W m$^{-2} \times (150/778)^2 = 50.6$ W m^{-2}, and $T_e = 101.7$ K. Since the observed emission from Jupiter is greater than this, it must have an internal heat source.
2. 233 K.
3. 271 K.
4. 255 K.
6. 42.8° summer, 81.3° winter at 47°N; 40.75° summer, 65.88° winter at 26°N.
7. $\theta_s = \varphi - \delta \pm 15°$ at noon: 1098 W m^{-2} (38.55°) North face, 1358 W m^{-2} (8.55°); South face in summer, 156 W m^{-2} (85.45°); North face, 814 W m^{-2} (55.45°); South face in winter.
8. 303 K conducting, 361 K nonconducting. For the nonconducting case, you need to move the ring out to a distance that is $\sqrt{4}$ times its current distance to get the same emission temperature of 255 K on the inside of the ring.

Global Physical Climatology. http://dx.doi.org/10.1016/B978-0-12-328531-7.00018-9

CHAPTER 3

1. 30 km.
2. 2.7 K day^{-1}, 5.1 K day^{-1} above, and 0.48 K day^{-1} below.
5. $T_s = \sqrt[4]{2.1}\, T_e$, $T_2 = \sqrt[4]{1.7}\, T_e$.
6. The model convective adjustment flux is 159 W m^{-2} from surface to the lower layer, 172 W m^{-2} from lower layer to the upper layer; net surface longwave loss is 80.5 W m^{-2}.
8. 194 K, 21 km.

CHAPTER 4

1. 270 W m^{-2}, 26 W m^{-2}.
2. For an isothermal hydrostatic atmosphere to warm at constant pressure requires $c_p \Delta T(p_s/g)$ J m^{-2}. To raise it up requires $R\Delta T(p_s/g)$, so the ratio is R/c_p. To show this, start from $PE = \int_0^\infty gz\,\rho\,dp$ and use the fact that the atmosphere is assumed isothermal and hydrostatic. You might also need to know that $\int_0^1 \ln p\,dp = -1$.
4. 5.4 W m^{-2} K^{-1} longwave; 12 W m^{-2} K^{-1} sensible flux.
5. 50.5°C asphalt, 43.5°C concrete.
6. 50.5°C dry asphalt; 33°C wet asphalt for $B_e = 0.25$.

CHAPTER 5

1. The average depth of groundwater on land is a whopping 52.7 m. For a land precipitation rate of 0.75 m year^{-1} it takes 69 years; if only 10% of the runoff is available (0.027 m year^{-1}) it takes 1952 years.
2. (a) 1.2 mm day^{-1}, (b) 0.3 mm day^{-1}, (c) 7.8 mm day^{-1}, (d) 1.7 mm day^{-1}
3. $B_e(0°C) = 0.48$, $B_e(15°C) = 0.20$, $B_e(30°C) = 0.07$.

CHAPTER 6

4. -54 m s^{-1} (Easterly).
5. OK you need to use the fact that the total amount of angular momentum transferred between the Earth and the atmosphere must be zero and that the moment arm is larger in the tropics. It is given that the area over which the wind stress is applied is the same for easterly and westerly and you can assume that the transfer coefficient is about the same.

6. $w = -2 \times 10^{-4}$ m s^{-1}.

7. To increase the net radiation from -20 to $+20$ W m^{-2} requires an albedo decrease of about 0.1 if the insolation is 400 W m^{-2}. If the surface warms up, you can use the Stefan–Boltzmann emission law to estimate how much more outgoing longwave radiation (OLR) you will lose, but if the air also moistens once you switch on the precipitation, you might get a stronger greenhouse effect and evaporation can help to cool the surface.

CHAPTER 7

1. Need $S \approx 35\%_0 \approx$ average.

2. ~13 m of sea ice if you choose an initial salinity of 30.5‰.

3. Would reverse the poleward heat transport.

4. Brings salty water (mostly below the surface where it is not freshened by rainfall) to a region where it can be cooled or where sea ice can form.

5. $\rho_t = 22.5$ becomes $\rho_t = 28.2$.

6. $c_w \Delta T = L(d/D)$ about 4.5% of the water mass would have to evaporate (or freeze), so the salinity would increase by 4.7, to be 36.6‰, for a potential density anomaly of about 30.

CHAPTER 8

1. In each case, the intermediate variance seems to peak on the eastern, downstream part of the high-frequency variance maximum.

2. It seems likely that this is related to the topographic barrier of the Andes and the Palmer Peninsula, or perhaps also the topography of Antarctica. The convective heat source in the tropical western Pacific also sends Rossby waves in this direction.

3. The Indian Ocean is the farthest from topographic barriers and the meridional gradient of sea surface temperature (SST) is larger there, too. The NH has two oceans and two major topographic barriers, Eurasia and North America.

4. This has to do with the vorticity balance of Rossby waves. The westward propagation associated with the beta effect becomes stronger relative to the eastward advection of relative vorticity as the zonal wavelength gets longer.

5. To explain why latitude shifts are more persistent than jet strengthening and weakening is an advanced topic in atmospheric dynamics, but it has to do with how the meridional propagation of Rossby waves depends on the jet structure.

6. The heat sources are over the warmest SST, but the waves can propagate freely over a greater range of longitudes, and the biggest amplitude waves tend to be the ones with the longest zonal wavelengths that span the globe.
7. Colder, denser water of the same depth would create larger pressure at depth, so the pressure is high in the east and low in the west. If the wind stress stopped, the water at depth would want to go west and the warmer surface water would go east to weaken the pressure gradient at depth.
8. A big MJO event would create equatorial westerly wind anomalies on the equator west of the OLR anomaly. These westerlies would push the warm surface water toward the east, like what happens in an El Niño.

CHAPTER 9

3. The decreased CO_2 and CH_4 during the ice ages explains about half of the cooling then.
5. It is about 0.13 Sverdrups, or about 10% of the deep-water formation rate.
6. If you mix 2 Sverdrups of 35‰ water with 0.13 Sverdrups of fresh water, you get a salinity of 32.9‰.
7. 38% of 10^{19} kg would have to melt in 100 years.

CHAPTER 10

3. $\lambda = \sqrt[4]{N+1}/4\sigma T_e^3$.
4. The static stability will increase more than the meridional gradient.
6. The more the heat transport weakens the temperature gradient, the more sensitive the ice line position is to global mean temperature change, and the more sensitive the climate.

CHAPTER 11

1. (a) $T_s = -2°C$, SH = 230 W m^{-2}, (b) $T_s = -23.9°C$,
 SH = 44 W m^{-2}, (c) $T_s = -27.6°C$, SH = 17 W m^{-2}.
 Derive $T_s = \left[(k_1 T_b/h_1) + CT\right]/\left[C + (k_1/h_1)\right]$, where $C = c_p \rho C_{DH} U_r$.
3. (a) 9.1 m, (b) 4.6 m.
7. Hudson's Bay is ice-free longer and stores more heat in the summer. In winter, this stored heat is released and keeps it warmer then. It is only about 100 m deep, but that is still a lot of stored heat.

CHAPTER 12

2. The emission temperature would have to be reduced by a factor of $\sqrt[4]{0.7}$ to 233 K, and the greenhouse effect increased to 222 W m^{-2}. The emission level would have to rise 3.7 km to make the emission temperature 22 K colder.

3. 2% annual mean, 40% at summer solstice.

4. The difference is about 110 W m^{-2} or 23%, with two-thirds from precession and one-third from obliquity.

CHAPTER 13

1. 1°C, 4°C, 25%.

3. About 4°C.

4. $\varepsilon = 10^{-4}$, $T_{\text{Strat}} = 275.2\,\text{K}$ when $\varepsilon = 1.12 \times 10^{-4}$.

5. Using the approximation (13.10), 32.5 years. If the forcing is held steady after that, the two cases equilibrate at 2.6 K and 1.3 K of warming.

Glossary

Absolute vorticity The sum of the planetary vorticity, which is associated with the rotation of Earth, and relative vorticity, which is associated with the fluid motion relative to the surface of Earth.

Absorption The annihilation of a photon and an equivalent release of energy.

Absorption line A discrete frequency corresponding to an energy transition of a molecule.

Aerosol A suspension of liquid or solid matter in air.

Albedo The fraction of incoming radiation that is reflected.

Aphelion The point on the orbit of Earth which is farthest from the Sun.

Atmospheric boundary layer The lowest part of the troposphere where the wind, temperature, and humidity are strongly influenced by the surface.

Back-scattering Scattering toward the direction of the source of the incident beam.

Biogeochemical Having to do with the interaction between chemical and biological processes on Earth.

Bioturbation The stirring of sediment by animal life.

Blackbody radiation The electromagnetic radiation emitted by an ideal blackbody. No actual substance behaves like a true blackbody, although platinum black and other soots come close. Hypothetically, it is a body that absorbs all of the electromagnetic radiation that strikes it, neither reflecting nor transmitting any of the incident radiation.

Bowen ratio The ratio of sensible to latent heating of the surface.

Bucket model A simple model for the soil–water budget in which the surface is represented by a field of buckets of specified depth.

Climate The synthesis of weather in a particular region.

Climate forcing An action external to the climate system that causes the climate to get warm or cool.

Climate sensitivity The relationship between the magnitude of the climate change and the measure of climate forcing (usually in K per W m^{-2}).

Cloud condensation nuclei Aerosol particles upon which molecules of vapor may condense.

Cloud microphysics The processes that control the formation, growth, and demise of individual cloud droplets or ice particles.

Conduction Heat transport in which a medium is required to transfer heat by collisions between atoms and molecules. No mass is exchanged.

Convection Heat transfer in which mass is exchanged. A net movement of mass may occur, but more commonly parcels with different energy amounts change places, so that energy is exchanged without a net movement in mass.

Convective adjustment A method used to simulate the effect of unresolved convective motions in a climate model.

Convective turbulence Generated when warm air parcels are accelerated upward by their buoyancy.

Coriolis parameter ($f = 2\Omega \sin \phi$) A measurement of twice the local vertical component of the rotation rate of a spherical planet.

Cryosphere Ice near the surface of Earth, including glaciers, snow cover, and sea ice.

Data assimilation Optimal updating of the four-dimensional state of a system (e.g., atmosphere or ocean) using a combination of a model forecast and data with known error characteristics.

Global Physical Climatology. http://dx.doi.org/10.1016/B978-0-12-328531-7.00019-0

Declination angle (see Season) The angle between the plane of the Equator and the Sun.

Dynamical Having to do with the equations of motion or variables contained therein.

Eccentricity The measure of how much the planetary orbit deviates from being perfectly circular. It controls the amount of variation of the solar-flux density at the planet as it moves through its orbit during the planetary year.

Eddy Deviations of flow from the time or zonal average.

Effective climate forcing The climate forcing that applies after the fast responses such as temperature change in the stratosphere and cloud changes in the troposphere have occurred, but before the surface temperature has changed. It is usually measured as the change in the top-of-atmosphere energy balance.

Ekman layer A layer of transition between the surface layer and the free atmosphere or ocean where Coriolis and frictional forces are both important in the momentum balance.

Ekman spiral An idealized mathematical description of the wind distribution in the boundary layer of the atmosphere or ocean, within which rotation and friction jointly determine the velocity profile.

El Niño A name given to the event when anomalously warm surface waters appear near the west coast of equatorial South America.

El Niño-Southern Oscillation (ENSO) phenomenon A joint reference made to the related oceanic and atmospheric variations that accompany warm and cold events in the equatorial Pacific.

Electronic excitation A process by which photons excite the electrons in the outer shells of an atom.

Emission The creation of a photon and a corresponding loss of energy by the emitting body.

Emission temperature The temperature at which a blackbody will emit energy at a specified rate, usually the blackbody emission temperature necessary to balance the absorption of solar radiation by a planet.

Energy balance A condition in which a system neither gains or loses energy.

Energy budget The measure of energy entering and leaving a system such as Earth's climate system.

Energy flux density The energy delivered per unit time per unit area ($W\,m^{-2}$).

Equilibration time The time required for a system to establish a new steady state after the application of a stimulus or a change in external conditions.

Evaporation A change of state from liquid to vapor.

Evapotranspiration The sum of evaporation and transpiration.

Extinction The sum of scattering and absorption of a direct beam of radiation.

Faculae Bright regions on the Sun.

Faint young Sun problem The combination of a less luminous Sun and a warm climate on Earth during the early history of the solar system.

Feedback mechanism A process that responds to a temperature change in such a way as to cause a further temperature change. Positive feedbacks amplify the temperature change and negative feedbacks reduce the magnitude of the temperature change.

Ferrel cell Weaker meridional cells in mid-latitudes which circulate in the opposite direction to the Hadley cell. Ferrel cells are thermodynamically indirect, as they transport energy from cold areas to warm areas.

Field capacity The maximum volume fraction of water that the soil can retain against gravity.

Forward scattering Scattering in the same direction as the incident beam.

Free atmosphere The portion of the atmosphere above the planetary boundary layer in which the effect of surface friction on the air motion is weak.

Gaia The concept of Earth as a single complex entity involving the biosphere, atmosphere, ocean, and land.

GCM (general circulation model) But it sometimes means global climate model.

General circulation The global system of atmospheric or oceanic motions.

Global warming potential A relative measure of the warming impact of releasing 1 kg of greenhouse gas into the atmosphere, which depends on its absorption efficiency, the amount of time it remains in the atmosphere after being released, and the time period over which the warming is integrated.

Greenhouse effect A condition in which the atmosphere warms the surface by being relatively transparent to solar radiation and absorbing and emitting terrestrial radiation very effectively.

Greenhouse gas A gas that efficiently absorbs and emits Earth's infrared emission – includes both naturally occurring gases and gases whose abundance is being changed by human activities.

Hadley cell A meridional cell in which air rises near the equator, flows toward the pole at upper levels, and sinks in the subtropical latitudes.

Halocarbon Compounds of carbon with halogens such as chlorine, bromine, and iodine.

Heat capacity The amount of energy that is required to raise the temperature by one degree.

Hour angle The longitude of the sub-solar point relative to its position at noon.

Hydrologic cycle The movement of water among the reservoirs of ocean, atmosphere, and land.

Hydrostatic balance A condition in which the gravity force that pulls atmospheric molecules toward the center of the planet is equal to the pressure-gradient force that pushes them out into space.

Ice nuclei Particles on which vapor can condense to make ice particles in air.

Infiltration The transfer of surface water to the soil.

Instrumental record The past history of climate as directly measured using modern instruments.

Internal energy Energy associated with the temperature of the atmosphere.

Intertropical convergence zone (ITCZ) The axis, or a portion thereof, of the broad trade-wind current of the tropics where the northerly and southerly trade winds meet, either in narrow bands or in the broader convergence zones over Indonesia, South America, and Africa.

Irradiance Radiant energy flux density ($W\,m^{-2}$).

Isothermal Of equal or constant temperature, with respect to either space or time. Isotope-forms of an element that have the same atomic number, but different atomic weights because they have more or less neutrons.

Isotropic scattering Scattering that is equally probable in all directions.

Jovian planets The planets of the solar system with physical characteristics similar to Jupiter, which include Jupiter, Saturn, Uranus, and Neptune.

K–T boundary The time period marking the end of the Cretaceous and the beginning of the Tertiary epochs about 65 million years ago.

Kelvin wave A type of wave that balances the Coriolis force against a topographic boundary or the equator. Plays a key role in the tropical atmosphere and ocean.

Lambert–Bouguer–Beer law of extinction Absorption is linear in the intensity of radiation and the absorber amount.

Lapse rate Rate of decline of temperature with height in the atmosphere, defined by $\Gamma = -\partial T/\partial z$.

Leaf area index (LAI) The ratio of the area of the horizontal projection of the top sides of all the leaves in the canopy to the surface area.

Line broadening An increase in the range of frequencies that can be absorbed or emitted during a particular molecular energy transition. The broadening mechanisms are natural, pressure (collision), and Doppler broadening.

Lithosphere The outer, solid portion of Earth, also known as Earth's crust.

Little Ice Age A period of expansion of mountain glaciers from about 1350 to 1800 in the Alps, Norway, Iceland, Alaska, and probably elsewhere.

Longwave radiation Thermal radiation emitted by Earth which has wavelengths between about 4 μm and 200 μm.

Luminosity The rate at which energy is released from the Sun by radiation in Watts.

Maunder minimum The period between 1645–1715 when sunspots were very few in number.

Meridional North–South.

Meridional wind The wind, or wind component, along the local North–South meridian, usually defined positive toward the North.

Mesosphere The atmospheric layer extending from the top of the stratosphere to the upper temperature minimum (the mesopause).

Monsoon A name for a wind system that changes in speed and direction with season.

Nuclear fusion The process whereby lighter elements combine to form heavier ones, releasing energy in the process.

Numerical integration The process of solving the equations of a numerical model on a computer.

Numerical model A model expressed in mathematical formulas and solved approximately on a computer.

Numerical simulation The sequence of states of the climate system as represented by a computer model, or the statistics of such a sequence.

Obliquity (or tilt) The angle between the axis of rotation of a planet and the normal to the plane of its orbit.

Oceanic mixed layer The top 20–200 m of water in contact with the atmosphere in which water properties are almost independent of depth because of rapid turbulent mixing.

Outgassing The process of releasing gases from the interior of a planet.

Paleoclimatic record Climate variables that are indirectly measured using physical, biological, and chemical information contained on land, in lake and ocean sediments, and in ice sheets.

Parameterization A process by which the effects of sub-grid–scale phenomena are specified from the knowledge of the variables at the grid scale of a climate model.

Photodissociation A process by which an energetic photon breaks the bond that holds together the atoms of a molecule.

Photoionization A process by which a photon removes electrons from the outer shells surrounding the nucleus, producing ionized atoms and free electrons.

Photolysis Chemical decomposition due to the interaction of a photon with a molecule.

Photosphere The region of the Sun from which most of its energy emission is released to space.

Planck's law of blackbody radiation An expression for the variation of monochromatic emissive power of a blackbody as a function of wavelength and temperature.

Planetary albedo The fraction of incoming solar energy reflected back to space.

Porosity of soil The volumetric fraction of the soil that can be occupied by air or water.

Potential density The density that sea water with a particular salinity and temperature would have at zero water pressure, or the density at surface air pressure.

Potential energy Energy associated with the gravitational potential of air some distance above the surface.

Potential evaporation The rate of evaporation that would occur if the surface was wet.

Precession parameter (e sinΛ) The critical parameter that describes the influence of eccentricity and longitude of perihelion on northern summer insolation.

Precipitation Any or all forms of water particles, whether liquid or solid, that fall from the atmosphere and reach the ground.

Primitive equations A simplified form of the equations of motion used by most global climate models.

Radiation Energy transport in which no medium is required and no mass is exchanged. Pure radiant energy moves at the speed of light.

Radiative cloud forcing Change of the radiative energy flux caused by the presence of liquid water and ice in the atmosphere.

Reanalysis A process that aims to assimilate historical observational data spanning an extended period, using a single consistent, state-of-the-art weather model and data-assimilation system to produce a consistent temporal record of the state of the atmosphere.

Relative humidity The dimensionless ratio of the actual water-vapor mixing ratio of the air to the saturation mixing ratio.

Richardson number A scaling parameter measuring the ratio of buoyancy to inertial forces in a fluid.

Rossby wave The basic type of wave whose theory describes the waves or eddies in the flow on a rotating sphere. The basic balance is between advection of the North–South gradient of planetary rotation and the East–West gradient of relative vorticity associated with the wave itself.

Rotational energy The energy associated with rotation (e.g., of a molecule).

Salinity Number of grams of dissolved salts in a kilogram of sea water.

Scattering An interaction between an object and radiation in which the radiation changes direction without a change in energy.

Season Expressed in terms of the declination angle of the Sun, which is the latitude of the point on the surface of the Earth directly under the Sun at noon.

Shadow area The area that a body sweeps out of a beam of parallel energy flux (see Figs 2.3 and 2.5).

Shortwave radiation Solar radiation which has wavelengths between about 0.2 and 4 μm.

Sink A point, line, or area where energy or mass is removed from a system.

Soil water zone The region which extends downward to the depth penetrated by the roots of the vegetation.

Solar constant The energy flux density of the solar emission at a particular distance. Now known as total solar irradiance.

Solar zenith angle The angle between the local normal to the Earth's surface and a line between a point on the Earth's surface and the Sun. It depends on latitude, season, and time of day.

Source A point, line, or area where mass or energy is added to a system, either instantaneously or continuously.

South Pacific convergence zone (SPCZ) The band of convection rooted near Indonesia and extending southeastward over the South Pacific Ocean that is most prominent in southern summer.

Specific humidity The dimensionless ratio of the mass of water vapor to the total mass of air, often given in units of grams of water vapor per kilogram of air.

SST Sea surface temperature.

Stationary eddy fluxes Fluxes of heat, mass, or momentum associated with stationary waves.

Stationary planetary waves East–West variations of the time-average state of the atmosphere.

Storm track The path followed by a center of low atmospheric pressure. Also the axis along which such mid-latitude systems frequently travel.

Stratosphere The atmospheric layer above the tropopause and below the mesosphere where the temperature generally increases with height. Most of the ozone is here.

Sub-grid–scale phenomena Phenomena that occur at scales smaller than the grid resolution of a climate model.

Sublimation The direct conversion of snow and ice to water vapor, without an intermediate liquid phase.

Subsolar point The point on the Earth's surface that falls on a line between the center of Earth and the center of the Sun.

Sunspots Dark spots seen on the Sun's photosphere.

Surface albedo The fraction of the downward solar flux density that is reflected by the surface.

Surface energy budget The energy flux per unit area passing through the surface boundary of the atmosphere, measured in watts per square meter, and apportioned between radiative, sensible, and latent fluxes, storage in the surface, and horizontal energy transport below the surface.

Surface layer The thin layer of air adjacent to the Earth's surface, where surface friction dominates the momentum balance and vertical fluxes are almost independent of height above the surface.

Synoptic-scale Of the spatial and temporal scale of the features typically seen on a weather map; a few days and a few thousand kilometers.

T-Tauri solar wind The intensified solar wind associated with a young star.

Temperature inversion A region of negative lapse rate where the temperature increases with altitude.

Terrestrial planets The planets of the solar system whose physical characteristics most resemble Earth, which include Mercury, Venus, Mars, and Earth.

Thermocline A layer of thermally stratified water about 1-km deep between the warmer surface layer of the ocean and the colder, deeper layer, both of which are of almost uniform temperature.

Thermosphere The atmospheric layer extending from the top of the mesosphere to outer limits of the atmosphere in which temperature increases with height.

Transient eddy fluxes Fluxes associated with rapidly developing weather disturbances especially in mid-latitudes.

Translational energy The energy associated with the movement of molecules or atoms through space (kinetic energy, temperature).

Transmission An interaction between an object and radiation in which the radiation passes the object without absorption or reflection.

Transpiration The release of moisture through the surface of leaves or other parts of plants.

Tropopause The altitude where the positive temperature lapse rate of the troposphere changes to the weak or negative lapse rate of the stratosphere.

Troposphere The atmospheric layer from Earth's surface to the tropopause; that is, the lowest 10–20 km of the atmosphere where the temperature generally decreases with altitude.

True anomaly (v) The angle formed at the center of the Sun between Earth and perihelion (see Fig. 12.11).

Turnover time The time required for the fluxes through a reservoir to completely replace the content of the reservoir.

Upwelling, downwelling Upward or downward mean vertical motion in the ocean.

Vibrational energy The energy associated with rapid variations of the interatomic distances within molecules.

Vorticity The curl of the velocity vector and a measure of the local rotation of a fluid.

Walker circulation The largest circulation cell oriented along the equator with rising in the Indonesian region and sinking to the East and West.

Wien's law of displacement The frequency of maximum emission is proportional to temperature.

Younger Dryas A cold event returning Europe to nearly glacial conditions about 11,000 years ago.

Zonal East–West, along a line of latitude.

Zonal wind The wind, or wind component, along the local parallel of latitude, usually defined positive toward the East.

References

Adler, R.F., et al., 2003. The version-2 global precipitation climatology project (GPCP) monthly precipitation analysis (1979–present). J. Hydrometeorol. 4, 1147–1167.

Alexander, M.A., Blade, I., Newman, M., Lanzante, J.R., Lau, N.C., Scott, J.D., 2002. The atmospheric bridge: the influence of ENSO teleconnections on air-sea interaction over the global oceans. J. Climate 15, 2205–2231.

Alley, R.B., 2000. The Younger Dryas cold interval as viewed from central Greenland. Quat. Sci. Rev. 19, 213–226.

Andersen, K.K., et al., 2004. High-resolution record of Northern Hemisphere climate extending into the last interglacial period. Nature 431, 147–151.

Andrews, T., Gregory, J.M., Webb, M.J., Taylor, K.E., 2012. Forcing, feedbacks and climate sensitivity in CMIP5 coupled atmosphere-ocean climate models. Geophys. Res. Lett. 39 doi: 10.1029/2012gl051607.

Arya, S.P., 1988. Introduction to Micrometeorology, vol. 42. Academic Press, San Diego, CA, p. 307.

Augustin, L., et al., 2004. Eight glacial cycles from an Antarctic ice core. Nature 429, 623–628.

Ball, W.T., Unruh, Y.C., Krivova, N.A., Solanki, S., Wenzler, T., Mortlock, D.J., Jaffe, A.H., 2012. Reconstruction of total solar irradiance 1974–2009. Astronom. Astrophys. 541, A27.

Barbante, C., et al., 2006. One-to-one coupling of glacial climate variability in Greenland and Antarctica. Nature 444, 195–198.

Barnes, E.A., Polvani, L., 2013. Response of the midlatitude jets, and of their variability, to increased greenhouse gases in the CMIP5 models. J. Climate 26, 7117–7135.

Barnett, T.P., et al., 2008. Human-induced changes in the hydrology of the western United States. Science 319, 1080–1083.

Barsugli, J.J., Battisti, D.S., 1997. The basic effects of atmosphere ocean thermal coupling on midlatitude variability. J. Atmos. Sci. 55, 477–493.

Battisti, D.S., Naylor, R.L., 2009. Historical warnings of future food insecurity with unprecedented seasonal heat. Science 323, 240–244.

Baumgartner, A., Reichel, E., 1975. The World Water Balance: Mean Annual Global, Continental and Maritime Precipitation and Run-Off. Elsevier Scientific Publishers, Amsterdam, The Netherlands, p. 179, 131 maps.

Bender, M.A., Knutson, T.R., Tuleya, R.E., Sirutis, J.J., Vecchi, G.A., Garner, S.T., Held, I.M., 2010. Modeled impact of anthropogenic warming on the frequency of intense Atlantic hurricanes. Science 327, 454–458.

Berger, A., Loutre, M.F., 1991. Insolation values for the climate of the last 10,000,000 years. Quat. Sci. Rev. 10, 297–317.

Berner, R.A., 1993. Paleozoic atmospheric CO_2: importance of solar radiation and plant evolution. Science 261, 68–70.

Berner, R.A., Kothavala, Z., 2001. GEOCARB III: a revised model of atmospheric CO_2 over phanerozoic time. American J. Science 301, 182–204.

Bindoff, N.L., et al., 2013. Detection and Attribution of Climate Change: from Global to Regional. In: Stocker, T.F. et al. (Ed.), Climate Change 2013: The Physical Science Basis. Contribution of Working Group I to the Fifth Assessment Report of the Intergovernmental Panel on Climate Change. Cambridge University Press, Cambridge, UK.

Global Physical Climatology. http://dx.doi.org/10.1016/B978-0-12-328531-7.00020-7

Bitz, C.M., Lipscomb, W.H., 1999. An energy-conserving thermodynamic model of sea ice. J. Geophys. Res. Oceans 104, 15669–15677.

Bladé, I., Liebmann, B., Fortuny, D., van Oldenborgh, G.J., 2012. Observed and simulated impacts of the summer NAO in Europe: implications for projected drying in the Mediterranean region. Climate Dyn. 39, 709–727.

Bond, T.C., et al., 2013. Bounding the role of black carbon in the climate system: a scientific assessment. J. Geophys. Res.-Atmos. 118, 5380–5552.

Bony, S., et al., 2006. How well do we understand and evaluate climate change feedback processes? J. Climate 19, 3445–3482.

Boucher, O., et al., 2013. Clouds and aerosols: In: Stocker, T.F., et al. (Eds.), Climate Change 2013 The Physical Science Basis. Contribution Fifth Assessment Report of Working Group I to the Intergovernmental Panel on Climate Change, Climate Change 2013, Cambridge University Press, Cambridge, UK 571–657.

Braconnot, P., et al., 2007a. Results of PMIP2 coupled simulations of the Mid-Holocene and Last Glacial Maximum – Part 2: feedbacks with emphasis on the location of the ITCZ and mid- and high latitudes heat budget. Clim. Past 3, 279–296.

Braconnot, P., et al., 2007b. Results of PMIP2 coupled simulations of the Mid-Holocene and Last Glacial Maximum – Part 1: experiments and large-scale features. Clim. Past 3, 261–277.

Bretherton, C.S., Smith, C., Wallace, J.M., 1992. An intercomparison of methods for finding coupled patterns in climate data sets. J. Climate 5, 541–560.

Brutsaert, W., 1982. Evaporation into the Atmosphere. D. Reidel Publishing Co., Dordrecht, Holland, p. 299.

Bryden, H.L., Hall, M.M., 1980. Heat transport by ocean currents across 25 N latitude in the Atlantic Ocean. Science 207, 884–886.

Bryden, H.L., Roemmich, D.H., Church, J.A., 1991. Ocean heat transport across 24°N in the Pacific. Deep Sea Res. 38, 297–324.

Budyko, M.I., 1969. The effects of solar radiation on the climate of the earth. Tellus 21, 611–619.

Cane, M.A., et al., 1997. Twentieth-century sea surface temperature trends. Science 275, 957–960.

Cavalieri, D.J., Parkinson, C.L., 2012. Arctic sea ice variability and trends, 1979–2010. Cryosphere 6, 881–889.

Ceppi, P., Hartmann, D.L., 2015. Connections between clouds, radiation, and midlatitude dynamics: a review. Curr. Clim. Change Rep. 1, 94–102.

Cess, R.D., 1975. Global climate change: an investigation of atmospheric feedback mechanisms. Tellus 27, 193–198.

Charlson, R.J., 1988. Have the concentrations of tropospheric aerosol particles changed? In: Rowland, F.S., Isaksen, I.S.A. (Eds.), Physical, Chemical and Earth Sciences Research Reports. Wiley-Interscience, Chichester, UK, pp. 79–90.

Charlson, R.J., Lovelock, J.E., Andreae, M.O., Warren, S.G., 1987. Atmospheric sulphur: geophysiology and climate. Nature 326, 655–661.

Charney, J.G., Fjortoft, R., von Neumann, J., 1950. Numerical integration of the barotropic vorticity equation. Tellus 2, 237–254.

Church, J.A., White, N.J., 2011. Sea-level rise from the late 19th to the early 21st century. Surveys Geophys. 32, 585–602.

Church, J.A., et al., 2013. Sea level change. In: Stocker, T.F., (Coauthors, Eds.), Climate Change 2013 The Physical Science Basis. Contribution of Working Group I to the Fifth Assessment Report of the Intergovernmental Panel on Climate Changes. Cambridge University Press, Cambridge, UK, 1137–1216.

Ciais, P, et al., 2013. Carbon and other biogeochemical cycles. The physical science basis. In: Stocker, T.F. et al. (Ed.), Contribution of Working Group I to the Fifth Assessment Report of the Intergovernmental Panel on Climate Change. Cambridge University Press, Cambridge, UK, pp. 465–570.

Clark, P.U., et al., 2012. Global climate evolution during the last deglaciation. Proc. Natl. Acad. Sci. USA 109, E1134–E1142.

CLIMAP Project Members, 1976. The surface of the ice age earth. Science 191, 1131–1137.

Collins, M., et al., 2010. The impact of global warming on the tropical Pacific Ocean and El Nino. Nature Geosci. 3, 391–397.

Compo, G.P., et al., 2011. The twentieth century reanalysis project. Quart. J. Royal Meteorol. Soc. 137, 1–28.

Crawford, K.C., Hudson, H.R., 1973. The diurnal wind variation in the lowest 1500 feet in central Oklahoma, June 1966–May 1967. J. Appl. Met. 12, 127–132.

Crowley, T.J., 1983. The geologic record of climate change. Rev. Geophys. Sp. Phys. 21, 828–877.

Dee, D.P., et al., 2011. The ERA-Interim reanalysis: configuration and performance of the data assimilation system. Quart. J. Royal Met. Soc. 137, 553–597.

Dickinson, R.E., 1984. Modeling evapotranspiration for three-dimensional global climate models. In: Hansen, J.E., Takahashi, T. (Eds.), Geophys. Monogr., American Geophysical Union, 58–72.

Donner, L.J., et al., 2011. The dynamical core, physical parameterizations, and basic simulation characteristics of the atmospheric component AM3 of the GFDL global coupled model CM3. J. Climate 24, 3484–3519.

Dreidonks, A.G.M., Tennekes, H., 1984. Entrainment effects in the well-mixed atmospheric boundary layer. Bound. Layer Meteor. 30, 75–105.

Durack, P.J., Wijffels, S.E., Matear, R.J., 2012. Ocean salinities reveal strong global water cycle intensification during 1950 to 2000. Science 336, 455–458.

Eagleman, J.R., 1976. The Visualization of Climate. Lexington Books, Lexington, MA, p. 227.

Eddy, J.A., 1976. The Maunder Minimum. Science 192, 1189–1202.

Etheridge, D.M., Steele, L.P., Langenfelds, R.L., Francey, R.J., Barnola, J.M., Morgan, V.I., 1996. Natural and anthropogenic changes in atmospheric CO_2 over the last 1000 years from air in Antarctic ice and firn. J Geophys. Res. Atmos. 101, 4115–4128.

Feng, S., Fu, Q., 2013. Expansion of global drylands under a warming climate. Atmos. Chem. Phys. 13, 10081–10094.

Fischer, E.M., Seneviratne, S.I., Vidale, P.L., Luethi, D., Schaer, C., 2007. Soil moisture – atmosphere interactions during the 2003 European summer heat wave. J. Climate 20, 5081–5099.

Flato, G., et al., 2013. Evaluation of climate models. In: Stocker, T.F. et al. (Ed.), Climate Change 2013: The Physical Science Basis. Contribution of Working Group I to the Fifth Assessment Report of the Intergovernmental Panel on Climate Change. Cambridge University Press, Lexington.

Flower, B.P., Hastings, D.W., Hill, H.W., Quinn, T.M., 2004. Phasing of deglacial warming and laurentide ice sheet meltwater in the Gulf of Mexico. Geology 32, 597–600.

Frierson, D.M.W., Hwang, Y.T., 2012. Extratropical influence on ITCZ shifts in slab ocean simulations of global warming. J. Climate 25, 720–733.

Fu, Q., Johanson, C.M., Wallace, J.M., Reichler, T., 2006. Enhanced mid-latitude tropospheric warming in satellite measurements. Science 312, 1179–11179.

Garfinkel, C.I., Hartmann, D.L., 2008. Different ENSO teleconnections and their effects on the stratospheric polar vortex. J. Geophys. Res. Atmos. 113, D18114.

Gillett, N.P., Thompson, D.W.J., 2003. Simulation of recent Southern Hemisphere climate change. Science 302, 273–275.

Goody, R.M., Yung, Y.L., 1989. Atmospheric Radiation, second ed. Oxford University Press, New York, p. 519.

Gregory, J., Webb, M., 2008. Tropospheric adjustment induces a cloud component in CO_2 forcing. J. Climate 21, 58–71.

Griffith, K.T., Cox, S.K., Knollenberg, R.G., 1980. Infrared radiative properties of tropical cirrus clouds inferred from aircraft measurements. J. Atmos. Sci. 37, 1077–1087.

Guilyardi, E., et al., 2009. Understanding El-Niño in ocean-atmosphere general circulation models: progress and challenges. Bull. Amer. Meteorolog. Soc. 90, 325–340.

Hahn, C.J., Warren, S.G., 1999. Extended edited synoptic cloud reports from ships and land stations over the globe, 1952–1996, NDP-026C, p. 71.

Hansen, J., Ruedy, R., Sato, M., Lo, K., 2010. Global surface temperature change. Rev. Geophys. 48, RG4004.

Hansen, J., et al., 2015. Ice melt, sea level rise and superstorms: evidence from paleoclimate data, climate modeling, and modern observations that 2°C global warming is highly dangerous. Atmos. Chem. Phys. Discuss. 15, 20059–20179.

Harrop, B.E., Hartmann, D.L., 2012. Testing the role of radiation in determining tropical cloud-top temperature. J. Climate 25, 5731–5747.

Hartmann, D.L., Larson, K., 2002. An important constraint on tropical cloud-climate feedback. Geophys. Res Lett. 29 doi: 10.1029/2002GL015835.

Hartmann, D.L., Ockert-Bell, M.E., Michelsen, M.L., 1992. The effect of cloud type on Earth's energy balance: global analysis. J. Climate 5, 1281–1304.

Hartmann, D.L., et al., 2013. Observations: atmosphere and surface. In: Stocker, T.F. et al., (Ed.), The Physical Science Basis. Contribution of Working Group I to the Fifth Assessment Report of the Intergovernmental Panel on Climate Change. Cambridge University Press, Cambridge, UK; New York, USA, pp. 159–254.

Held, I.M., 1978. Vertical scale of an unstable baroclinic wave and its importance for Eddy heat-flux parameterizations. J. Atmos. Sci. 35, 572–576.

Held, I.M., 1982. Climate models and the astronomical theory of the ice ages. Icarus 50, 449–461.

Held, I.M., Suarez, M.J., 1974. Simple albedo feedback models of ice-caps. Tellus 26, 613–629.

Held, I.M., Winton, M., Takahashi, K., Delworth, T., Zeng, F.R., Vallis, G.K., 2010. Probing the fast and slow components of global warming by returning abruptly to preindustrial forcing. J. Climate 23, 2418–2427.

Hemming, S.R., 2004. Heinrich events: massive late pleistocene detritus layers of the North Atlantic and their global climate imprint. Rev. Geophys. 42, RG1005.

Herzberg, G., Herzberg, L., 1957. Constants of polyatomic molecules. In: Gray, D.E. (Ed.), American Institute of Physics Handbook. McGraw-Hill, New York, 7-145-147-161.

Hewitt, H.T., et al., 2011. Design and implementation of the infrastructure of HadGEM3: the next-generation Met Office climate modelling system. Geosci. Model Dev. 4, 223–253.

Holton, J.R., Hakim, G.J., 2012. An Introduction to Dynamical Meteorology, fifth ed. Academic Press, San Diego, CA, p. 533.

Horel, J.D., Wallace, J.M., 1981. Planetary-scale atmospheric phenomena associated with the Southern Oscillation. Mon. Wea. Rev. 109, 813–829.

Hoskins, B.J., Karoly, D.J., 1981. The steady linear response of a spherical atmosphere to thermal and orographic forcing. J. Atmos. Sci. 38, 1179–1196.

Hoskins, B.J., James, I.N., White, G.H., 1983. The shape, propagation and mean-flow interaction of large-scale weather systems. J. Atmos. Sci. 40, 1595–1612.

Houze, R.A., 2009. Cloud Dynamics, second ed. Academic Press, San Diego, CA, p. 432.

Huang, S.P., Pollack, H.N., Shen, P.Y., 2000. Temperature trends ever the past five centuries reconstructed from borehole temperatures. Nature 403, 756–758.

Hunt, G.E., 1973. Radiative properties of terrestrial clouds at visible and infrared thermal window wavelengths. Quart. J. Roy. Meteor. Soc. 99, 346–369.

Imbrie, J., Imbrie, K.P., 1979. Ice Ages: Solving the Mystery. Enslow Publishers, Short Hills, NJ, p. 224.

Imbrie, J., Imbrie, J.Z., 1980. Modeling the climatic response to orbital variations. Science 207, 943–953.

Joughin, I., Smith, B.E., Medley, B., 2014. Marine ice sheet collapse potentially under way for the Thwaites Glacier basin. West Antarctica. Science 344, 735–738.

Jouzel, J., et al., 2007. Orbital and millennial Antarctic climate variability over the past 800,000 years. Science 317, 793–796.

Kalnay, E., et al., 1996. The NCEP/NCAR 40-year Reanalysis Project. Bull. Amer. Meteor. Soc. 77, 437–471.

Kao, H.-Y., Yu, J.-Y., 2009. Contrasting Eastern-Pacific and Central-Pacific Types of ENSO. J. Climate 22, 615–632.

Kiehl, J.T., Trenberth, K.E., 1997. Earth's annual global mean energy budget. Bull. Amer. Meteorol. Soc. 78, 197–208.

Klein, S.A., Soden, B.J., Ngar, C.L., 1999. Remote sea surface temperature variations during ENSO: evidence for a tropical atmospheric bridge. J. Climate 12, 917–932.

Kopp, G., Lean, J.L., 2011. A new, lower value of total solar irradiance: evidence and climate significance. Geophys. Res. Lett. 38, L01706.

Kopp, G., Lawrence, G., Rottman, G., 2005. The total irradiance monitor (TIM): science results. Solar Phys. 230, 129–139.

Kraucunas, I.P., Hartmann, D.L., 2005. Equatorial superrotation and the factors controlling the zonal-mean zonal winds in the tropical upper troposphere. J. Atmos. Sci. 62, 371–389.

Krivova, N.A., Vieira, L.E.A., Solanki, S.K., 2010. Reconstruction of solar spectral irradiance since the Maunder minimum. J. Geophys. Res. Space Phys. 115, A12112.

Kubar, T.L., Hartmann, D.L., Wood, R., 2008. Radiative and convective driving of tropical high clouds. J. Climate 21, 5510–5526.

Lacis, A., Hansen, J., Sato, M., 1992. Climate forcing by Stratospheric Aerosols. Geophys. Res. Lett. 19, 1607–1610.

Lambert, F., et al., 2008. Dust-climate couplings over the past 800,000 years from the EPICA Dome C ice core. Nature 452, 616–619.

Laskar, J., Robutel, P., Joutel, F., Gastineau, M., Correia, A.C.M., Levrard, B., 2004. A long-term numerical solution for the insolation quantities of the Earth. Astronom. Astrophys. 428, 261–285.

Lau, N.C., Nath, M.J., 1996. The role of the "atmospheric bridge" in linking tropical Pacific ENSO events to extratropical SST anomalies. J. Climate 9, 2036–2057.

Lea, D.W., Pak, D.K., Peterson, L.C., Hughen, K.A., 2003. Synchroneity of tropical and high-latitude Atlantic temperatures over the last glacial termination. Science 301, 1361–1364.

Lettau, H.H., Davidson, B., 1957. Exploring the Atmosphere's First Mile, vol. 2. Pergamon Press, London, p. 578.

Leventer, A., Williams, D.F., Kennett, J.P., 1982. Dynamics of the Laruentide ice sheet during the last deglaciation – Evidence from the Gulf of Mexico. Earth Planet. Sci. Lett. 59, 11–17.

Liebmann, B., Smith, C.A., 1996. Description of a complete (interpolated) outgoing longwave radiation dataset. Bull. Am. Meteorol. Soc. 77, 1275–1277.

Liebmann, B., et al., 2012. Seasonality of African Precipitation from 1996 to 2009. J. Clim. 25, 4304–4322.

Limpasuvan, V., Hartmann, D.L., 2000. Wave-maintained annular modes of climate variability. J. Clim. 13, 4414–4429.

Lin, B., et al., 2008. Assessment of global annual atmospheric energy balance from satellite observations. J. Geophys. Res. Atmos. 113, D16114.

Lisiecki, L.E., Raymo, M.E., 2005. A Pliocene-Pleistocene stack of 57 globally distributed benthic delta O-18 records. Paleoceanography 20, doi: 10.1029/2004pa001071.

Loeb, N.G., et al., 2009. Toward optimal closure of the Earth's top-of-atmosphere radiation budget. J. Clim. 22, 748–766.

Lohmann, U., Feichter, J., 2005. Global indirect aerosol effects: a review. Atmos. Chem. Phys. 5, 715–737.

Lorenz, E.N., 1963. Deterministic nonperiodic flow. J. Atmos. Sci. 20, 130–141.

Lorenz, E.N., 1967. The Nature and Theory of the General Circulation of the Atmosphere. World Meteorological Organization, Washington, DC, p. 161.

Lorenz, D.J., 2014. Understanding midlatitude jet variability and change using Rossby wave chromatography: poleward-shifted jets in response to external forcing. J. Atmos. Sci. 71, 2370–2389.

Loulergue, L., et al., 2008. Orbital and millennial-scale features of atmospheric CH_4 over the past 800,000 years. Nature 453, 383–386.

Lovelock, J.E., 1979. Gaia: A New Look at Life on Earth. Oxford University Press, Oxford, England, p. 157.

Luthi, D., et al., 2008. High-resolution carbon dioxide concentration record 650,000–800,000 years before present. Nature 453, 379–382.

MacFarling Meure, C., et al., 2006. Law Dome CO_2, CH_4 and N_2O ice core records extended to 2000 years BP. Geophys. Res. Lett. 33, L14810.

Madden, R.A., Julian, P.R., 1971. Detection of a 40–50 day oscillation in the zonal wind in the tropical Pacific. J. Atmos. Sci. 28, 702–708.

Madden, R.A., Julian, P.R., 1972. Description of global-scale circulation cells in the tropics with a 40–50 day period. J. Atmos. Sci. 29, 1109–1123.

Mahrt, L., Heald, R.C., Lenschow, D.H., Stankov, B.B., 1979. An observational study of the structure of the nocturnal boundary layer. Bound. Layer. Meteor. 17, 247–264.

Maloney, E.D., Hartmann, D.L., 1998. Frictional moisture convergence in a composite life cycle of the Madden-Julian oscillation. J. Clim. 11, 2387–2403.

Manabe, S., Wetherald, R.T., 1967. Thermal equilibrium of the atmosphere with a given distribution of relative humidity. J. Atmos. Sci. 24, 241–259.

Manabe, S., Smagorinsky, J., Strickler, R.F., 1965. Simulated climatology of a general circulation model with a hydrologic cycle. Mon. Wea. Rev. 93, 769–798.

Manley, G., 1974. Central England temperatures: monthly means 1659–1973. Quart. J. Roy. Meteor. Soc. 100, 389–405.

Mantua, N.J., Hare, S.R., Zhang, Y., Wallace, J.M., Francis, R.C., 1997. A Pacific interdecadal climate oscillation with impacts on salmon production. Bull. Amer. Meteorolog. Soc. 78, 1069–1079.

Marshall, J., Speer, K., 2012. Closure of the meridional overturning circulation through Southern Ocean upwelling. Nature Geoscience 5, 171–180.

Mayer, M., Haimberger, L., 2012. Poleward atmospheric energy transports and their variability as evaluated from ECMWF reanalysis data. J. Clim. 25, 734–752.

McCartney, M.S., 1982. The subtropical recirculation of mode waters. J. Marine Res. 40, 427–464.

McCartney, E.J., 1983. Absorption and Emission by Atmospheric Gases: The Physical Processes. Wiley, New York, p. 320.

McCoy, D.T., et al., 2015. Natural aerosols explain seasonal and spatial patterns of Southern Ocean cloud albedo. Sci. Adv. 1, e1500157.

Meehl, G.A., Hu, A., Arblaster, J.M., Fasullo, J., Trenberth, K.E., 2013. Externally forced and internally generated decadal climate variability associated with the interdecadal Pacific Oscillation. J. Clim. 26, 7298–7310.

Milankovitch, M., 1941. Canon of Insolation and the Ice Age Problem. Israel Program for Scientific Translations TT 67-51410.

Mirinova, Z.F., 1973. Albedo of the Earth's surface and clouds. In: Kondratev, I.A. (Ed.), Radiation Characteristics of the Atmosphere and the Earth's Surface. NASA, USA, p. 580.

Monin, A.S., Obukhov, A.M., 1954. Osnovnye zakonomernosti turbulentnogo permeshivaniia v prizemnom sloe atmosfery (basic laws of turbulent mixing in the atmosphere near the ground). Akademiia Nauk SSSR, Geolfizicheskii Institut, Trudy 24, 163–187.

Morice, C.P., Kennedy, J.J., Rayner, N.A., Jones, P.D., 2012. Quantifying uncertainties in global and regional temperature change using an ensemble of observational estimates: The HadCRUT4 data set. J. Geophys. Res. Atmos. 117 doi: 10.1029/2011jd017187.

Myhre, G., et al., 2013. Anthropogenic and natural radiative forcing. In: Stocker, T.F. et al., (Ed.), Climate Change 2013: The Physical Science Basis. Contribution of Working Group I to the Fifth Assessment Report of the Intergovernmental Panel on Climate Change. Cambridge University Press, Cambridge, UK; New York, USA, pp. 659–740.

National Research Council, 1975. Understanding Climate Change: A Program for Action. 239 pp.

Nerem, R.S., Chambers, D.P., Choe, C., Mitchum, G.T., 2010. Estimating mean sea level change from the TOPEX and Jason altimeter missions. Mar. Geodesy 33, 435–446.

North, G.R., 1975. Theory of energy balance climate models. J. Atmos. Sci. 32, 2033–2043.

O'Gorman, P.A., Schneider, T., 2009. The physical basis for increases in precipitation extremes in simulations of 21st-century climate change. Proc. Nati. Acad. Sci. USA 106, 14773–14777.

Pacala, S., Socolow, R., 2004. Stabilization wedges: solving the climate problem for the next 50 years with current technologies. Science 305, 968–972.

Parkinson, C.L., Cavalieri, D.J., 2012. Antarctic sea ice variability and trends, 1979–2010. Cryosphere 6, 871–880.

Pavlova, T.V., Kattsov, V.M., Govorkova, V.A., 2011. Sea ice in CMIP5 models: closer to reality? Trudy GGO (in Russian) 564, 7–18.

Peixoto, J.P., Oort, A.H., 1984. Physics of climate. Rev. Mod. Phys. 56, 365–429.

Peltier, W.R., 1994. Ice Age paleotopography. Science 265, 195–201.

Pendergrass, A.G., Hartmann, D.L., 2014. The atmospheric energy constraint on global-mean precipitation change. J. Clim. 27, 757–768.

Penman, H.L., 1948. Natural evaporation from open water, bare soil and grass. Proc. Roy. Soc. London 193, 120–145.

Phillips, N.A., 1956. The general circulation of the atmosphere: a numerical experiment. Quart. J. Royal. Meteor. Soc. 82, 123–164.

Pierrehumbert, R.T., Abbot, D.S., Voigt, A., Koll, D., 2011. Climate of the neoproterozoic. In: Jeanloz, R., Freeman, K.H. (Eds.), Annual Review of Earth and Planetary Sciences, vol. 39. pp. 417–460.

Pitman, A.J., 2003. The evolution of, and revolution in, land surface schemes designed for climate models. Int. J. Climatol. 23, 479–510.

Pollack, J.B., Toon, O.B., Ackerman, T.P., McKay, C.P., Turco, R.P., 1983. Environmental effects of an impact-generated dust cloud: implications for the Cretaceous-Tertiary extinctions. Science 219, 287–289.

Ramanathan, V., Carmichael, G., 2008. Global and regional climate changes due to black carbon. Nature Geoscience 1, 221–227.

Rayner, N.A., et al., 2003. Global analyses of sea surface temperature, sea ice, and night marine air temperature since the late nineteenth century. J. Geophys. Res. Atmos. 108, 4407.

Reichler, T., Kim, J., 2008. How well do coupled models simulate today's climate? Bull. Amer. Meteorolog. Soc. 89, 303–311.

Richardson, L.F., 1922. Weather Prediction by Numerical Process. Reprinted by Dover Publications in 1965, p. 236.

Rignot, E., Mouginot, J., Morlighem, M., Seroussi, H., Scheuchl, B., 2014. Widespread, rapid grounding line retreat of Pine Island, Thwaites, Smith, and Kohler glaciers, West Antarctica, from 1992 to 2011. Geophys. Res. Lett. 41, 3502–3509.

Ripley, E.A., Redmann, R.E., 1976. Grassland. In: Monteith, J.L. (Ed.), Vegetation and the Atmosphere. Academic Press, New York, pp. 351–398.

Rosenfeld, D., et al., 2008. Flood or drought: How do aerosols affect precipitation? Science 321, 1309–1313.

Rossow, W.B., Schiffer, R.A., 1991. ISCCP cloud data products. Bull. Amer. Meteor. Soc. 72, 2–20.

Rossow, W.B., Schiffer, R.A., 1999. Advances in understanding clouds from ISCCP. Bull. Amer. Meteor. Soc. 80, 2261–2287.

Santer, B.D., et al., 2013. Identifying human influences on atmospheric temperature. Proc. Natl Acad. Sci. USA 110, 26–33.

Sardeshmukh, P.D., Hoskins, B.J., 1988. The generation of global rotational flow by steady idealized tropical divergence. J. Atmos. Sci. 45, 1228–1251.

Sato, M., Hansen, J.E., McCormick, M.P., Pollack, J.B., 1993. Stratospheric aerosol optical depths, 1850–1990. J. Geophys. Res. 98, 22987–22994.

Scheff, J., Frierson, D.M.W., 2014. Scaling potential evapotranspiration with greenhouse warming. J. Climate 27, 1539–1558.

Schlesinger, M.E., Ramankutty, N., 1994. An oscillation in the global climate system of period 65–70 years. Nature 367, 723–726.

Schmidtko, S., Johnson, G.C., Lyman, J.M., 2013. MIMOC: a global monthly isopycnal upper-ocean climatology with mixed layers. J. Geophys. Res. Oceans 118, 1658–1672.

Scotese, C.R., 2001. Atlas of Earth History, Volume 1, Paleogeography, PALEOMAP Project, p. 52.

Seager, R., Hoerling, M., 2014. Atmosphere and ocean origins of North American droughts. J. Climate 27, 4581–4606.

Seager, R., Vecchi, G.A., 2010. Greenhouse warming and the 21st century hydroclimate of southwestern North America. Proc. Natl. Acad. Sci. USA 107, 21277–21282.

Sellers, W.D., 1965. Physical Climatology. University of Chicago Press, Chicago, p. 272.

Sellers, W.D., 1969. A climate model based on the energy balance of the earth-atmosphere system. J. Appl. Meteor. 8, 392–400.

Sellers, P.J., et al., 1997. Modeling the exchanges of energy, water, and carbon between continents and the atmosphere. Science 275, 502–509.

Shimanouchi, T., 1967a. Tables of Molecular Vibrational Frequencies, Part I.

Shimanouchi, T., 1967b. Tables of Molecular Vibrational Frequencies, Part 2.

Shimanouchi, T., 1968. Tables of Molecular Vibrational Frequencies, Part 3.

Slingo, A., Schrecker, H.M., 1982. On the shortwave radiative properties of stratiform water clouds. Quart. J. Roy. Meteor. Soc. 108, 407–426.

Slingo, A., Nicholls, S., Schmetz, J., 1982. Aircraft observations of marine stratocumulus during JASIN. Q.J.R.M.S. 108, 833–856.

Smagorinsky, J., 1963. General circulation experiments with the primitive equations I. The basic experiment. Mon. Wea. Rev. 91, 99–164.

Smagorinsky, J., 1974. Global atmospheric modeling and numerical simulation of climate. In: Hess, W.N. (Ed.), Weather and Climate Modification. Wiley and Sons, New York, pp. 633–686.

Soden, B.J., Held, I.M., 2006. An assessment of climate feedbacks in coupled ocean-atmosphere models. J. Clim. 19, 3354–3360.

Soden, B.J., Wetherald, R.T., Stenchikov, G.L., Robock, A., 2002. Global cooling after the eruption of Mount Pinatubo: a test of climate feedback by water vapor. Science 296, 727–730.

Soden, B.J., Held, I.M., Colman, R., Shell, K.M., Kiehl, J.T., Shields, C.A., 2008. Quantifying climate feedbacks using radiative kernels. J. Clim. 21, 3504–3520.

Solomon, S., et al., 2007. Climate change 2007. The Physical Science BasisCambridge University Press, 800 pp.

Spencer, J.W., 1971. Fourier series representation of the position of the sun. Search 2, 172.

Stephens, G.L., et al., 2002. The cloudsat mission and the A-train – a new dimension of space-based observations of clouds and precipitation. Bull. Amer. Meteor. Soc. 83, 1771–1790.

Stevens, B., Feingold, G., 2009. Untangling aerosol effects on clouds and precipitation in a buffered system. Nature 461, 607–613.

Stocker, T. F., et al., 2013. Climate Change 2013 The Physical Science Basis. In: Stocker, T.F. (Coauthors, Eds.), Contribution Fifth Assessment Report of Working Group I to the Intergovernmental Panel on Climate Change, Cambridge University Press, Cambridge, UK, 1533.

Stull, R.B., 1988. Introduction to Boundary Layer Meteorology. Kluwer Academic Publishers, Dordrecht, The Netherlands, p. 666.

Tanner, C.B., 1960. Energy balance approach to evapotranspiration from crops. Soil Sci. Soc. Am. Proc. 24, 1–9.

Taylor, K.E., Stouffer, R.J., Meehl, G.A., 2012. An overview of the CMIP5 and the experiment design. Bull. Amer. Meteorolog. Soc. 93, 485–498.

Thompson, D.W.J., Wallace, J.M., 2000. Annular modes in the extratropical circulation. Part I: month-to-month variability. J. Clim. 13, 1000–1016.

Thorne, P.W., Lanzante, J.R., Peterson, T.C., Seidel, D.J., Shine, K.P., 2011. Tropospheric temperature trends: history of an ongoing controversy. Wiley Interdisciplinary Rev. Clim. Change 2, 66–88.

Thornthwaite, C.W., 1948. An approach toward a rational classification of climate. Am. Geogr. Rev. 38, 55–94.

Trenberth, K.E., Caron, J.M., Stepaniak, D.P., 2001. The atmospheric energy budget and implications for surface fluxes and ocean heat transports. Clim. Dyn. 17, 259–276.

Trenberth, K.E., Fasullo, J.T., Balmaseda, M.A., 2014. Earth's energy imbalance. J. Clim. 27, 3129–3144.

Untersteiner, N., 1984. The cryosphere. In: Houghton, J.T. (Ed.), The Global Climate. Cambridge University Press, Cambridge, UK, pp. 121–140.

U.S. Standard Atmosphere Supplements, 1966. Environmental Science Services Administration. 289 pp.

Valley, S.L., 1965. Handbook of Geophysics and Space Environments. McGraw-Hill, New York, p. 683.

Vallis, G.K., 2006. Atmospheric and Oceanic Fluid Dynamics. Cambridge University Press, Cambridge, UK, p. 745.

van Vuuren, D.P., et al., 2011. The representative concentration pathways: an overview. Clim. Change 109, 5–31.

Vehrencamp, J.E., 1953. Experimental investigation of heat transfer at an air-surface interface. Trans. Am. Geophys. Umis. 34, 22–30.

Wallace, J.M., Gutzler, D.S., 1981. Teleconnections in the geopotential height field during the Northern Hemisphere winter. Mon. Wea. Rev. 109, 784–812.

Wallace, J.M., Hobbs, P.V., 2006. Atmospheric Science: An Introductory Survey, second ed. Academic Press, New York, p. 464.

Warren, S.G., Hahn, C.J., London, J., Chervin, R.M., Jenne, R.J., 1986. Global Distribution of Total Cover and Cloud Type Amounts Over Land. NCAR TN-317+STR.

Warren, S.G., Hahn, C.J., London, J., Chervin, R.M., Jenne, R.J., 1988. Global Distribution of Total Cloud Cover and Cloud Type Amounts Over the Ocean. NCAR TN-317 +STR.

Watanabe, M., et al., 2010. Improved climate simulation by MIROC5. Mean states, variability, and climate sensitivity. J. Climate 23, 6312–6335.

Watson, A.J., Lovelock, J.E., 1983. Biological homeostasis of the global environment: the parable of Daisyworld. Tellus 35B, 284–289.

Webster, P.J., Holland, G.J., Curry, J.A., Chang, H.R., 2005. Changes in tropical cyclone number, duration, and intensity in a warming environment. Science 309, 1844–1846.

Wheeler, M.C., Hendon, H.H., 2004. An all-season real-time multivariate MJO index: Development of an index for monitoring and prediction. Monthly Weather Rev. 132, 1917–1932.

Wielicki, B.A., Barkstrom, B.R., Harrison, E.F., Lee, R.B.I., Smith, G.L., Cooper, J.E., 1996. Clouds and the Earth's Radiant Energy System (CERES): an earth observing system experiment. Bull. Amer. Meteor. Soc. 77, 853–868.

Wilks, D.S., 2011. Statistical Methods in the Atmospheric Sciences, vol. 100. Academic Press, Oxford; Waltham, MA, p. 676.

Yapp, C.J., Poths, H., 1992. Ancient atmospheric CO_2 pressures inferred from natural geothites. Nature 355, 342–343.

Yeo, K.L., Krivova, N.A., Solanki, S.K., Glassmeier, K.H., 2014. Reconstruction of total and spectral solar irradiance from 1974 to 2013 based on KPVT, SoHO/MDI, and SDO/HMI observations. Astronom. Astrophys. 570, doi: 10.1051/0004-6361/201423628.

Zachos, J., Pagani, M., Sloan, L., Thomas, E., Billups, K., 2001. Trends, rhythms, and aberrations in global climate 65 Ma to present. Science 292, 686–693.

Zdunkowski, W.G., Crandall, W.K., 1971. Radiative transfer of infrared radiation in model clouds. Tellus 23, 517–527.

Zebiak, S.E., Cane, M.A., 1987. A model El Niño-Southern Oscillation. Mon. Wea. Rev. 115, 2262–2278.

Zelinka, M.D., Klein, S.A., Hartmann, D.L., 2012a. Computing and partitioning cloud feedbacks using cloud property histograms. Part I: cloud radiative kernels. J. Clim. 25, 3715–3735.

Zelinka, M.D., Klein, S.A., Hartmann, D.L., 2012b. Computing and partitioning cloud feedbacks using cloud property histograms. Part II: attribution to changes in cloud amount, altitude, and optical depth. J. Clim. 25, 3736–3754.

Zelinka, M.D., Klein, S.A., Taylor, K.E., Andrews, T., Webb, M.J., Gregory, J.M., Forster, P.M., 2013. Contributions of different cloud types to feedbacks and rapid adjustments in CMIP5. J. Clim. 26, 5007–5027.

Zhang, C., 2013. Madden-Julian Oscillation: bridging weather and climate. Bull. Amer. Meteorol. Soc. 94, 1849–1870.

Zhang, Y., Wallace, J.M., Battisti, D.S., 1997. ENSO-like interdecadal variability: 1900–93. J. Clim. 10, 1004–1020.

Zhou, J., Tung, K.-K., 2013. Observed tropospheric temperature response to 11-yr solar cycle and what it reveals about mechanisms. J. Atmos. Sci. 70, 9–14.

Subject Index